Essentials of
Functional MRI

Essentials of
Functional MRI

Patrick W. Stroman

CRC Press
Taylor & Francis Group
Boca Raton London New York

CRC Press is an imprint of the
Taylor & Francis Group, an **informa** business

CRC Press
Taylor & Francis Group
6000 Broken Sound Parkway NW, Suite 300
Boca Raton, FL 33487-2742

© 2011 by Taylor & Francis Group, LLC
CRC Press is an imprint of Taylor & Francis Group, an Informa business

No claim to original U.S. Government works

ISBN: 978-1-4398-1878-7 (Pbk)

Library of Congress Cataloging-in-Publication Data

Stroman, Patrick W.
    Essentials of functional MRI / Patrick W. Stroman.
        p. ; cm.
    Includes bibliographical references and index.
    ISBN 978-1-4398-1878-7 (hardcover : alk. paper)
    1. Magnetic resonance imaging. I. Title.
    [DNLM: 1. Magnetic Resonance Imaging--methods. WN 185]

RC78.7.N83S77 2011
616.07'57--dc23   2011024101

Visit the Taylor & Francis Web site at
http://www.taylorandfrancis.com

and the CRC Press Web site at
http://www.crcpress.com

This book is dedicated to
Janet,
Felicity, Liam, and Clara.

Contents

Contents

Preface

In 1990, when I was a graduate student, I heard someone with a long, prestigious career of developing and using magnetic resonance methods make an offhand comment that "everything worth doing in MRI has been done." Functional magnetic resonance imaging (fMRI) was first reported not long after. I have not forgotten the comment because it always makes me wonder what new development will be reported tomorrow and what we will be able to do with fMRI, 10 years from now. At the very least, I believe fMRI will become a very powerful and commonly used tool for providing doctors with information about their patients' neural function anywhere in the central nervous system—how it has been affected by trauma, disease, congenital effects, and other causes. I also believe this will lead to new treatments, improved outcomes, and better medical research.

This book is intended for students, researchers, and clinicians who want to understand the theory and practice of fMRI in sufficient detail to use it for neuroscience research, clinical research, and (eventually) for clinical practice, such as for diagnosis, monitoring of treatment outcomes, and treatment planning.

The chapters are organized so that the more fundamental principles come first and then are built upon to explain how magnetic resonance (MR) images are created, and finally how MR imaging is used to produce maps of neural function with fMRI. It is not expected that the reader will read this book from beginning to end, or will even start at the beginning. Sometimes you may find it useful to jump ahead to understand why more basic concepts are necessary or to put them into context. Chapter 9 describes current clinical applications of fMRI, of which there are only a few, and a much larger number of studies that have been done with patient groups, which—rather than being used for diagnosis—demonstrate how fMRI can be used as a valuable clinical tool. This chapter might be a useful starting point, depending on your interests, to show you the value of reading the chapters that precede it. To help those who just want to get to "the answer," major sections of Chapters 2 through 8 end with "Key Points" boxes that highlight the most important concepts to take away from these chapters. A useful "Glossary of Terms" and an "Index" are included at the end of the book.

You may notice that, probably unlike most other books on fMRI, this book contains quite a lot of discussion about alternative fMRI methods, other than the most widely used method, based on BOLD (blood oxygenation–level dependent) contrast. It also includes a discussion of fMRI of areas of the central nervous system other than the cortex, such as the brainstem and spinal cord. Although the vast majority of fMRI studies will use the BOLD method to study brain function, I have included these other extensions of fMRI, not just because of my research interests, but because they can help to further the understanding of key concepts as well as strengths and limitations of fMRI. It is also quite possible that some of these alternative methods will prove to be extremely important in the future, or may be the basis of even better functional imaging methods.

As the author, I wrote this book from the point of view of someone who has worked on fMRI for many years (although not since its inception), and with a particular interest in spinal cord function. This interest required reinvestigation of many of the concepts that had already been developed for fMRI of the brain, including imaging methods, the underlying basis of neuronal activity–related signal changes, and data analysis methods. Although my experience with the bulk of this work was like reinventing the wheel, it provided a solid grounding in the underlying concepts of fMRI. I have been motivated by a very strong interest in understanding how to implement fMRI as a clinical tool (in particular, for clinical assessment of spinal cord function). The process of preparing this book has contributed to this goal, and I hope it will contribute to a broader understanding of the technological, conceptual, and practical barriers to clinical use of fMRI.

MATLAB™ is a registered trademark of The MathWorks, Inc. For product information, please contact:

The MathWorks, Inc.
3 Apple Hill Drive
Natick, MA 01760-2098
USA
Tel: 508-647-7000
Fax: 508-647-7001
E-mail: info@mathworks.com
Web: www.mathworks.com

Acknowledgments

I am very grateful to Chase Figley, Jordan Leitch, Rachael Bosma, and Chris Kidd for the helpful feedback and advice they gave me while I was writing this book; and I thank Massimo Filippi and Alicia Peltsch-Williams for providing the figures for Chapter 9. I would also like to acknowledge the help of the following reviewers who generously read through a complete draft of the manuscript and provided valuable comments: Dr. Robert Savoy at Harvard Medical School, Professor Scott Huettel at Duke University, Professor Christopher Rorden at Georgia Institute of Technology, and Dr. W. Einar Mencl at Yale University.

In addition, I would like to thank my many research collaborators and my colleagues at the Centre for Neuroscience Studies at Queen's University (Kingston, Ontario, Canada), who, probably without knowing it, helped me to see fMRI from other points of view.

About the Author

Patrick. W. Stroman, Ph.D., is Canadian and was born in Lethbridge, Alberta, but grew up in Penticton, British Columbia. He received a B.Sc. with honors (co-op program) in physics at the University of Victoria from 1983 to 1988, and then a Ph.D. in applied sciences in medicine, with a focus on magnetic resonance imaging, at the University of Alberta in Edmonton from 1988 to 1993. In 1997, after 4 years as a postdoctoral fellow at the Quebec Biomaterials Institute, Laval University, in Quebec City, he moved to Winnipeg, Manitoba, to take a position as a research officer at the Institute for Biodiagnostics (IBD), National Research Council of Canada. While at the IBD he began developing functional magnetic resonance imaging (fMRI) of the spinal cord (spinal fMRI). In 2004, he moved to Kingston, Ontario, to join the Centre for Neuroscience Studies at Queen's University, with cross-appointments in the Departments of Physics and Diagnostic Radiology, and to be the director of the new Queen's MRI facility. He is now an associate professor, holds a Canada research chair in imaging physics, and leads a research program that remains focused on the development of spinal fMRI as a tool for clinical assessments and spinal cord research.

1

Introduction

Functional magnetic resonance imaging (fMRI) of neural function has been in existence only as of about 1992 (1–4), and since that time its use for neuroscience research has expanded rapidly. The basic method has remained the same: Magnetic resonance images of the brain, brainstem, or spinal cord are acquired repeatedly for several minutes to detect changes in the images over time. Whenever neuronal activity in a region changes as part of a cognitive process, sensory stimulus, motor task, and so forth, the appearance of the tissues in that region changes subtly in the magnetic resonance (MR) images and can be detected if the changes occur consistently each time the same function is performed. Clinical uses of fMRI for diagnosis and monitoring have been proposed since the method was first developed, and yet this potential remains largely untapped so far for clinical practice, although fMRI studies of many different neurological disorders have been carried out. Important questions need to be answered. For example, is fMRI reliable and sensitive enough to guide clinical decisions? Are the results worth the time and effort that could be spent on other tests or patients?

There is no single answer to these questions; rather, the answers depend on the clinical information needed, how the fMRI data are acquired, and what MRI equipment is available. Functional MRI is not a single method—it has many variations, and there is no standard method that is optimal for every situation. The translation of research methods to standardized and validated clinical methods is also confused by the many points of view that are presented by researchers from different disciplines. Researchers differ as to which acquisition and analysis methods are optimal and how the results can be interpreted, and also use their own terminology, resulting in extensive jargon associated with fMRI. However, as is evident in examples of fMRI applications in Chapter 9 and in explanations of methods in other chapters, there is actually a considerable amount of consistency in the fMRI acquisition methods that are currently used for a wide range of applications. Moreover, although methods for fMRI analysis have a fairly wide range of options, ultimately, as long as the results accurately reflect the neural function that occurred during an fMRI study, the analysis method is irrelevant.

Thus, as our understanding of the physiology underlying the neuronal-activity–related MR signal changes improves, there appears to be greater agreement about the correct interpretation of fMRI results and their reliability. With a good understanding of its underlying theory, fMRI can be used to study a wide range of neurological disorders, and most of the effort during the design process can be put into choosing the timing and method of tasks or stimulation to elicit the neural activity of interest. This understanding of the theory will also enable users of fMRI to incorporate future advances in MRI hardware and software. Based on the vast number of fMRI studies that have been published showing clinically relevant information about neurological disorders, and despite its many challenges, the clinical use of fMRI appears to be inevitable. It also appears to be inevitable that the methods used will continue to develop and evolve. All users

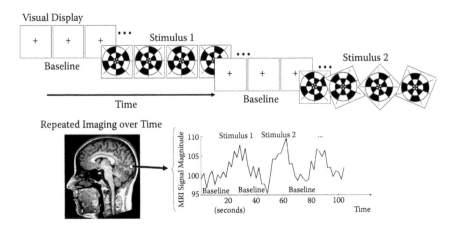

FIGURE 1.1 Summary of the basic process used for fMRI, with the example of a visual stimulus that is varied in time. MR images are acquired quickly and repeatedly to describe a time series, while the task or stimulus condition is varied systematically to elicit systematic changes in neural activity in the regions of interest. Subtle changes in the tissues, corresponding to changes in neural activity, can be detected in the time series of images.

of fMRI would therefore benefit from a better understanding of the underlying theory of MRI, fMRI study design, data analysis, and the interpretation of fMRI results.

This book is intended to describe the essential theory and practice of fMRI for clinicians and researchers who are not experts at MRI. No specialized background in physics and math is assumed. Each underlying basic idea, such as how the MR signal is detected, builds upon previous concepts (e.g., what the MR signal is and where it originates from). The chapters are therefore organized in a logical order, starting with the fundamentals of MRI; how an image is formed; how we use those images to get information related to neural function; and finally, how we can use MRI as a tool in the form of fMRI. Again, Chapter 9 describes current clinical applications and others that are expected to be precursors of future clinical applications. Each section in Chapters 2 through 8 ends with a summary of "Key Points"; a "Glossary of Terms" is also provided at the end of the book.

The basic method for functional MRI is depicted in Figure 1.1 and shows the key points that MR images are acquired repeatedly over time, while the person being studied changes tasks or cognitive processes in order to elicit a change in neural activity. The acquired images are based on the same MRI methods used for conventional anatomical imaging (see Chapter 4). Functional MRI is just a specific use of conventional MRI, and almost any modern MRI system can be used for fMRI. As indicated above, there are tissue changes that occur when the neural activity and related metabolic demand change, and these tissue changes are reflected in the MR signal (as discussed in Chapter 6). However, the changes that must be detected are subtle and cannot be seen by eye simply by comparing two images. Other effects such as physiological motion and random noise can also cause subtle changes in images, so every difference between two images cannot be attributed to neural activity.

In order to support the conclusion that image changes are caused by a change in neural activity, we systematically change between two (or more) tasks or conditions repeatedly. Any image changes that consistently occur, corresponding to the change in task or condition, are most likely related to differences in neural activity. As a result, we need to acquire images of the brain (or other region of interest) repeatedly, and as fast as possible. However, there are trade-offs between image quality, speed, and resolution, as discussed in Chapter 5. This point relates to a number of questions that arise. For example, why don't we just image with high enough resolution to see individual neurons, in order to be certain that we are detecting neural responses? One reason is that it takes time to measure and record the data that compose an image. If we image the entire three-dimensional brain, with one point of data representing each 25 μm × 25 μm × 25 μm cubic volume (the rough average size of a nerve cell body, or *soma*), we would need about 100 billion data points to cover an average brain with a volume of 1500 cm^3. Even sampling the data at a rate of 10 billion points per second would take too long for the purposes of fMRI. This sampling rate is currently impossible for reasons discussed in Chapter 3, and it would require data transfer rates and writing to disc at more than 20 GB/sec (which has become possible only recently). As discussed in Chapters 6 and 7, we want to image the brain about every 3 sec (or faster). Each individual axial brain image would have to be about 8000 × 6400 points, and we would need about 5600 images to span the whole brain. Compare this to current high-definition televisions (HDTVs) with 1080p format, which display 1920 × 1080 pixels. That is, we could not even display one entire axial slice at the full resolution of the data. We would need to arrange 24 HDTVs in a 4 × 6 grid to display the whole slice. While perhaps a fun exercise, it is not at all practical.

Another factor affecting the spatial resolution is that the signal we detect is predominantly from water and lipids. The signal from each 25 μm cubic volume would be very low compared with the background noise, as discussed in Chapters 3 and 5. As a result, we are limited to working with image data with a resolution of a few millimeters in each direction, and taking a few seconds to image the brain for each time point. Nonetheless, with good choice of imaging methods (Chapter 6), efficient fMRI study design (Chapter 7), and effective data analysis (Chapter 8), we can achieve high sensitivity to neural activity and reliably characterize changes in brain function that result from neurological disorders or trauma (Chapter 9). A summary of the concepts discussed in Chapters 6, 7, and 8 is also provided in the Appendix.

References

1. Ogawa S, Lee TM, Kay AR, Tank DW. Brain magnetic resonance imaging with contrast dependent on blood oxygenation. *Proc Natl Acad Sci USA* 1990;87(24):9868–9872.
2. Ogawa S, Lee TM, Nayak AS, Glynn P. Oxygenation-sensitive contrast in magnetic resonance image of rodent brain at high magnetic fields. *Magn Reson Med* 1990;14(1):68–78.
3. Kwong KK, Belliveau JW, Chesler DA, Goldberg IE, Weisskoff RM, Poncelet BP, Kennedy DN, Hoppel BE, Cohen MS, Turner R. Dynamic magnetic resonance imaging of human brain activity during primary sensory stimulation. *Proc Natl Acad Sci USA* 1992;89(12):5675–5679.
4. Menon RS, Ogawa S, Kim SG, Ellermann JM, Merkle H, Tank DW, Ugurbil K. Functional brain mapping using magnetic resonance imaging. Signal changes accompanying visual stimulation. *Invest Radiol* 1992;27 Suppl 2:S47–S53.

2

Basic Concepts

A few key concepts are useful to understand before delving into explanations of how magnetic resonance (MR) images are constructed and how we can use them to map neural function, because these concepts come up repeatedly or have an overall influence over the theory and practice of functional magnetic resonance imaging (fMRI). One very influential factor is the construction of the MRI system itself, because it determines the limited space and environment that we have to work within. Another important concept to understand is how numerical data can be represented as images, since all of the data used for fMRI are in the form of images. Furthermore, important mathematical concepts that occur throughout the theory of how images are constructed, and how they are analyzed, are based on the common idea of representing data as a sum of meaningful components. These concepts include separating data into components (such as with simple linear fitting), the general linear model (GLM), and the Fourier transform. Although the level of detail presented in this chapter might go beyond what is needed to use these ideas for fMRI in practice, some readers may want to know the details, and so they are included here.

2.1 Basic Anatomy of an MRI System

A good starting point is to understand the basic features of an MRI system, especially with regard to how its shape and operation will affect how fMRI studies are done. Some people who want to use fMRI may not have had the opportunity to look in detail at all of the components of an MRI system. This section briefly describes an MRI system's key elements that are in the MR "magnet room" and are seen by people who participate in fMRI studies. This section also describes the various magnetic fields, which typically cannot be sensed. The descriptions include how the various magnetic fields are created and basic principles of MR signal detection.

2.1.1 The Static Magnetic Field

One of the biggest factors that influences how functional MRI studies are carried out and what functions are studied is the shape and size of the MRI system itself (Figure 2.1). Magnetic resonance imaging requires a very strong magnetic field that has precisely the same magnitude (i.e., strength) and direction everywhere in the region we want to image. One of the key properties used to describe the quality of an MRI system is the uniformity, or homogeneity, of its magnetic field. For example, today's high-quality MRI systems made for clinical use in hospitals will have magnetic fields that vary by less than 5 parts per million (ppm) over a 40 cm diameter spherical volume in the region to be used for imaging (based on a manufacturer's minimum

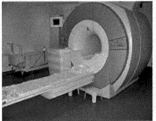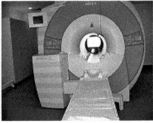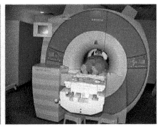

FIGURE 2.1 A typical clinical MRI system, which is also used for fMRI.

specifications for 3 tesla MRI systems in 2004). The strength of the magnetic field can be difficult to understand because we rarely encounter strong magnetic fields in everyday life. Unless you have some metal in your body from surgery or an injury, or keys or coins in your pockets, you may not be aware that you are in a magnetic field at all. To describe the strength of the field, we use units of gauss (G) or tesla (T), where 10,000 gauss is equal to 1 tesla. Some examples to put the magnetic field strength of a typical MRI system into perspective are as follows:

Earth's magnetic field	0.3 to 0.6 gauss, depending on your location
Limit considered safe for people with implanted pacemakers or neurostimulators	5 gauss
Magnetic strip used to hold refrigerator doors shut	~20 gauss at the surface
Electromagnets used on cranes for lifting scrap metal, such as cars	Around 10,000 gauss or 1 tesla
Common MRI systems currently used in hospitals	1.5 tesla to 3 tesla
Research MRI systems for humans (so far)	Up to 9.4 tesla for whole-body systems Up to 11.7 tesla for head-only systems
Highest magnetic field (used for research, not for MRI)	45 tesla (National High Magnetic Field Laboratory at Florida State University, Gainesville)

The purpose of having MRI systems with higher fields is to get a stronger signal, as described in Chapter 3. The downside of higher magnetic fields is that they take a lot of energy to create, both in terms of electrical energy and in terms of the work to design and build the magnet. This observation is especially true if we want the magnetic field to be uniform over a volume large enough for imaging a human body.

The basic concept used both for creating the magnetic fields we need for MRI and for detecting the MR signal is that if we run an electrical current through a wire, then a magnetic field is created around the wire (Figure 2.2). If we change the direction of the current, then we also change the direction of the magnetic field everywhere. Instead of using a straight wire, we can bend it into a loop and create a region in the middle of the loop with a more uniform magnetic field. For the design of MRI systems, this idea is extended further by wrapping wires in repeated loops along the surface of a cylinder, which makes the magnetic field stronger and more uniform over a larger volume at the center of the cylindrical coil of wire (Figure 2.3).

In practice, the coils of wire used to create the magnetic field for an MRI system are not wrapped in evenly spaced loops, but the spacing and density of the windings are varied to create

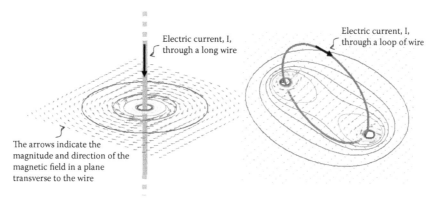

Electric current, I, through a long wire

Electric current, I, through a loop of wire

The arrows indicate the magnitude and direction of the magnetic field in a plane transverse to the wire

FIGURE 2.2 Magnetic fields created by passing an electrical current, I, through a straight conducting wire (left) or through a circular loop of wire (right). The arrows indicate the magnitude (length of the arrow) and the direction of the magnetic field at each position. The contour lines run along positions where the magnetic field magnitude is constant. The contour lines are further apart where the magnetic field is more uniform.

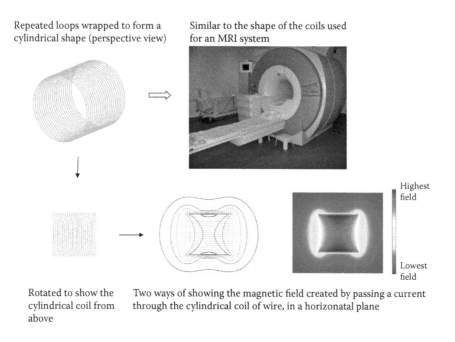

Repeated loops wrapped to form a cylindrical shape (perspective view)

Similar to the shape of the coils used for an MRI system

Highest field

Lowest field

Rotated to show the cylindrical coil from above

Two ways of showing the magnetic field created by passing a current through the cylindrical coil of wire, in a horizonatal plane

FIGURE 2.3 The magnetic field created by passing an electrical current through a continuous conducting wire wrapped on the surface of a cylinder. The shape of this cylindrical coil of wire is the basis for the general shape of most modern MRI systems.

the optimally uniform field at the center of the magnet. This field is called B_0, and its direction is through the long axis of the cylinder (i.e., the north pole of the magnet is at one end of the cylinder, and the south pole is at the other end). To have consistency across descriptions of MR systems, we define the z-axis of our coordinate system to be parallel to B_0. If we have a person laying supine in the magnet, then the z-axis is in the head-foot direction, the x-axis is right–left, and the y-axis is anterior–posterior.

The wires used to create B_0 are also made of special alloys, which are superconducting (meaning they have no electrical resistance) at very low temperatures. The wires are contained in a cryostat filled with liquid helium at roughly –270°C. The specifics of the designs are proprietary information held by magnet manufacturers. The key features of the magnet design, as they impact on fMRI, are that the person being imaged must be placed at the center of the MR magnet, which is typically around 2 meters (79 inches) long and typically has an inner opening (called the *bore*) with a diameter of 65 cm (25.6 inches). All fMRI experiments must be performed within this confined space. In addition, since the magnet is superconducting, once it is powered up to produce the desired magnetic field, it stays on without any additional electrical power. This is an important safety consideration because the magnet is generally never turned off.

2.1.2 Magnetic Field Gradients

As mentioned earlier, a key property of the static magnetic field of the MR system is its uniformity, or homogeneity. But, anything we place inside the magnetic field, including a person, tends to change the magnetic field slightly. To make the magnetic field as uniform as possible, and to compensate for changes caused by putting different objects or people in the field, we *shim* the field. Shimming is typically handled in one of two ways. One is done at the time the magnet is first installed and powered up, by placing small amounts of iron at specific locations within long trays that line the cylindrical magnetic field coil. Iron is magnetic, so the pieces of iron alter the magnetic field around them. This is called *passive* shimming and is usually a time-consuming, iterative process. Once the shimming is complete, though, the magnetic field is highly uniform over the center region where the imaging takes place. The second method of shimming uses several more sets of wire coils, like the one used to create the main magnetic field (B_0), except these are not superconducting and are designed to change the spatial distribution of the field in a number of different patterns. Each time we place a person inside the magnet, the MR system can quickly map the magnetic field and compute the electrical current needed in each shim coil to make the magnetic field as uniform as possible. This is called *active* shimming.

After all of this effort to make a highly uniform magnetic field, during imaging we alter the magnetic field by turning on magnetic field *gradients*. These are magnetic fields that vary linearly with position, across the region we want to image, to get spatial information from the MR signal. Exactly how and why we perform this procedure is explained in the next chapter. For now, we want to focus on the equipment used to make these field gradients.

The basic principles used to produce the magnetic field gradients are exactly the same as those used to create the main magnetic field. The important difference is that now, instead of creating a uniform magnetic field, we want to create one that changes perfectly linearly in strength across the region we use for imaging. To create a gradient along the long axis of the magnet (defined as the z-axis), we can use two simple loops as shown in Figure 2.4. If we place them some distance apart and pass equal electrical currents through the two loops in opposite directions, then they create two equal magnetic fields but with opposite directions. At the mid-point between the coils their fields cancel out, and as you get closer to either loop the field gets larger in magnitude, but the direction depends on which coil you are closest to, as shown in Figure 2.4. In the middle region between these two loops the field varies almost perfectly linearly with position, creating the gradient we want. Closer to either loop, the variation with position is no longer linear, and we cannot use this region for imaging, as explained in Chapter 4.

Producing gradients in the other directions, perpendicular to the z-axis, such as horizontally right–left (x-direction) or vertically (y-direction), is done in the same way only with differently shaped loops of wire (called *gradient coils*). Still, the most important features are that the magnetic field produced by the gradient coils must be parallel to B_0, and the *gradient direction* is the direction you would move along to have the field strength change. With gradient coils to

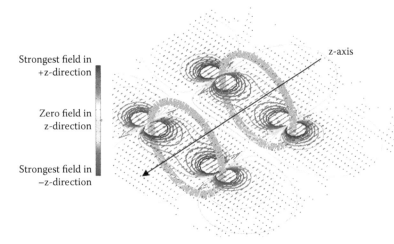

Strongest field in
+z-direction

Zero field in
z-direction

Strongest field in
−z-direction

z-axis

Example of a magnetic field gradient of
+1 gauss/cm in the z-direction

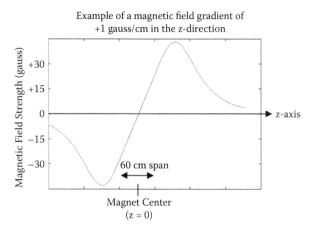

Magnetic Field Strength (gauss)

+30

+15

0

−15

−30

60 cm span

z-axis

Magnet Center
(z = 0)

FIGURE 2.4 The magnetic field produced by passing an electrical current in opposite directions through two loops, spaced some distance apart. This is the basic idea used to create a gradient through the long axis of the MR magnet (the z-axis, parallel to B_0). The plot on the right shows the field variation through the pair of loops, with the numbers selected to show the example of a gradient of 1 gauss/cm. This example shows the linear region of the gradient has a span of about 60 cm, in this case. This relationship does not hold for every MR system because it depends on the actual gradient coil design.

produce linear gradients in each of the x, y, and z directions, we can produce a net gradient in any direction we want simply by turning on two or more gradients at a time. For example, if we put electrical currents through the x- and y-gradient coils to produce gradients of 1 gauss/cm in each direction, the fields sum to produce a net gradient of 1.41 gauss/cm in the x–y plane, exactly midway between x and y.

The direct impact of the gradient coils on fMRI, in addition to producing the gradient fields we need for imaging, is that they decrease the space available in the MR system. The coils used to create the main magnetic field, B_0, are often around 1 meter in diameter for whole-body MRI systems. The three gradient coils need to be positioned within the main magnet, and the center points of the three gradient fields and of the main field must be at the same position. This is the *isocenter* of the MR system and is the point where the magnetic field does not change, regardless of which magnetic field gradients are turned on. The space taken up by these coils reduces the available space to a diameter of about 65 cm in many current MR systems, as mentioned earlier. Another significant impact of the gradient fields is that turning them on and off produces sounds because of the considerable forces between the static field and the gradient fields. The stronger the gradient, the louder the sound tends to be, and gradients that are turned on and off more quickly produce sounds with a higher pitch. The sounds produced are loud enough that people being imaged generally need to wear hearing protection. Depending on the nature of the fMRI study, these sounds can have a significant impact on the brain function being studied if they cause distraction or interfere with audio cues or stimuli.

2.1.3 Radio-Frequency Magnetic Fields

Yet another form of magnetic fields needed for MRI is one that oscillates or rotates at high frequencies for *excitation*, as described in Chapter 3. Again, these magnetic fields are produced by passing electrical currents through loops of wire. By rapidly changing the direction of the current, the magnetic fields produced also oscillate at the same frequency. In practice these fields are made to oscillate in what is termed the *radio-frequency* range of the electromagnetic spectrum, so they are called radio-frequency, or RF, fields. The loops of wire are referred to as RF coils. As shown in Figure 2.2, the direction of the magnetic field is determined by the direction of the current as well as the shape and orientation of the loop of wire. By using two loops of wire oriented 90° to each other, we can produce a total magnetic field in any direction we choose, and can also make this field rotate around in time (Figure 2.5).

This concept can be extended by using combinations of more coils, and the designs are not limited to circular loops. The number and shape of loops can be designed as needed to produce the desired volume with uniform magnetic field intensity. However, the combination of coils also must be electrically *tuned* to function at the desired frequency, just like a radio antenna. How to do so is beyond the scope of this description, but it is worth pointing out that this factor limits the possible shape and size of the coils.

Another fairly obvious point is that the RF coils must be placed around the body part to be imaged. For reasons to be explained later, we need the magnetic field to be as uniform as possible across the region to be imaged. As in Figure 2.5, the magnetic field is weaker further from the coils, but more uniform. In theory, we could just increase the current through the coils to make the magnetic field stronger, if needed. However, there are limits to how much current we can use without causing the coils to heat up and without reaching the limits of our power supply.

An important factor that influences the coil design is that we need to use RF coils for detecting, or receiving, the MR signal as well. Just as we can pass a current through a loop of wire to make a magnetic field, any magnetic field that varies in time induces an electrical current in a loop of wire. The theorem of reciprocity states that the spatial distribution of the magnetic field created by passing a current through a coil is the same as the distribution of its sensitivity when used as a receiver (1). In other words, in order for the RF coils to be sensitive when detecting the MR signal, they need to be as close as possible to the body part to be imaged. But, we need the RF coils further away to produce uniform magnetic fields for excitation. In many cases the RF coils are designed to balance these two requirements as much as possible (Figure 2.6), or in some cases RF coils are designed for excitation-only or for receive-only, so that there is no need to compromise the design of either. In the latter case, the result is that we need at least two different RF coils around the area to be imaged.

For the purposes of fMRI, the necessity of the RF coils places additional restrictions on the tasks that can be performed or the environments that can be presented for studying some neural functions. For example, RF coils placed around the head for fMRI of the brain can affect the

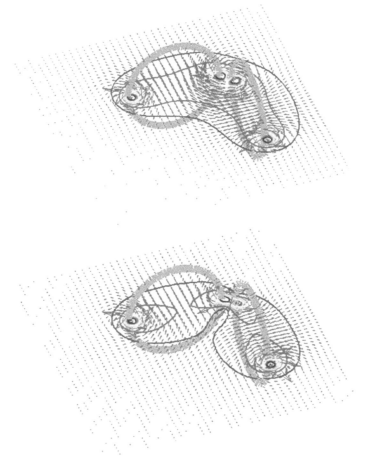

FIGURE 2.5 Magnetic fields created by passing electrical currents through two loops of wire. The direction of the resulting magnetic field in the region between the loops can be varied by adjusting the magnitude and direction of the current in each coil. The change in field between the left and right figures is caused by reversing the current direction in one of the coils.

visibility of visual displays and can make participants feel more claustrophobic. However, in many cases it is possible to use mirrors so that the participant can see outside of the magnet, or visual displays may be built into goggles placed in front of the eyes so that the presence of the RF coil is less noticeable. It is also an option on some MRI systems to use only the posterior half of head or neck coils, as receivers only, to provide unobstructed vision, or to enable the head to be in a slightly raised or turned position if needed, at the expense of some reduction in sensitivity (2).

The remaining components of a complete MRI system are typically computers, electronics, and amplifiers, which are hidden away in an equipment room. Although they are essential to the operation of the MRI system, they do not impact on how fMRI studies are done, are not visible to a person participating in an fMRI study, and are not described in detail here. However, they are shown schematically in Figure 2.7.

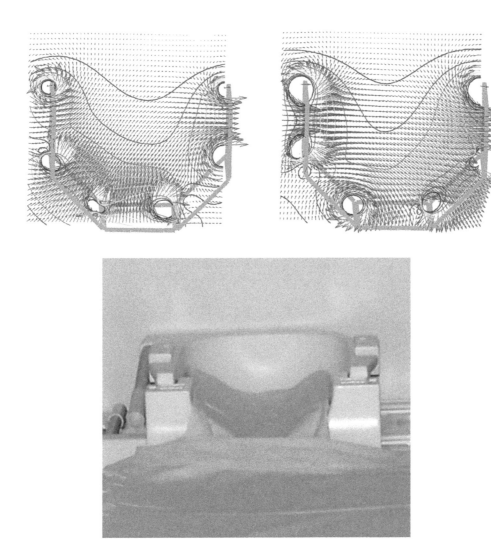

FIGURE 2.6 Magnetic fields created by passing electrical currents through five square loops of wire, arranged to fit around the back of a person's head, similar to what might be used in a posterior head coil (shown on the right). The direction of the resulting magnetic field in the region between the loops can be varied by adjusting the magnitude and direction of the current in each coil, as can be seen by comparing the two magnetic field plots (left and middle frames).

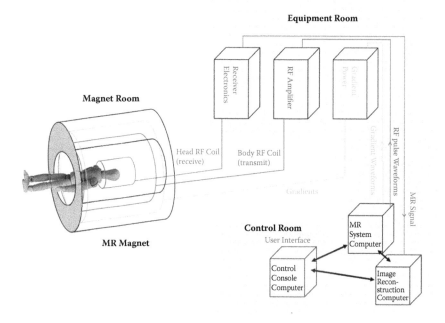

FIGURE 2.7 Schematic of a complete typical MRI system, including the magnet components and the electronic components and computers used to operate the system.

Key Points

1. A magnetic field is created when an electrical current is passed through a conductor, such as a wire.
2. If the wire is curved to form a closed loop (square, circular, oval, etc.), the magnetic field created will be directed through the loop at the center.
3. A shape similar to a long wrap of wire around the surface of a cylinder is used to make the very large static magnetic field of the MRI system, with typical strengths of 15,000 to 70,400 gauss (equal to 1.5 tesla to 7.04 tesla).
4. Other shapes of loops are used to make magnetic fields with linear gradients, which in most MRI systems can be up to around 4.5 gauss per cm. These gradient fields are typically switched on and off quickly during imaging.
5. Smaller loops are used to make magnetic fields that oscillate rapidly in the radio-frequency range (around 63 MHz to 300 MHz). These fields are pulsed briefly (for a few msec) during imaging.

2.2 Representing Images with Numbers and Vice Versa

While we are familiar with looking at pictures on a page or canvas, or images on a computer monitor or television, it is much less common to think about the numbers that an image represents. What is the connection between an image and numbers? All images that can be displayed on a computer monitor or TV screen or stored in computer memory or on a disc are digital images (Figure 2.8). In other words, they are represented as discrete points as opposed to a

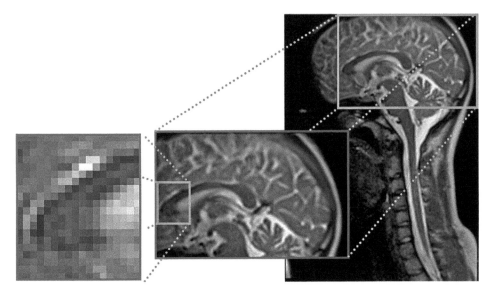

FIGURE 2.8 Examples of digital images (MRI) with progressively greater magnification to show details, and the limits of spatial resolution. Note that magnifying the image does not tend to show greater detail because we cannot know the finer details that are not represented in the original image data.

continuous surface such as you would see by looking closely at paint on a canvas. If you look very closely at some types of computer monitors or television screens, you may see that the display is composed a large number of small dots of color. Each dot is called a picture element, or pixel. It is also now common to hear about digital cameras being characterized by a number of megapixels. This refers to how many million pixels produced in the pictures you would get with this camera. For example, if a camera produces digital photos that are 2048 pixels by 1536 pixels, then the total number of pixels used is 2048 × 1536 = 3,145,728, and it would be called a 3.1 megapixel camera. Each pixel within any digital image, regardless of how it is produced, is assigned a color or shade of gray, which is indicated by one or more numbers.

The colors used to represent MR images typically indicate only the strength of the MR signal from each point in a region of the body or an object. By displaying the signal in a two-dimensional plane it looks like a picture of the anatomy, but we have to keep in mind that it is not. There are ways in which the pictures we obtain with MRI may not accurately represent the anatomy because of spatial distortions or limits of the fine details we can detect. The pictures can also just as easily show more details than we see with our eyes when looking at the anatomy. Their phenomenon is explained in more detail in Chapter 5. In a digital photograph the measurement method detects colors, so they can be reflected accurately in the image. With MRI we create what are termed *false-color* images because we don't measure a visible quality of tissues in the body—we measure a physical quality of the tissues and represent it visually. We can therefore choose to assign colors to each value in an image in any way. If we assign the colors randomly, then the resulting image tends to look like a random mess; instead, colors are typically applied in a smooth progression over the range of values, which is called a *color scale* (see examples in Figure 2.9). We can apply any color scale as shown in Figure 2.10 without changing the image information, but some features can be emphasized preferentially. However, in most cases, especially when looking at the anatomy, colors can be misleading or distracting, and it is best to stick with a gray scale so that images are displayed consistently.

The color scales used in the examples in Figure 2.10 are only used when the image information does not already indicate a specific color. To specify a color, we typically need three

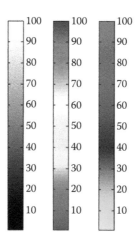

FIGURE 2.9 Examples of color scales used to label image values from 0 to 100.

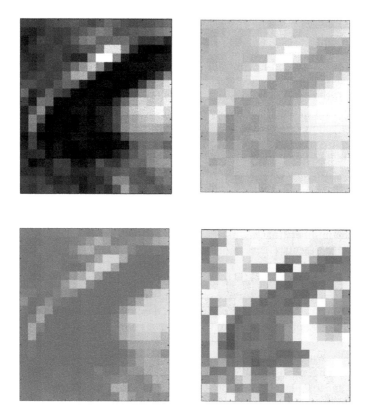

FIGURE 2.10 Examples of the same magnified portion of the image shown in Figure 2.8 except that here it is not smoothed and the individual square pixels are visible (this is called *pixelated*). The same image is shown in gray scale on the left, and with three different color scales.

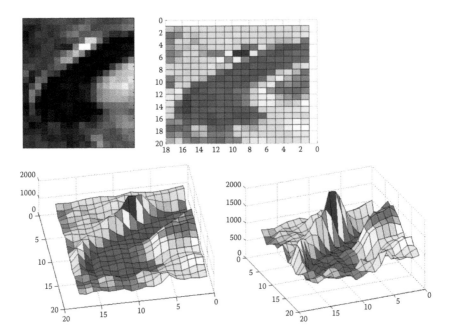

FIGURE 2.11 Examples of an image shown with a color scale and as a surface, with the height of the surface used to indicate the value of each pixel.

or four values per pixel, depending on the color-coding format we use. Some common color-coding schemes are RGB for red-green-blue and CMYK for cyan-magenta-yellow-black. The RGB scheme uses three values for each pixel to indicate the amount of red, green, and blue, respectively, to show for that pixel; and they combine by adding the wavelengths. For example, with the RGB scale, yellow is created by mixing red and green, because yellow is between red and green in the visible spectrum. However, we assume the color scale is cyclic (i.e., it wraps around), and purple is made by adding blue and red (even though red is the longest visible wavelength and purple is one of the shortest). Shades of gray are produced anytime all three color scales have equal values, and brighter colors have higher values. In other words, black is created when all three colors are set to zero; and to indicate white, we need to set all three color values to the maximum value.

Returning to the numbers that are used to create digital images, we can see that each pixel in an image really represents one or more numbers, and we can show these many different ways, not just as images. For example, Figure 2.11 shows how a grayscale image could instead be shown as a grid of colored rectangles or as a surface plot where the height of the surface reflects the value of each pixel. The surface plot of the image data more dramatically shows the changes in pixel values between certain regions, but obscures some areas behind the peaks. It also shows how we could plot the change in pixel values along selected lines to see structural details in another way.

Each black line drawn across the surface plots in Figure 2.12 shows the variation of pixel values across a specific range of positions. We could choose a single line, as in Figure 2.6, and extract the data to show it as a line plot. The purpose of doing this is to demonstrate how there is nothing particularly special about data in the form of an image—it just tends to have more dimensions than other forms of data, such as physiological measurements. Image data, just like any other data, can be characterized by applying statistical tests or can be modified by smoothing, filtering, and so forth. Finally, an important idea to point out is that we can also acquire image data repeatedly over time to add yet another dimension. Pixel values can just as easily be plotted as a function of any component of spatial position or as a function of time.

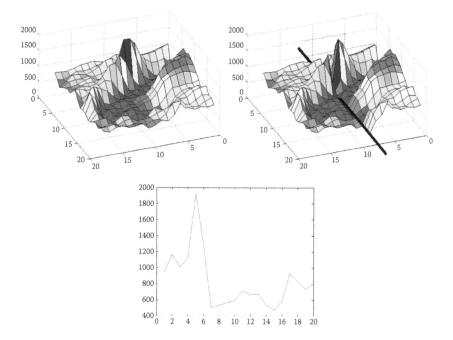

FIGURE 2.12 Example of extracting a single line of data from a surface plot, and showing it as a line plot of the pixel values as a function of position.

Key Points

6. A digital image, such as an MR image, is simply a two-dimensional grid of numbers.
7. Each point in the grid is a picture element, or pixel.
8. If there is one number per pixel, the numeric data can be shown as an image, by assigning a color or gray scale to the range of values. These are displayed in place of the numeric values.
9. A true-color image has 3 numeric values per pixel, indicating the red, green, and blue (RGB) components to display, or 4 values if the cyan, magenta, yellow, black (CMYK) format is used.

2.2.1 MR Image Data Formats

Because images come in many different shapes and sizes, we need to get some information about an image before we can read the data or load them into a computer program for display. This information includes the number of pixels spanned by the data in each direction and the format in which the numbers are stored. Additional information might include the size of the field-of-view (i.e., the distance spanned by the image data in each direction), the position coordinates for the center or corners of the image so that it can be positioned in relation to other images, details of how the image was acquired, and information about the person who was imaged. To provide all this information and more, standard formats have been developed so images can be shared, and software for reading and displaying images can be made for widespread use. The standard used for medical imaging is the DICOM standard, which stands for Digital Imaging and Communications in Medicine. This standard includes definitions for data file formats as well as for how to store, print, and transmit images across computer networks. While most MRI systems will output the image data in DICOM format, this is rarely used for fMRI data analysis, and so it is common to convert to some other format that is more convenient to work with. As a result, it is useful to know some of the details of the common file formats because reading the data is typically the first step in analyzing fMRI data (discussed in Chapter 8).

A typical feature of commonly used file formats, including the DICOM standard, is that information about the image is appended to the beginning of the image data, in what is called a *header*. Another option that is occasionally used is to divide the image data and header data into two separate files, which must be assigned file names that reflect their link. The data file itself is really just a series of binary numbers, being zeros and ones, which are each called *bits*. Binary numbers are used because the data are stored electronically, and each bit can only be set as being *on* (one) or *off* (zero). Similarly, there is no way to record number symbols or otherwise indicate larger numbers with a single piece of information. To read the data and to represent higher numbers, it is necessary to know how many bits are used to represent each number. For example, if we were able to read ordinary base-10 numbers, we would read digits between 0 and 9. Reading a string of numbers such as:

```
012345678901234567890123456789012
```

```
might mean:     01 23 45 67 89 01 23 45 67 89 01 23 45 67 89 01
or possibly:    0123 4567 8901 2345 6789 0123 4567 8901.
```

But, if we knew the numbers were defined in four-digit groups, we would know we could read numbers between 0 and 9999, or –999 to +999 if one digit was used to indicate whether the number was positive or negative. In the same way, a 4-bit binary number can be between 0 and 1111, or from –111 to +111. The meaning of each bit is similar to the meaning of each digit in a base-10 number. For example, we know that with a four-digit base-10 number, the digits represent {1000 100 10 1} in order from left to right. In binary, or base 2, the bits represent {8 4 2 1}. Picking 101 as an arbitrary example of a binary number, we can see that this would equal 5 in base-10 numbers, since it is equal to 4 + 1. In other words, an 8-bit binary number can be used to represent numbers from 0 to 255, because the bits would represent {128 64 32 16 8 4 2 1}, or could be –127 to 127 if we use the first bit to indicate positive or negative, that is, {± 64 32 16 8 4 2 1}. This relationship does not give much range for numbers to represent images; instead, we can use 16-bit or 32-bit numbers to represent much larger numbers. For reasons of computer architecture, numbers are typically defined in groups of 8, 16, 32, 64, … bits because these numbers are also powers of 2, and any group of 8 bits is called a *byte*. We can represent characters of the alphabet as well, because each character of the alphabet and several symbols have been assigned a number in the coding standard known as ASCII (American Standard Code for Information Interchange). In fact, there are 95 ASCII characters defined, and these are assigned numbers 32 to 126. So, we can read in a long series of bits and convert them to a series of numbers or characters, as long as we know in which form they are written.

We return now to the DICOM standard mentioned earlier. A DICOM image contains data defined in blocks as follows:

Group	Element	VR	Length in Bytes	Data
2 bytes	2 bytes	2 bytes	2 bytes	Variable length

The value representation (VR) values are formatting codes consisting of two characters that, together with the length of the data in bytes, indicate exactly how to read the data. The meaning of the formatting codes, and also other implicit formats that are occasionally used, are defined in the DICOM standard. Each Group and Element combination also has an explicitly defined name. For example, Element numbers 16 and 17 of Group 40 are 2-byte integers and are defined as *xsize* and *ysize*, which are the two dimensions of the image data. In this way, all the details of how, where, and when the image was acquired, and details about the person who was imaged, are included in the image header. Even if some elements are missing or out of order, each entry in the header is identified by its Group and Element number that precedes it. The last entry of the DICOM data contains the size and format of the image data, followed by the image data itself.

While this format is very flexible and contains a great deal of information, it means that a lot of header information is repeated with every image. Most MR image data contains multiple two-dimensional image slices to span a range of anatomy, such as a brain. Each of the images would be stored with its own header information. Functional MRI data requires images to be acquired repeatedly over time, in order to describe a time course. So, with DICOM format images, we might have 30 images to span the brain, each repeated 100 times to describe a time course, for a total of 3000 images in one fMRI data set. Alternative formats for storing images have been devised, such as the Analyze format, which is used by the ANALYZE image processing program (The Mayo Clinic, Rochester, New York) and was adopted by Statistical Parametric Mapping (SPM, Wellcome Trust Centre for Neuroimaging). This format uses one header file that is 348 bytes long and a separate file containing only the image data for all of the slices that were acquired at one time. For the example above, the fMRI data set would consist of 100 image data files and the associated header files. The trade-off is that the header cannot contain as much information about how the image data were acquired or the person who was imaged as with the DICOM format.

However, a more recent development is an extension of the Analyze format called NIfTI-1. The Data Format Working Group (DFWG) of the Neuroimaging Informatics Technology Initiative (NIfTI) (http://nifti.nimh.nih.gov/) developed this data format specification to make it easier to use different fMRI data analysis software packages with the same data. For example, it may be desirable to use a feature that is available in one software package, and continue the analysis or display with features from another software package. Many of the widely used software packages (as discussed in Chapter 8) support the NIfTI-1 format. The basic structure is similar to the Analyze format described briefly above but includes more information in the data header.

In practice, when doing fMRI, probably the only time we need to think about the image format is when trying to determine if we have the right format for a particular analysis software. It is important to understand that all formats are not the same, and some conversion to another format may be needed. Fortunately, many widely used fMRI analysis packages are able to read DICOM format and can write output in NIfTI-1 format. Also, software options for converting between various formats are available from sources such as NITRC (Neuroimaging Informatics Tools and Resources Clearinghouse, http://www.nitrc.org/).

Key Points

10. MR images (like any digital images) must be stored in clearly defined formats so that the image data can be read and loaded into computer software, for display and analysis.
11. The image data must be accompanied by information about the dimensions of the image, in pixels, and the format and number of bits used to represent the value of each pixel in the image.
12. The information accompanying the image can also include details about how the image was acquired, the person imaged, and so forth, and is called the image header.
13. The header information can be stored before the image data in the same file or can be stored as a separate, accompanying file.
14. The standard format for digital medical images, as output by most MRI systems, is the DICOM standard.
15. MR image data, such as for fMRI, are typically converted from DICOM to other formats for convenience.

2.3 Recurring Math Concepts: Representing Data as Sums of Meaningful Components

Throughout the explanations of how MRI data are acquired, how images are constructed, and how fMRI data are analyzed, are the math concepts described in the following sections. These include (1) breaking down any data set into components, such as with curve fitting; (2) the general linear model (GLM); and (3) the Fourier transform. While these might seem like separate topics that are unrelated, the basic concept they have in common is very important: that is, the idea that any data set can be represented as a sum of components, or data subsets, that can explain features of the data or provide meaningful information.

2.3.1 Decomposing Signals or Images into Simpler Components

A concept that is used repeatedly in MRI for constructing images and for fMRI data analysis is that data with any number of dimensions can be represented as a sum of simpler patterns. In the case of MRI data, we use the idea of summing patterns to create the total image because the data are actually acquired as a set of patterns, which are summed to create an image. While this idea is very vague for now, it is described in detail in Chapter 5, and for now we want to just focus on how signals or images in general can be represented as a sum of simpler components or patterns. In the case of fMRI data analysis, dividing the data into simpler patterns can make it easier to understand. But, there are many examples of this idea throughout all forms of data analysis methods.

One very simple and perhaps familiar example of decomposing data into simpler components is fitting a line or a curve to values that are measured as a function of some independent variable, such as time, position, and so forth. For example, if we have the following set of data:

X:	1	2	3	4	5	6	7	8	9	10
Y:	13.60	17.24	19.22	20.67	22.94	23.20	24.99	22.20	24.57	24.76

then simply looking at the progression of the values of Y, or plotting Y versus X in a graph, we can see an increasing trend. To describe how Y changes, though, we need to fit some curve to the data. A linear fit works quite well, as can be seen in Figure 2.13. This graph shows that we can describe the data as a sum of a constant function with Y = 15.34 at all values of X, and a linear ramp function with Y increasing by 1.09 each time X increased by 1. However, we would obtain a better match to the measured data if we used a curve of the form $Y = a_1 + a_2X + a_3X^2$. This is a second-order polynomial, and the curve-fitting results would tell us the best values of a_1, a_2, and a_3. We can also think of it as a sum of three patterns: (1) a constant function with a value of 11.2, (2) a linear ramp with Y increasing by 3.1 each time X is increased by 1, and (3) an X^2 curve with a scale factor of –0.1861. These variations are shown in Figure 2.14. Depending on the values being measured, the separate patterns that fit the data may have some physical meaning. For example, they could each represent a specific effect that contributes to the measured values. For the purposes of this discussion, this example only serves to illustrate the basic concept of decomposing a set of measurements into sums of simpler patterns (or lines, or functions, or whatever you prefer to call them).

Key Points

16. A sequence of measured values (i.e., a data set) can be expressed as a sum of patterns that are easier to understand or interpret.
17. Examples of decomposing data into simpler patterns are very common, such as fitting a line or a curve to data.

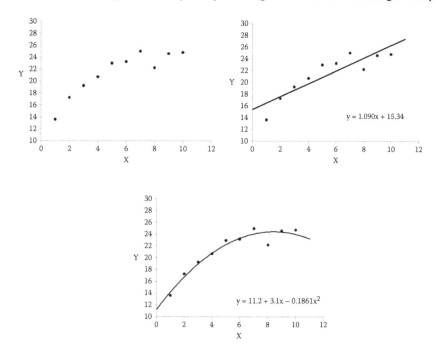

FIGURE 2.13 A sample data set, with fits to two different functions, as examples of decomposing data into simpler shapes or functions.

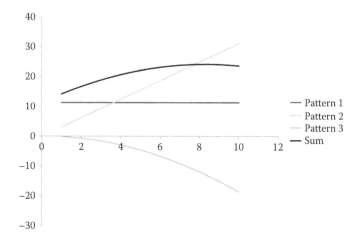

FIGURE 2.14 Example of decomposing a pattern of measured data points into three simpler components, which are the elements of a second-order polynomial: (1) a constant function, (2) a linear function, and (3) a parabolic function (X2). The measured data (black line) is the sum of these three patterns.

2.3.2 General Linear Model

In many cases it may be desirable to determine how, or if, a sequence of numbers or measurements can be expressed as sums of other functions, such as other changing values in time, for example. With the GLM (3) to decompose our measured data, $S(t)$ we use the same idea as above:

$$S(t) = a_1P_1 + a_2P_2 + a_3P_3 + e$$

In this formula, P_1, P_2, and P_3 represent a set of predefined patterns of our choosing, and are called the *basis functions*. In this example, three patterns were chosen, but any number can be used. Together they form our *basis set*. These patterns must have the same number of points as our measured data and are typically defined with a mean value of zero. As a result of setting mean values to zero, the basis set must either include one constant function with a value of 1, or S must be corrected to have a mean value of zero as well. For example, the GLM could be defined as $S(t) = a_1 + a_2P_2 + a_3P_3 + e$, and then the average value of S does not need to be zero. The values a_1, a_2, and a_3 are the values that will be determined and are the output of the GLM in order to give the best fit. The value "e" is simply whatever is left over, called the *residual*, and the best fit gives the values of a_1, a_2, and a_3 that minimize the sum of the squares of the values in "e." That is, the best fit values of a_1, a_2, and a_3 make the values in "e" as small as possible.

For example, say we have measured some physiological value such as a person's heart rate while exercising on a treadmill. It might be more useful in this case to see if the heart rate is related to other changing patterns in time, such as the rate of breathing, the amount of work the person is doing, and the slope of the treadmill. Fictional data are shown in Figure 2.15. Right away it is clear that these patterns might also be related to each other. It wouldn't make sense to fit our measured values to all three of these patterns. Also, each of these patterns has a different unit (beats/min, calories/min, etc.). Without showing yet how the GLM is done, it is useful to look at a few examples of possible applications and what problems may occur. First, if we try to express the heart rate as a sum of the other three patterns, we get this result:

heart rate = 2.55 × breathing rate + 2.66 × calorie burn rate – 0.33 × treadmill slope

Figure 2.16 shows that the GLM fit to the heart rate appears to be quite good, but the results do not make much physical sense. The results of the fit indicate that the heart rate decreases with increasing treadmill slope. If we omit the treadmill slope from the GLM, then we get this result (Figure 2.17):

heart rate = 2.99 × breathing rate + 1.86 × calorie burn rate

The result of this fit appears to be just as good as the previous one. So, how do we know which results to trust? Two key points that have been overlooked so far are that (1) we should include

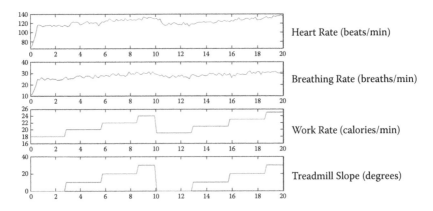

FIGURE 2.15 Sample data for use in an example of a GLM.

a constant function to account for the average value of the measured data, and (2) we need to choose uncorrelated basis functions, with average values of zero (Figure 2.18).

$$\text{heart rate} = 3.27 \times (\text{calorie burn rate} - \text{avg. rate}) + 122.3$$

(Here, *avg. rate* is used to indicate the average calorie burn rate, which we want to subtract to make the basis function have a mean value of zero.)

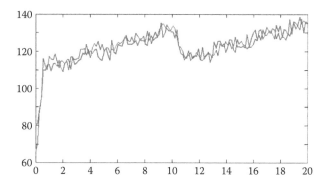

FIGURE 2.16 The red line shows the GLM fit to the measured heart rate, which is shown in blue.

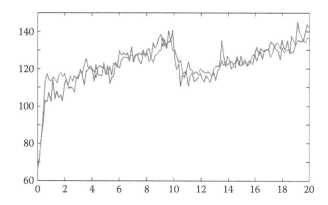

FIGURE 2.17 The red line shows the second GLM fit to the measured heart rate (plotted in blue).

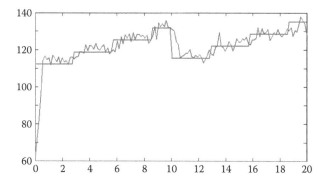

FIGURE 2.18 The GLM fit this time with linearly independent functions, with mean values of zero, plus a constant function.

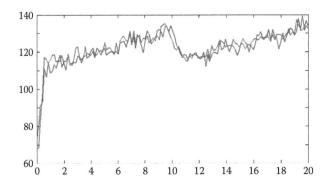

FIGURE 2.19 Results of the GLM fit (red line) to the original data (blue line), with a basis set composed of the four linearly independent functions in Figure 2.21.

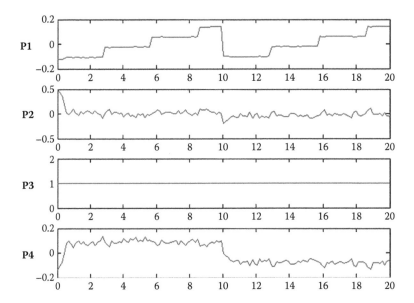

FIGURE 2.20 Linearly independent basis functions for the GLM.

This fit to the example data demonstrates that the heart rate appears to be related to the calorie burn rate, and that it changes by 3.27 beats/min for each change of 1 calorie/min. We can also see that the average heart rate over the 20 min of exercise was 122.3 beats/min. Finally, we can go further and use a principal components analysis (which we will not go into here, see instead Section 8.2.4) to make four basis functions that are completely linearly independent, based on the original three functions plus a constant function (Figures 2.19 and 2.20).

$$\text{heart rate} = 80.0 \times P_1 - 79.2 \times P_2 + 122.3 \times P_3 + 1.1 \times P_4$$

These results again show that the measured heart rate values have a strong dependence on a function matching the pattern of the calorie burn rate, and an average value of 122.3. The other two functions have relatively little contribution (such as for function P_4), or the functions show relatively little information (function P_2).

Clearly, there are many more ways we could analyze these data if we wanted, and we have not yet looked at the statistical significance of the fit values (i.e., how confident we can be that they are not equal to zero). But, the purpose here is just to show the idea of how we could use the GLM and how the results of the GLM can give information about our data. This example also serves to demonstrate the importance of a good choice of basis functions.

The GLM is frequently used for fMRI data analysis, in much the same way as shown in the previous example, and is described in detail in Chapter 8. For fMRI, we use basis functions that match the timing of when participants are resting or performing various tasks, so that we can detect related signal changes in the MR image data. This is the basis of many popular fMRI analysis software packages such as Statistical Parametric Mapping (SPM), Brain Voyager, AFNI, and others, also discussed in Chapter 8.

In most cases, people using fMRI do not need to explicitly perform the math involved with a GLM, since the GLM is incorporated into the analysis software and it is sufficient just to understand the meaning of the results. Nonetheless, it is useful to understand the basic ideas of the underlying math because it is necessary to define the basis functions to use for fMRI analysis, even if an advanced analysis software package is used. The method for using the GLM to calculate the weighting of each function in the basis set is relatively easy if you are very familiar with matrix algebra. The basic idea was mentioned above:

$$S(t) = a_1 P_1 + a_2 P_2 + a_3 P_3 + e$$

This relationship can be expressed in matrix form:

$$S = AP + e$$

which means:

$$S(t) = \begin{bmatrix} a_1 & a_2 & a_3 \end{bmatrix} \begin{matrix} P_1(1) & P_1(2) & \dots \\ P_2(1) & P_2(2) & \dots \\ P_3(1) & P_3(2) & \dots \end{matrix} + e$$

Here $P_1(1)$ means the first point in P_1, $P_1(2)$ means the second point, and so forth. In these equations, we know $S(t)$ because it is our measured data, and we know the basis functions because we choose them. The only values we do not know are a_1, a_2, a_3, and e. In the most basic form, we can determine A with:

$$A = S/P$$

and we can determine e with $e = S - AP$. However, dividing by matrices is not always straightforward, and it is usually more accurate to multiply by the inverse:

$$A = SP^{-1}$$

There may be other mathematical manipulations we need to perform to get the inverse, but for the purposes of this description, the basic idea is sufficient. The main point is that while the GLM may seem quite complicated on the surface, the basic underlying ideas are quite easy to understand.

Key Points

18. The general linear model (GLM) is another method for decomposing data into sums of other patterns, but for the GLM we use functions of our choosing, such as:

$$S(t) = a_1 P_1 + a_2 P_2 + a_3 P_3 + e$$

19. The functions we use (P_1, P_2, and P_3, in the equation above) are called the basis functions, which together form the basis set.

20. For the GLM to have a unique solution, the basis functions must be linearly independent.

21. The value of e is the pattern of whatever is leftover, because the fit to the basis functions is not expected to be perfect.

2.3.3 The Fourier Transform

A special method of decomposing data into sums of other patterns is the use of sine and cosine functions as the basis functions. This approach is used frequently in MRI and fMRI. At first it might seem like a strange choice, but as shown below, sine and cosine functions can actually be summed to create almost any pattern of values.

Sine and cosine functions have values that oscillate between 1 and –1 and can be made to have any time interval between the peaks. The values can be viewed as the two-dimensional coordinates of a point on a circle, with the specific point on the circle indicated by the angle from the horizontal axis, specified by ωt in Figure 2.21. Here, ω is used to indicate the rate at which the angle is changing. The horizontal coordinate is the value of $\cos(\omega t)$, and the vertical coordinate is the value of $\sin(\omega t)$. If the angle ωt increases in time, the point sweeps around the circle and the values of $\cos(\omega t)$ and $\sin(\omega t)$ oscillate as shown in Figure 2.21. The position of the point on the circle in Figure 2.21 can be expressed as a complex number, $\cos(\omega t) + i \sin(\omega t)$, where i is the imaginary number (i.e., the nonreal part of a complex number) equal to $\sqrt{-1}$, and indicates the imaginary axis, or vertical axis in the figure. The other axis is called the *real* axis (horizontal axis in the figure). The sum of $\cos(\omega t) + i \sin(\omega t)$ can also be written as $e^{i\omega t}$, or $\exp(i\omega t)$, which means the same thing. (This value e is the constant equal to approximately 2.71828 and is not the same as the e used in the previous section.) One complete cycle has the angle ωt sweeping through values spanning 360° or equivalently, 2π radians. So, we can define a rate of oscillation by setting ωt to go through a complete cycle in whatever time interval we want, such as $\omega = 2\pi f$, where f is the frequency of oscillation, in cycles per second (π is the constant 3.141592654).

Sine and cosine functions have two very important properties that make them particularly useful for decomposing patterns into linear combinations of these functions. The first is that these functions are *linearly independent* if they are oscillating at different frequencies. In other words, $e^{i\omega t}$ at one frequency cannot be expressed as a linear combination of functions at other frequencies. If two sequences of numbers are "orthogonal," say $a_1, a_2, \ldots a_n$ and $b_1, b_2, \ldots b_n$, then $a_1 b_1 + a_2 b_2 + \ldots + a_n b_n = 0$, and they are also linearly independent. In the case of sine and cosine functions, or $e^{i\omega t}$, if $a(t) = e^{i\omega_1 t}$ and $b(t) = e^{i\omega_2 t}$, then $a(t)b(t) = e^{i(\omega_1 + \omega_2)t}$. This product is simply an oscillating function at a different frequency than the original $a(t)$ or $b(t)$ and is symmetric around zero as shown in Figure 2.22. There are as many positive values as negative

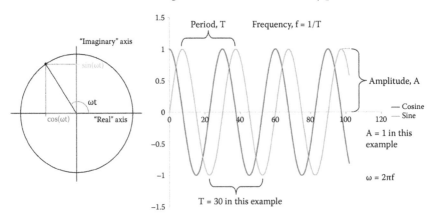

FIGURE 2.21 Sine and cosine functions.

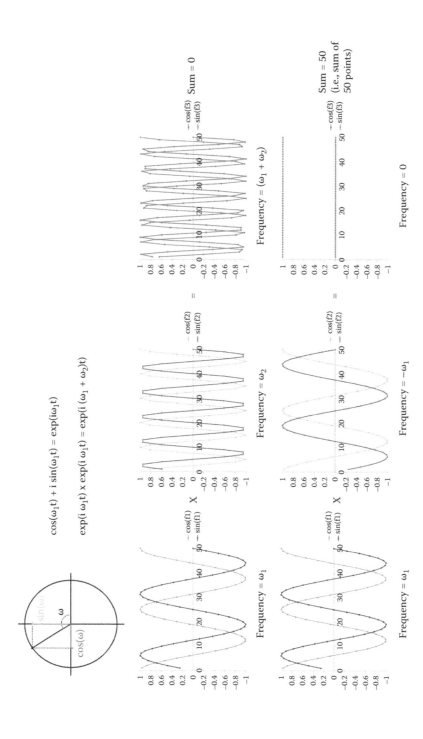

FIGURE 2.22 Examples of multiplying rotating functions, exp(−iωt) at different frequencies. The sum of the values of all points in one or more complete cycles is equal to zero, except in the special case when the two frequencies are exactly opposite, meaning $\omega_1 = -\omega_2$. In this special case, the product of the two frequencies is equal to 1 at every point, and the sum of all points is equal to the number of points.

values in a complete cycle, so summing points spanning one or more complete cycles gives a net total of zero, showing that $a(t)$ and $b(t)$ are linearly independent. There is only one special case of this, when $a(t) = e^{i\omega_1 t}$ and $b(t) = e^{-i\omega_1 t}$, so $a(t)b(t) = e^0 = 1$ at every value of t, and the sum of N values is equal to N.

The second very useful property of sine and cosine functions is that they form a *complete* set. Therefore, they can describe any sequence of numbers, as long as the sequence is continuous and has a single value at each measured point. For example, a pattern such as the one shown in Figure 2.21 is suitable, whereas a function such as $1/x$ for values of x from -10 to 10 is not, because of the undefined value at $x = 0$. Accepting that within these limitations any sequence of numbers, S, measured at N different points in time can be described as a sum of sine and cosine functions, then this relationship can be written as:

$$S(t) = a_1 \, e^{i\omega_1 t} + \ldots + a_n e^{i\omega_n t}$$

This equation may include any frequencies, ω_n, as needed. We could also write it out in long form, listing terms for all frequencies, and allow for the fact that many of the values of a_n may be equal to zero.

So, making use of the two important properties mentioned above, if we multiply $S(t)$ by the function $f(t) = e^{-i\omega_k t}$ at each point in time, and sum the N values, then all of the components with frequencies not equal to ω_k sum to zero, and the one with a frequency equal to ω_k sums to Na_k. This outcome demonstrates how we can determine the value of a_k, simply by dividing the result by N. We could do the same thing at all frequencies and determine all of the values of a_1 to a_n. Once we know these values, then we can describe our measured values as a sum of $e^{i\omega t}$ functions over a range of frequencies. We can also describe the sequence of values and characterize some of its properties by listing or plotting the values of a_1 to a_n. This is called the *Fourier transform*.

In the preceding description, there is a serious practical problem with attempting to test for values of a_k at all frequencies, one at a time. Since there is a continuum of frequencies, testing would take forever. Also, the measured values themselves are not continuous, but sampled at discrete points, so we do not know the values between the measured points or outside of the measured range. It is worth noting that sine and cosine functions extend forever in both directions (i.e., positive and negative values of time). So, what values should we model outside of the measured range? To deal with these questions, we need to use the *discrete* form of the Fourier transform.

The Fourier transform can be made discrete by treating the data as though they repeat forever in both directions, as shown in Figure 2.23. A pattern of numbers that repeats cyclically forever is much better suited to being described with functions that also repeat cyclically forever (i.e., sines and cosines). With the data repeating forever, a finite number of frequencies are sufficient to completely describe the measured points at discrete values. Mathematically, this relationship is written:

$$a_k = \frac{1}{N} \sum_{t=1}^{N} S(t) e^{-i\omega_k t}$$

Here, the symbol Σ indicates the sum over N points at different values of t ranging from 1 to N. We can see that it is still the same idea as before—we multiply our series of numbers, $S(t)$, by a function $e^{i\omega_k t}$, add up the N values, and then divide by N.

To know the frequency of oscillation of any component of $S(t)$, we need to have at least two measured points within each complete cycle. So, if we have measured values every Δt seconds, then the highest frequency that we can determine is $1/(2\Delta t)$ in cycles per second (i.e., hertz, Hz), which is called the *Nyquist* frequency. With N measured points, the largest number of frequency components we can determine is also equal to N. The slowest oscillation we can determine is 0 (i.e., a constant function), which means the spacing between

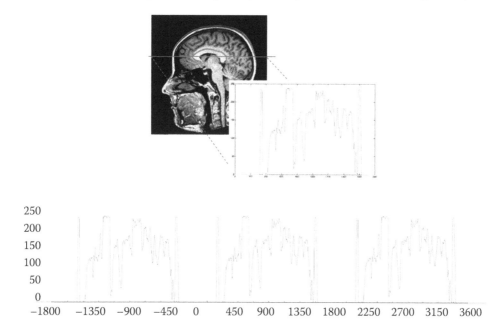

FIGURE 2.23 Replication of a pattern of numbers in order to make the Fourier transform discrete.

frequencies (i.e., the resolution) is $1/(N\Delta t)$, with frequencies spanning from $-(N-1)/(2N\Delta t)$ to $1/(2\Delta t)$. If we try to determine the frequency of a function that oscillates at a frequency higher than $1/(2N\Delta t)$, it would actually appear to be at a lower frequency, because we are not sampling fast enough. In fact, the frequencies $-1/(2\Delta t)$ and $1/(2\Delta t)$ are indistinguishable. When a frequency is outside the range we can determine, and as a result appears to be at a different frequency, it is called *aliasing* (see Figure 2.24).

It is important to understand the Fourier transform (i.e., the decomposition of a set of data into sums of oscillating functions) because it is a key principle involved in the construction of MR images and in determining how the data must be sampled. The effects of aliasing are often visible in MR images as they produce a number of different image artifacts.

The relationship between the sampled data, $S(t)$, and its Fourier transform, $s(\omega)$, (i.e., the list of a_k values described above) is shown in Figure 2.25. Since the frequency components are cyclic, with $\omega_{N/2}$ equivalent to $-\omega_{N/2}$ we can represent the range of frequencies equally from $-\omega_{N/2-1}$ to $\omega_{N/2}$, or 0 to ω_{N-1}. When looking at a plot of the results of the Fourier transform, it is necessary to know where the $\omega = 0$ value is plotted.

An example of how the Fourier transform might be applied to one line of data through an MR image is shown in Figure 2.26. For most data, such as for a line of MRI data, the lower frequencies always tend to have much higher amplitudes than the higher frequency components. The relative amplitudes of the oscillating functions are not shown to scale in Figure 2.26. If they were drawn to scale, the higher frequency oscillations would be too small to be visible. The actual magnitudes of each frequency component are shown in Figure 2.27, and the much lower amplitudes of the high-frequency components can be seen. This figure also shows again how the amplitudes of specific frequency components can be determined.

The same concepts described above can be applied to two-dimensional image data as well, simply by applying the Fourier transform line by line in one direction and then line by line in the other direction. An example of the two-dimensional Fourier transform is shown in Figure 2.28. Because the Fourier transform is a summation of components, the order of summation does not matter. In other words, it does not matter in which direction we apply it first, when working with

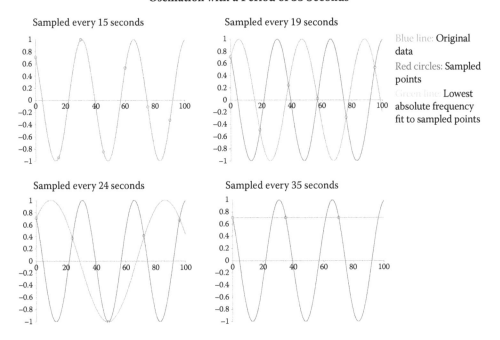

Oscillation with a Period of 35 Seconds

FIGURE 2.24 Examples of *aliasing* caused by sampling an oscillating function too slowly to be able to accurately determine the frequency of oscillation. Blue lines show the actual functions, and green lines show the frequency that would be determined by fitting to the points sampled at the red circles.

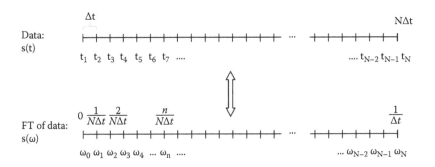

FIGURE 2.25 Relationship between the timing interval and number of sampled points, with the frequency interval and span of the Fourier transform of the sampled points.

two-dimensional data. In exactly the same way, the Fourier transform can be applied to more dimensions as well (3-D, 4-D, etc.).

Applications of the Fourier transform for MR image construction and for use in fMRI data analysis arise repeatedly in later chapters. Also, important limitations of MRI capabilities and sources of image artifacts and distortions are related to the Fourier transform and are discussed in later chapters. For now, the main concept to understand is that the Fourier transform is a useful way of decomposing any set of data, whether one-dimensional as in the examples above, two-dimensional such as an image, or even higher, into sums of oscillating functions.

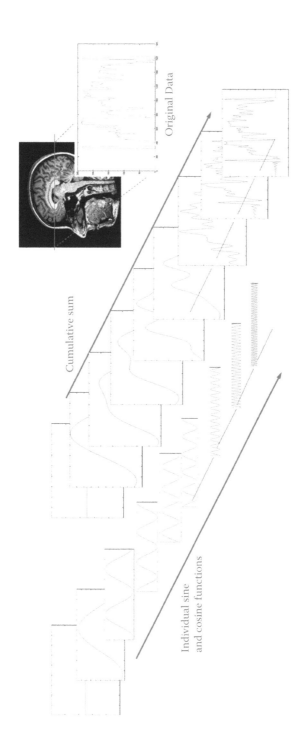

FIGURE 2.26 Example of decomposing a set of sample data into a sum of sine and cosine functions. The amplitudes of the individual sine and cosine functions are not shown to scale so that they are large enough to be visible. Also, not all frequencies are shown, for clarity.

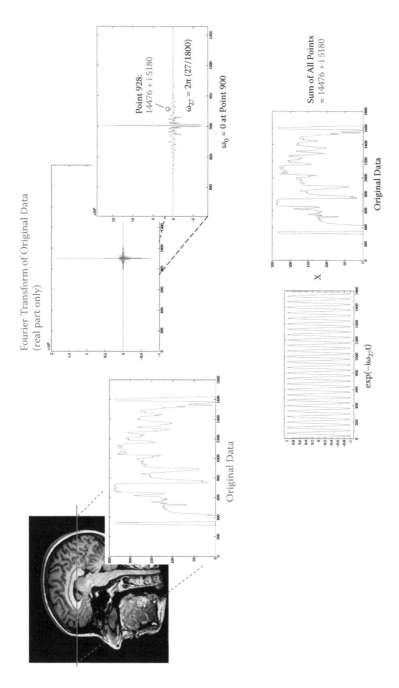

FIGURE 2.27 Example of the Fourier transform of one line of image data. The method for extracting the magnitude of a particular frequency component is also demonstrated schematically in the lower half of the figure.

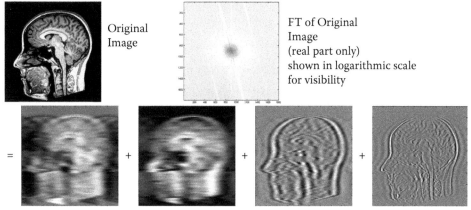

Original Image

FT of Original Image (real part only) shown in logarithmic scale for visibility

Four selected groupings of frequency components of the original image

FIGURE 2.28 Example of the two-dimensional Fourier transform of an MR image. The bottom row shows four images reconstructed from selected portions of the Fourier transform to demonstrate how the frequency components can be summed to create the original image.

Key Points

22. A special case of decomposing data into simpler patterns is to represent the data as a weighted sum of sine and cosine functions, or more specifically $e^{i\omega t}$.

$$S(t) = a_1 e^{i\omega_1 t} + \dots + a_n e^{i\omega_n t}$$

23. This function might be more easily recognizable as $e^{i\omega t} = cos(\omega t) + i\,sin(\omega t)$.

24. If functions are linearly independent, then one function cannot be represented as a sum of the others. More specifically, if two functions are orthogonal, then if we multiply them point-by-point, the resulting values sum to zero and they also form a linearly independent set.

25. A complete set of functions is one that can be used to represent any set of data.

26. The set of functions composed of all $e^{i\omega t}$ functions (that is, at all values of ω) is linearly independent and complete.

27. This decomposition of data into oscillating functions is called the Fourier transform and comes up repeatedly in MRI and fMRI.

28. The outputs of the Fourier transform are the values of $a_1 \dots a_n$ mentioned above.

29. The discrete Fourier transform can be applied to data sampled at discrete points (as opposed to a continuum of values), by treating the data as though it repeats forever outside of the measured range.

2.3.4 Useful Properties of the Fourier Transform

The Fourier transform is used so frequently in MRI that it is helpful to be familiar with its properties. Certain forms of functions also appear frequently and so are useful to know. These are (1) rectangular functions, meaning those with constant values over a range and some other baseline value otherwise; (2) *sinc* functions, which are of the form $sin(x)/x$; (3) exponentially decaying functions; and finally (4) Gaussian functions. These functions are used as examples below to illustrate the properties of the Fourier transform. These will also be used as shortcuts in later descriptions for approximating how various effects will influence the appearance of MR images and causes of image artifacts.

The key relationships between these functions are their Fourier transforms, which can be summarized as follows and are demonstrated in Figure 2.29:

1. The Fourier transform of a rectangular function is a sinc function.

2. The Fourier transform of a sinc function is a rectangular function.

3. The Fourier transform of an exponential function is a Lorentzian function. (A Lorentzian function of time, t, has the form $1/(1 + t^2/\sigma^2)$ with the value σ being a scaling factor.)

4. The Fourier transform of a Gaussian function is also a Gaussian function (has the form e^{-t^2/σ^2}).

The functions and their Fourier transforms shown in Figure 2.29 show the following properties:

1. The center point of the Fourier transform (i.e., the point at $\omega = 0$) is equal to the sum of all of the values in the input data.

2. Input data with broad distributions have Fourier transforms with narrow distributions, and vice versa.

3. Shifting the position of the data in the input function results in a change in only the phase of the Fourier transform of the input data, not the magnitude.

Another very useful property that is not obvious by looking at Figure 2.29 is known as the *Fourier convolution theorem*. This rule states that the Fourier transform of two functions when multiplied together is equal to the convolution (Figure 2.30) of the Fourier transforms of the two functions, and can be written mathematically as:

$$FT[AB] = FT[A] \otimes FT[B]$$

Here, the symbol $\otimes$ indicates convolution, and FT indicates the Fourier transform operation. Another consequence of this theorem is that the following is also true:

$$A \otimes B = FT^{-1}[FT[A]\,FT[B]]$$

In this equation, FT^{-1} indicates the inverse Fourier transform operation.

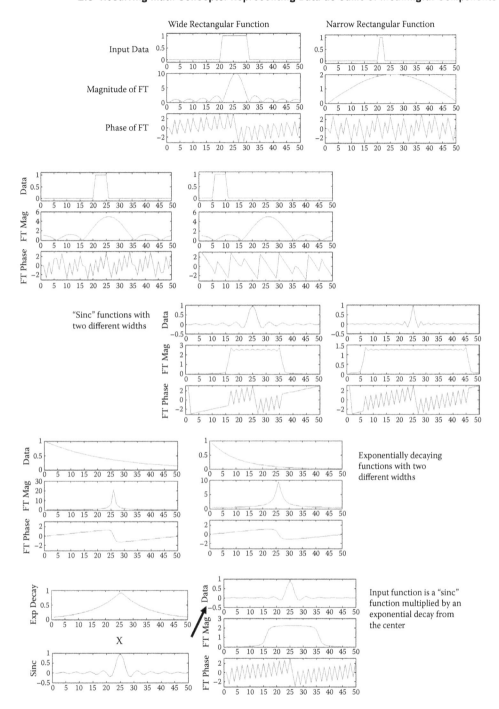

FIGURE 2.29 Examples of some key properties of the Fourier transform and transforms of functions that come up often in MRI, including rectangular functions, sinc functions, and exponentially decaying functions.

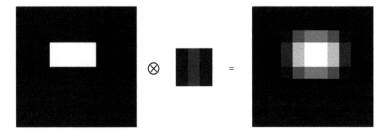

$$0 \times 1/8 + 0 \times 1/4 + 0 \times 1/8$$
$$+ 0 \times 1/8 + 1 \times 1/4 + 1 \times 1/8$$
$$+ 0 \times 1/8 + 1 \times 1/4 + 1 \times 1/8 = 3/4$$

FIGURE 2.30 An example of the convolution operation between two 2-D grids of numbers (top), or alternatively, the same example shown as images (bottom). In this example the convolution results in a spatial smoothing or blurring of the original image.

References

1. Hoult DI, Bhakar B. NMR signal reception: virtual photons and coherent spontaneous emission. *Concepts in Mag Reson* 1997;9(5):277–297.
2. Gallivan JP, Cavina-Pratesi C, Culham JC. Is that within reach? fMRI reveals that the human superior parieto-occipital cortex encodes objects reachable by the hand. *J Neurosci* 2009;29(14):4381–4391.
3. Chatfield C, Collins AJ. *Introduction to Multivariate Analysis.* London: Chapman & Hall; 1980.

Source of the MR Signal and Its Properties

The first step toward a complete understanding of how to apply and interpret functional magnetic resonance imaging (fMRI) data is to understand what the fMRI data really are. That is, where does the magnetic resonance (MR) signal come from, and how does it relate to anatomy and physiology? The concepts introduced in this chapter are essential to every MRI method, whether for imaging structure, function, or any other physiological process. These concepts include the origin of the MR signal, how it is detected, and how we use it to get physiologically relevant information based on the *relaxation times*.

3.1 Origins of the MR Signal

This section describes the origins of the signal that we use to create MR images and how we detect the signal and also get information about the chemical environment of the fluids or tissues in the body. At the very heart of all MRI methods is the source of the magnetic resonance signal—the nucleus of the hydrogen atom, ^1H—owing to its magnetic properties and its great abundance in the body (1). In biological tissues hydrogen is mostly in water (H_2O) (Figure 3.1) and lipids ($CH_3(CH_2)_nCOOH$, for example), both of which are important components of neural tissues. There are other sources of hydrogen, such as neurotransmitters and metabolites, but these signal sources are miniscule compared with those from water and lipids (estimated at ~0.03%) (2) and cannot be detected without specialized methods for MR spectroscopy. The discussions that follow will focus on the signal from hydrogen nuclei in water and lipids at body temperature, in order to lead up to applications of functional MRI in humans.

The hydrogen nucleus has two particular properties that determine all of the MRI methods to follow. The first is that it is magnetic, with a north and a south pole like any magnet, and the second is that it spins on its axis (3). These properties are actually related, and the axis the nucleus spins around is the same as the magnetic field axis, which runs between its north and south magnetic poles. For MRI, we are interested in the total net magnetic field that is produced by all of the hydrogen nuclei in a volume, since even 1 μL of water contains roughly 10^{20} hydrogen nuclei. This net magnetization is what we detect and record as the magnetic resonance signal. Given that the hydrogen nucleus is composed of a single proton, one of the main factors determining the strength of the MR signal is referred to as the *proton density*, meaning the number of hydrogen nuclei within a given volume to produce the MR signal. It is still necessary though to begin by looking at the behavior of the individual hydrogen nuclei to lead up to describing the signal we actually observe.

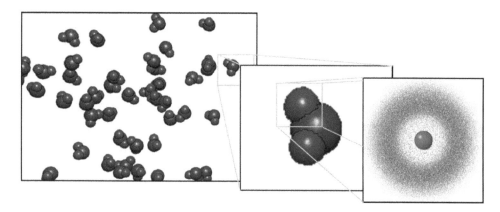

FIGURE 3.1 A representation of a small sample of water (left), an individual water molecule with the oxygen atom depicted in red and hydrogen in blue (middle), and a single hydrogen atom with the nucleus (a single proton) depicted in blue and the distribution of the single electron indicated by the gray cloud (right).

In order to describe the magnetic properties of a hydrogen nucleus, we refer to its *magnetic moment*, which is an indication of how strong a magnet it is. Every hydrogen nucleus in the universe (as far as we know) has the same magnetic moment and spins with the same speed and direction. Typically in water, or lipids, or in any form, the hydrogen nuclei are randomly oriented and their individual magnetic moments cancel each other out. However, when placed inside the strong magnetic field of an MRI system (which we will call B_0), the nuclei tend to align with the magnetic field. Now, it is important to keep in mind that this is only the hydrogen nuclei aligning; the orientation of chemical bonds and entire molecules are not affected, nor is the motion of the molecules. A good analogy for this situation is a compass needle, which tends to point north (Figure 3.2). The compass itself is not pulled north; just the needle orientation is affected.

The hydrogen nuclei are a little more complicated though, and they actually have two possible orientations, being either parallel or antiparallel to the magnetic field of the MRI system. This would be like your compass needle being able to point either north or south. What determines the direction of orientation of the hydrogen nuclei is their energy state, or the energy of the interaction between the nucleus and the magnetic field. Think of a magnet on the door of your refrigerator. When you bring the magnet very close to the door, you can feel it pull and then it sticks to the door. The force you feel, and that keeps the magnet from falling to the ground, has a certain amount of energy.

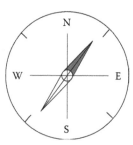

FIGURE 3.2 A typical compass as an analogy of how hydrogen nuclei align with the magnetic field of an MRI system.

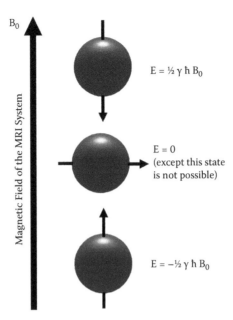

FIGURE 3.3 Energy of the interaction between a hydrogen nucleus and the magnetic field of an MRI system, with strength B_0. The values γ and $\hbar$ are physical constants: the proton gyromagnetic ratio and Planck's constant divided by 2π, respectively.

The two possible orientations of a nucleus in a magnetic field do not have the same energy though. The energy (specifically the magnetic energy—called the *Zeeman* energy) depends on the product of the magnetic moment and the magnetic field of the MRI system, as shown in Figure 3.3. Since both of these quantities each have a magnitude and a direction, the energy depends on the relative directions of the two. When the two are pointed the same direction, we refer to them as being parallel, whereas when they are pointed in exactly opposite directions, we call this antiparallel. When the magnetic moment is aligned parallel to B_0, then the energy is at its minimum; and when they are antiparallel, the energy is at its maximum. It is the energy difference between the two possible states that will become important later, as we have to deal with transitions from one state to another. The energy of the two states depends on the strength of the magnetic field of the MRI system, B_0, and on physical constants that we cannot control. As indicated in Figure 3.3, the physical constants that play a role are the proton gyromagnetic ratio, γ, and $\hbar$ which is Planck's constant divided by 2π. The proton gyromagnetic ratio is a constant for all hydrogen nuclei, and this constant will come up repeatedly in the sections to follow, so it is worth mentioning here. Its value depends on the units used, but it is commonly represented as 42.6 MHz/T, where T indicates the units tesla (for a full description of the units and their meaning, refer to the "Glossary of Terms" at the end of the book).

Key Points

1. The source of the signal used for MRI is the nucleus of the hydrogen atom, which is abundant in water and lipids in biological tissues.
2. In a strong magnetic field the hydrogen nucleus will align parallel or antiparallel to the magnetic field, because the nucleus itself is magnetic.
3. The two possible orientations of alignment have different energies, which depend on the direction of alignment and the strength of the magnetic field.

3.2 The Equilibrium State—Magnetization in Tissues

An important job now is to determine just how many of the hydrogen nuclei will be in the higher energy state and how many will be in the lower energy state. Since there is always a tendency to move toward greater disorder or randomness (second law of thermodynamics, which we won't go into here), the lower energy state is preferred. That is, there will be more hydrogen nuclei in the lower energy state than in the higher state. The difference between the numbers of nuclei in the two states depends on the energy difference between the states and on the temperature. The temperature matters because the hydrogen nuclei (and of course the water or lipids they are in) also have thermal energy, and the random thermal motion tends to push the nuclei out of alignment. When all of these forces are balanced, the hydrogen nuclei are in the *equilibrium state*. The actual difference between the two states, for water at body temperature, is only about 10 parts per million (ppm) in a 1.5 tesla MRI system, 20 ppm in a 3 tesla system, or 46 ppm at 7 tesla (according to Maxwell-Boltzmann statistics) (3). This means that more are aligned in one direction than the other, and the water in the body is magnetized, albeit weakly, when inside an MRI system. *Magnetization* refers to the net magnetic moment per unit volume and is the sum of all of the individual magnetic moments of the hydrogen nuclei. The resulting magnetization is parallel to the magnetic field of the MRI system, and we will refer to its magnitude as M_0. Even though the tissues are magnetized, there are no net forces on the tissues; and blood flow, water diffusion, and so forth, are all unaffected. Therefore, there are no physiological effects (that we know of), and a person can be in a strong magnetic field and not feel it.

Before going further, it is useful to point out that even though the nuclei are aligned either parallel or antiparallel to the magnetic field of the MRI system, this does not mean they are fixed in one orientation or the other. It is equally appropriate to think in terms of how many nuclei are in each state at some instant in time as to think of what proportion of time one nucleus will spend in each state. The nuclei can jump between states, as long as the total energy of all of the nuclei together remains constant. In addition, each individual nucleus has a component of its magnetic moment that is transverse (i.e., at a 90° angle) to B_0. That is, the axis of the nucleus does not align perfectly parallel to B_0. The transverse component of the magnetization can also point in any direction within a 360° circle around B_0; there is no preferred transverse direction. As mentioned before, even a very small volume of water, such as 1 μL, will contain almost 10^{20} hydrogen nuclei; and the total transverse component of the magnetization, in the equilibrium state, is exactly zero. In practice, for understanding the theory behind MR imaging, it is often easier to think in terms of the net magnetization of a very small volume of water, as opposed to thinking about how each individual nucleus behaves.

Key Points

4. When placed in the strong magnetic field of an MRI system, the tissues in the body become weakly magnetized.
5. Once at equilibrium, the magnetization of the tissues depends on the strength of the magnetic field, the temperature, and the number of hydrogen nuclei.
6. This equilibrium magnetization has a magnitude M_0, parallel to the magnetic field of the MRI system, and has no transverse component.

3.3 Behavior of the Magnetization When Not at Equilibrium

So far, we have worked through how the tissues become weakly magnetized when placed in the strong magnetic field of the MRI system, as a result of the magnetic properties of the hydrogen nuclei. This does not explain how we measure the MR signal, though. To do this,

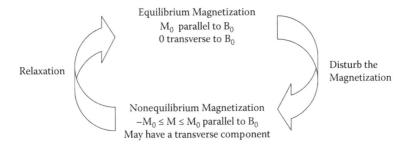

Equilibrium Magnetization
M_0 parallel to B_0
0 transverse to B_0

Relaxation

Disturb the
Magnetization

Nonequilibrium Magnetization
$-M_0 \leq M \leq M_0$ parallel to B_0
May have a transverse component

FIGURE 3.4 Schematic of the process of excitation from equilibrium and the return to equilibrium by means of relaxation.

it is necessary to understand another important concept, that is, how the magnetic moments behave when they are not parallel to the magnetic field. This might seem to contradict the last section, because the magnetization was described as having a particular direction and magnitude. But, a key point is that the tissues in the body *become* magnetized when placed inside an MRI system (Figure 3.4)—we haven't talked about how long this takes. Actually, it only takes a few seconds, but still it is not an instantaneous process. The process of getting the hydrogen nuclei into the equilibrium state is called *relaxation* and is the topic of Section 3.6.

For now, let's just imagine we have a hydrogen nucleus that is inside the magnetic field, B_0, of an MRI system. But, the nucleus is not yet in equilibrium, so it is not aligned with B_0. Even though there are forces tending to pull the two into alignment, the nucleus does not just snap into place. This is prevented by the fact that it is spinning and so it has angular momentum. Angular momentum is much like the more familiar linear momentum, such as that which makes it difficult to stop your car quickly when driving, or what you feel when standing on a bus or subway that changes speed quickly. Angular momentum, however, is the momentum carried by a rotating object. You can feel the effects of angular momentum any time you ride a bicycle, because the momentum of the wheels keeps you stable and upright. You can also see its effects when you spin a top and it does not fall over; instead it wobbles around, as shown in Figure 3.5.

In the same manner as the spinning top in Figure 3.5, the angular momentum of the hydrogen nucleus causes it to wobble or *precess* around the direction of B_0 when it is placed inside the MRI system (Figure 3.6). This is because the angular momentum of the spinning top would remain constant if there were no friction, and the angular momentum of the hydrogen nucleus does remain constant. Instead of causing the hydrogen nucleus to rotate into alignment with B_0,

FIGURE 3.5 A gyroscope with its heavy red disc at the center spinning rapidly on its axis. The sequence of pictures shows the gyroscope wobbling, or *precessing*, due to its angular momentum, instead of simply falling over under the force of gravity.

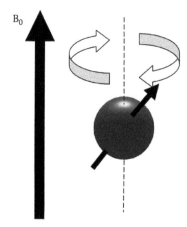

FIGURE 3.6 Precession of a hydrogen nucleus under the influence of a magnetic field, B_0.

the force causes it to precess around the direction of B_0 and its total angular momentum stays constant. For the purposes of MRI, a property that is very useful is that the speed of the precession is known and depends on the strength of the magnetic field. The frequency of rotation, ω, is equal to γB, where γ is the gyromagnetic ratio discussed above and B is the magnetic field. This is the *Larmor equation* and is probably the most useful equation to remember for understanding MRI theory. The frequency of precession is also called the *Larmor frequency*. So far we have only talked about the magnetic field being B_0, but in later chapters methods for spatial encoding will use additional magnetic fields, so it is useful to leave this equation in its general form.

The precessing magnetization can induce an electrical current in a receiver coil, as discussed in Chapter 2. This is the MR signal that we actually measure. But, there are still a few important steps to understand before the description of the MR signal is complete. In its simplest form, the signal, S, can be described (as introduced in Chapter 2) with

$$S = S_0\, e^{-i\omega t}$$

or equivalently,

$$S = S_0\, [\cos(\omega t) + i\, \sin(\omega t)]$$

where S_0 is the magnitude of the signal, ω is the frequency of rotation, and t represents time. The exponential term ($e^{-i\omega t}$) describes the rotation of the signal.

Key Points

7. When placed inside a magnetic field, a hydrogen nucleus does not snap into alignment with B_0; instead it precesses around the direction of B_0 at a fixed frequency.
8. The Larmor equation, $\omega = \gamma B$, gives the frequency of precession ω, in any magnetic field, B. The value γ is the gyromagnetic ratio, which is a constant.

3.4 Pushing the Magnetization Away from Equilibrium—The RF Pulse

The descriptions to this point have covered how the hydrogen nuclei behave at equilibrium and when not at equilibrium, but have not yet explained how the nuclei move between these two conditions—that is, the transitions indicated by the arrows in Figure 3.4.

First, the transition from equilibrium to another condition can be accomplished by applying a second magnetic field in addition to B_0. However, this second magnetic field must oscillate, or rotate, in time. One way of describing the effect of this second magnetic field is to think about how much energy is needed to have a single hydrogen nucleus transition from the lower energy state to the higher energy state. The energy difference between the two states is described above and is equal to $\gamma\hbar B_0$. The energy of a magnetic field oscillating at a frequency, ω, is equal to $\hbar\omega$. This means that an oscillating magnetic field at the frequency $\omega = \gamma B_0$ has the exact amount of energy needed to allow the transition from the low state to the high state. This equation might look familiar because it is also exactly the Larmor frequency at which the hydrogen nuclei precess when in a magnetic field of B_0.

Another way of thinking of how we can push the magnetization out of equilibrium is to look at the total effect of adding a second magnetic field to B_0 on the total magnetization from a very small volume of tissue. Starting at equilibrium, the magnetization is M_0 directed parallel to B_0, and so to simplify the following discussion we will call this the z-axis of our three-dimensional (3D) coordinate system. Initially, when we add the second magnetic field, which we will call B_1, we can see two things right away. The first is that B_1 has to be directed 90° to B_0. Since we need to turn B_1 on and off quickly, there is no practical way we can make it as strong as B_0. Most MRI systems today have magnetic field strengths between 1.5 T and 7 T, whereas the B_1 field we can turn on and off quickly is only around 1 gauss, which is equal to 0.0001 T. If we applied B_1 parallel to B_0, then the two would just sum and the result would be essentially the same as B_0. If, on the other hand, B_1 is 90° to B_0, as shown in Figure 3.7, then the sum of the two is at a very slight angle to B_0 and to M_0. Immediately after B_1 is applied, M_0 will therefore precess around the sum of $B_1 + B_0$ and will move away from alignment with B_0.

Adding a constant B_1 field to B_0 will not have much effect if we did nothing else but just let M_0 precess around the net magnetic field that is ever so slightly at an angle to B_0. The trick is to make B_1 rotate around B_0 at the Larmor frequency. A rotating B_1 field also meets the requirement of providing the exact energy needed to make the hydrogen nuclei transition from the lower energy state to the higher one. If B_1 oscillates or rotates at the frequency ω_0, it has energy equal to $\hbar\omega_0$, which is the same as the energy difference between the states that the nuclei can be in. The total magnetic field, the sum of $B_1 + B_0$, will change in time as B_1 rotates, and M_0 will precess around this moving target. As M_0 moves around the total magnetic field, the angle between M_0 and the total field continually increases. In effect, the total field keeps moving away from M_0. This effect makes it possible to rotate M_0 a large angle away from its original position. In fact, if B_1 rotates at the Larmor frequency, then we could rotate the magnetization a full 180° and invert it. It is easiest to see how M_0 will behave with B_1 applied if we use a frame of reference that also

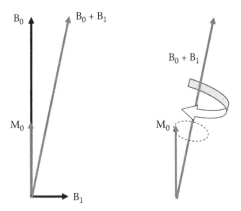

FIGURE 3.7 Initial effect of adding a second magnetic field, B_1, to magnetization already in equilibrium with a field, B_0.

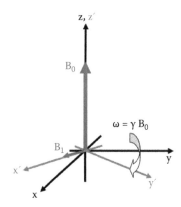

FIGURE 3.8 Rotating frame of reference for describing the effects of a B_1 field rotating at the Larmor frequency.

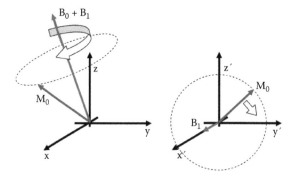

FIGURE 3.9 Precession of M_0 around the sum of B_0 and B_1 in the stationary frame (left) and in the rotating frame (right).

rotates at the Larmor frequency, as in Figure 3.8. If this rotating frame and the stationary frame have the same z-axis, but the x- and y-axes rotate around z at the Larmor frequency, then our B_1 magnetic field will look stationary in this rotating frame.

In the rotating frame of reference, M_0 does not appear to be rotating around B_0. This is similar to the situation when you are riding in a car and you look at another person in the car with you. Compared with yourself, this person does not appear to be moving. A more dramatic example is to look at the room around you right now. It probably looks stationary, even though you, it, and the rest of the earth are traveling at 107,300 km/h around the sun. Similarly, in the rotating frame we see only M_0, looking stationary and initially along the z-axis, and we see B_1, somewhere in the x′-y′ plane. For now we can just choose to have B_1 to be along the x′-axis as we continue this example. In any frame of reference the behavior has to be consistent, so we will see M_0 precess around B_1 at the frequency $\omega = \gamma B_1$ (this is the Larmor equation again) (Figure 3.9).

As a specific example, we could choose a B_1 field with a magnitude of 0.5 gauss, or 5×10^{-5} T. Even though the unit of tesla is a bit cumbersome in this case, we need it because earlier we had our gyromagnetic ratio expressed as 42.6 MHz/T. This means that M_0 will precess around B_1 at 2130 Hz, and to rotate the magnetization 90° from B_0, then we have to turn off B_1 after only 0.12 msec.

This is a good time to review what we have so far. If we turn on a magnetic field, B_1, that is directed 90° to B_0, and if B_1 rotates at the Larmor frequency, then we can rotate M_0 away from alignment with B_0. This occurs even if B_1 is tiny compared with B_0, and this is the *resonance* condition that gives rise to the name *magnetic resonance*. In magnetic fields of 1.5 tesla to 7 tesla, the

Larmor frequency is approximately 64 MHz to 300 MHz, respectively. A reference from every-day life is when you turn on a radio and choose a station at around 100 FM, you are detecting a radio signal at 100 MHz. These frequencies are in what is called the *radio-frequency* range of the electromagnetic spectrum, and since we apply B_1 only briefly, we called it a radio-frequency, or RF, pulse. We can also see from the description above that B_1 has to rotate at very close to the Larmor frequency or it will have no effect. If M_0 is able to precess all the way around the sum of $B_1 + B_0$, then it will start to move back toward where it started. With the example above, this would mean M_0 would have to rotate only one extra revolution in 0.12 msec. This could occur with a frequency difference of 8333 Hz between B_1 and the Larmor frequency, which, as mentioned above, is typically between 64 MHz and 300 MHz. This is only an estimate of how close the frequencies need to be, but it shows that if the frequencies are not quite similar, then the RF pulse will have no effect. This turns out to be a very useful property that we will use later for spatially selective RF pulses. Another useful point is that we can rotate M_0 by whatever angle we choose from alignment with B_0, called the *flip angle*, simply by applying B_1 for just the right duration. How this B_1 pulse is created in practice in an MRI system is described in Chapter 2 (Section 2.1.3).

Key Points

9. The magnetization can be pushed away from equilibrium by a brief pulse of a small magnetic field, B_1, which rotates at the Larmor frequency.
10. B_1 must be oriented 90° degrees to B_0, it will rotate the magnetization at the frequency γB_1, and its duration can be set to rotate the magnetization as needed.
11. The angle the magnetization is rotated away from alignment with B_0 is called the *flip angle*.
12. Since this magnetic field rotates at the Larmor frequency and is applied only briefly, it is called a *radio-frequency*, or RF, pulse.

3.5 Detecting the MR Signal

After the RF pulse has been applied and turned off, we now have the situation that the magnetization has been tipped some angle away from B_0, and this magnetization will now precess around B_0 at the Larmor frequency. It is worth noting that all of the hydrogen nuclei would have been tipped the same way if they all experienced the same B_1 field, and so, at least initially, they are all pointing the same direction. The total magnetization has the same magnitude that it did at equilibrium, namely M_0, but it is now at an angle to B_0. This rotating magnetization can induce an electrical signal in a receiver coil, as described in Chapter 2, and this is the MR signal that we record. The signal will oscillate at the same frequency that the magnetization rotates, as mentioned in Section 3.3, and its strength depends on the angle of the magnetization from B_0. The angle matters because the signal depends on how much the magnetization oscillates. For example, if the magnetization was tipped only 5° from B_0, then its precession around B_0 would not cause much of a changing magnetic field. It would just have a slight wobble. If the magnetization was tipped by 45° though, then the magnetization would wobble in a large cone. If it was tipped by 90°, the magnetization would lie completely in the x–y plane (i.e., the transverse plane) and it would sweep around in a flat circle and would produce the largest possible fluctuation in the magnetic field and the largest MR signal. Tipping the magnetization by 180° would again result in no signal at all, because the magnetization would be back on the z-axis, just inverted from its original position. The strength of the signal detected therefore depends on the equilibrium magnetization (which depends on the field strength and the number of hydrogen nuclei) and on the sine of the flip angle (Figure 3.10).

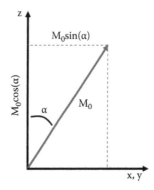

FIGURE 3.10 Transverse and longitudinal components of the magnetization after being tipped an angle α from B_0.

It is worth noting that the MR signal induced in a receiver coil is caused by the rotation of the magnetization. It is not caused by the hydrogen nuclei changing to the lower energy state and emitting energy as they relax back to equilibrium (4). The energy lost during relaxation is transformed into thermal energy, and the rate of relaxation does not determine the strength of the signal that is detected. It is the magnitude of the component of magnetization that is transverse to B_0 at any instant in time that determines the strength of the MR signal.

Key Points

13. The MR signal is the electrical signal induced in a receiver coil by the rotating magnetization.
14. The MR signal strength depends on the magnitude of M_0 and on the flip angle.

3.6 Relaxation Back to Equilibrium

When a body is first put inside an MRI system, the tissues become weakly magnetized, as described earlier. This means there has to be some process for the hydrogen nuclei to change energy and settle into the equilibrium state. The same process also has to return the magnetization to equilibrium after it has been disturbed by an RF pulse. Since the RF pulse puts energy into the hydrogen nuclei, this return to equilibrium involves the hydrogen nuclei losing energy to *relax* back to equilibrium, and so it is called *relaxation*.

An important concept to understand is that relaxation is not a rotation of the magnetization back to equilibrium. It is not the opposite of the effect of the RF pulse, with the magnetization rotating through a given angle. Instead, keep in mind that even 1 µL of water contains roughly 10^{20} hydrogen nuclei, and they all act independently, and we observe only the sum of all of their magnetic moments. It is necessary to describe the relaxation as two components, one being parallel to B_0, and the other transverse to B_0. The component parallel to B_0, called the longitudinal relaxation, will cause the magnetization to grow to its equilibrium value of M_0, and it is characterized by the time T_1. The transverse relaxation will cause the transverse magnetization to decay to its equilibrium value of zero and has the characteristic time T_2. The two components of relaxation do not occur at the same rate; the transverse (T_2) relaxation always occurs more quickly than the longitudinal (T_1) relaxation.

The most common mechanism for relaxation is through interactions between the hydrogen nuclei. As a side note, since each nucleus is a magnetic dipole (i.e., has two poles, north and south), these are called *dipole–dipole* interactions (3). When two hydrogen nuclei move close together, the total magnetic field around each one is affected slightly. The field experienced by one nucleus as it moves near another one could be higher or lower, depending on how the two are orientated. If you have ever had a chance to play with two magnets, you might have noticed that they will attract each other when aligned one way and repel each other when aligned the opposite way. This is because the north pole of a magnet will repel the north pole of another magnet but will be attracted to its south pole. All molecules move around and also rotate because of their thermal energy, and so there are a number of ways that hydrogen nuclei, such as those in water, can move around relative to each other. As a result, each nucleus experiences a total magnetic field that has a small random fluctuation in time, and this is the key to relaxation.

In the last section, we saw how a magnetic field rotating at the Larmor frequency can have a large effect on the magnetization, even if the magnetic field is quite weak (as with the RF pulse). In the same way we will see that if these small field fluctuations are at the Larmor frequency, they can cause the hydrogen nuclei to change energy. To see how this is possible, we first need to look at the different ways that two interacting hydrogen nuclei can change their magnetic energy, as shown in Figure 3.11.

The energy changes for the possible transitions shown in Figure 3.11 are 0, $\hbar\gamma B_0$, or $2\hbar\gamma B_0$. This is the exact amount of energy that must be gained or lost by this pair of hydrogen nuclei for each transition. One big question is then, where does the energy go, or where does it come from? The answer is that the magnetic energy is lost to, or provided by, the thermal energy of the system. Thermal energy is the kinetic energy of the randomly moving molecules, and the link between the thermal energy and the magnetic energy is that the random movement provides the necessary fluctuations in magnetic fields that are needed to cause the nuclei to change magnetic energy states. The energy changes of 0, $\hbar\gamma B_0$, or $2\hbar\gamma B_0$ mentioned above correspond to oscillating magnetic fields (or to be more general we could say electromagnetic fields) at frequencies of 0, γB_0, or $2\gamma B_0$. In terms of the Larmor frequencies these are oscillations of 0, ω_0, and $2\omega_0$. An example of this can be visualized by imagining two nuclei coming close together so that their

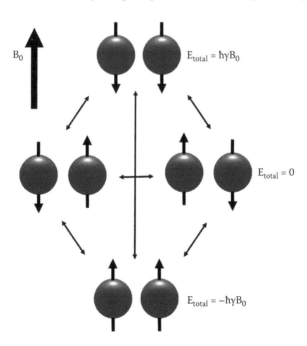

FIGURE 3.11 Possible energy levels, and transitions between them, for two interacting hydrogen nuclei in an MRI system with a magnetic field of B_0.

magnetic fields repel, resulting in one or both moving away more quickly than they were moving before, but one having flipped to the lower magnetic energy state. So, the total energy has not changed, it has just changed form from magnetic energy to thermal energy. The amount of magnetic energy is very small compared with the total thermal energy though, and the change in temperature of the hydrogen nuclei tends to be negligible.

Among the possible energy transitions shown in Figure 3.11, there are two that would not cause a net change in energy. These are shown in the middle row of Figure 3.11, with the two hydrogen nuclei simply exchanging energy. There are also two transitions that would have either both nuclei gaining energy or both losing energy, with a jump in either direction between the top and bottom rows of Figure 3.11. But, there are eight possible transitions that have only one nucleus gaining or losing energy (5). The higher number of possible interactions means that the single-level transitions are more likely to happen than the other possible transitions. Also, downward transitions are more likely to occur than upward transitions, and the nuclei will tend to lose magnetic energy. This is how they settle back to the equilibrium state.

Molecules are continually moving and interacting, and this is characterized by the temperature, which is a measure of the average kinetic energy, as mentioned above. Even though this movement is random, it is nonetheless limited by the physical properties of the liquid, tissues, and so forth. Somewhat similar to what we call *white noise*, which has equal components at all frequencies, the random movements of molecules have equal components at all frequencies, but only below a certain threshold frequency. There are physical limits to how often the molecules will interact with their surroundings to change direction of travel or rotation. This means that above a certain frequency the amplitude of the motion components drops off quickly, and there are no components of the motion at higher frequencies. The frequency where the motion contribution drops off, that is, the threshold frequency, can be used to characterize the random motion. If you are going to study this in more detail in more advanced papers or books, it is useful to know that this threshold frequency is $1/\tau_c$, where τ_c is the *correlation time* that is used to characterize the random movements. For now, we can just use a qualitative description and compare two types of molecules with different movements. Molecules with relatively unrestricted movement will have motion with contributions from higher frequency components. Because the motion is made up of a wide range of frequencies, the contribution from each frequency is relatively small. In contrast, molecules with restricted movement will have a lower threshold frequency, meaning the movement will have little to no contribution from higher frequency components, but the amplitude of motion will be larger in each of the frequency components that contribute.

For relaxation of magnetization, this means that it is possible that molecules with unrestricted motion (short correlation time) will have frequency contributions well above the Larmor frequency. Since the motion includes all frequencies below the threshold, it will also have a component at exactly the Larmor frequency. However, molecules with slower motion (more restricted) and a lower threshold frequency, although still above the Larmor frequency, would have an even larger component of motion at exactly the Larmor frequency. In this case the more restricted movement would cause relaxation to occur more quickly. Taking this example further, if the motion was even more restricted, so that the threshold frequency was below the Larmor frequency, the motion might have a much smaller or even zero component of movement at the Larmor frequency. In this case the relaxation due to transitions at the Larmor frequency might not be able to occur. Real-life examples of these three situations are intracellular water, water in blood plasma, and water in cerebrospinal fluid (CSF). In these examples the movement of the water would be affected by the organelles, structural elements, proteins, and so on inside the cell; in the case of blood plasma would be affected to a lesser degree by the plasma proteins; or finally in the CSF water would have much lower concentrations of macromolecules to interact with. The water in each of these three situations has different rates of relaxation, and we will look at these in more detail below.

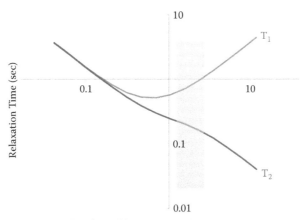

FIGURE 3.12 Comparisons of T_1- and T_2-values at a given Larmor frequency (ω) for hydrogen nuclei with different rates of movement, as characterized by the correlation time, τ_c. Movement with more frequent interactions is represented by shorter correlation times, toward the left of the horizontal axis, and movement with less frequent interactions has longer correlation times and Is represented further to the right on the horizontal axis. The yellow band indicates the typical ranges of movement encountered in biological tissues.

At this point it is necessary to understand how the different frequency components of motion can affect the relaxation times. When a hydrogen nucleus flips its orientation and changes its energy state, both its transverse and longitudinal magnetic components are flipped. This might seem obvious, but it is worth noticing. The transition in the middle row of Figure 3.11, with the two nuclei simply exchanging energy, has a total change in magnetic energy of zero. This transition does not result in a change in the total longitudinal magnetization, but there is a change in the transverse magnetization. Why? Because the longitudinal components are only up or down, and the two nuclei simply trade orientations; but the transverse magnetization components can be distributed in all directions in the transverse plane and are unlikely to have been in exactly opposite directions before the two nuclei flipped. Only the interactions at frequencies of ω_0 and $2\omega_0$ contribute to the longitudinal relaxation, but interactions at 0, ω_0, and $2\omega_0$ contribute to transverse relaxation. As a result, the transverse relaxation is always faster than longitudinal relaxation, as shown in Figure 3.12. Another consequence is that the longitudinal relaxation is most efficient, or in other words, occurs most quickly, when the threshold frequency of the movement (i.e., $1/\tau_c$) is similar to the Larmor frequency (the actual minimum is when $\omega_0\tau_c$ equals approximately 0.62). If the movement of the hydrogen nuclei is either more restricted or less restricted, the longitudinal relaxation tends to happen more slowly. The trade-off is between (1) having lower amplitude contributions at higher frequencies with less restricted motion, and (2) having higher amplitude motion contributions at lower frequencies with more restricted motion. Similarly, longitudinal relaxation tends to occur more slowly in biological tissues at higher magnetic fields simply because the Larmor frequency is higher. Transverse relaxation behaves differently, and tends to be faster for hydrogen nuclei with more restricted movement. It also tends to occur a little more slowly at higher fields, but in practice the transverse relaxation rates can be considered to be fairly consistent across the range of magnetic fields from 1.5 T to 7 T.

Approximate expressions for T_1 and T_2 in water were first developed by Bloembergen et al. (5) (and a brief account of the work appeared in *Nature* 1947;160:475). The expression for T_1 is therefore known as the BPP expression. Another derivation of equations describing T_1 and T_2 is given by Solomon (6). These are good resources for anyone wanting to have a deeper understanding of the underlying relaxation theory.

Another important factor that can influence how two hydrogen nuclei will interact is the average distance between them, and this is demonstrated in the practical examples that follow. If the nuclei tend to spend more time in close proximity, they will have more frequent and/or stronger interactions. This will facilitate the exchange of energy. In biological tissues this is a significant effect; generally, more mobile water environments tends to have longer T_1- and T_2-values, and more restricted water environments tend to have shorter values. Similarly, lipids tend to have shorter T_1- and T_2-values.

Key Points

15. *Relaxation* refers to the return of the hydrogen nuclei to the equilibrium state.
16. There are two separate components of relaxation: *longitudinal*—the recovery of the magnetization parallel to B_0, characterized by the time T_1; and *transverse*—the decay of the transverse magnetization to zero, characterized by the time T_2.
17. Relaxation occurs as a result of magnetic interactions between hydrogen nuclei, enabling them to lose energy.
18. Random thermal motion causes fluctuations in the magnetic field experienced by each nucleus.
19. The random motion can have the specific frequency components that permit the exchange of energy between nuclei and from magnetic energy to thermal energy.

Actual T_1- and T_2-values measured in 3 T and 1.5 T MRI systems are listed in Table 3.1. The trends in these values reflect the influence of the mobility of the hydrogen nuclei on the relaxation times, as well as the influence of being bound into lipids. The key difference between white matter and gray matter is the myelin sheaths surrounding axons in white matter, and these myelin layers are composed largely of lipids. For imaging the brain and spinal cord, the differences in relaxation properties between white matter and gray matter are extremely useful, both for anatomical imaging, and for detecting de-myelinating diseases such as multiple sclerosis.

Table 3.1 Relaxation Times Measured at 1.5 T and 3 T for Various Tissues				
	T_1-Values (msec) 1.5 T	T_2-Values (msec) 1.5 T	T_1-Values (msec) 3 T	T_2-Values (msec) 3 T
Adipose tissue	259	84	282	84
White matter				
Frontal lobe	556 ± 20	79 ± 2	699 ± 38	69 ± 2
Occipital lobe	616 ± 32	92 ± 4	758 ± 49	81 ± 3
Gray matter				
Frontal lobe	1048 ± 61	99 ± 4	1209 ± 109	88 ± 3
Occipital lobe	989 ± 44	90 ± 4	1122 ± 117	79 ± 5
Cerebrospinal fluid	4300	1442	4300	1442

Sources: Bottomley, P.A. et al., *Med Phys* 11, 4, 425–448, 1984 (8); Donahue, M.J. et al., *Magn Reson Med* 56, 6, 1261–1273, 2006 (9); Lu, H. et al., *J Magn Reson Imaging* 22, 1, 13–22, 2005 (7).

The idea of having more than one relaxation environment, such as in the myelin sheaths and in the intracellular and extracellular water, is extremely important for understanding relaxation in biological tissues (10–12). In fact, all tissues have a considerable number of different relaxation environments. Water will adsorb to the surfaces of many molecules in solution, forming a hydration layer around each. The mobility of the water in a hydration layer is determined by the molecule that the hydration layer is formed around, thereby creating a different relaxation environment for the water. However, water molecules exchange rapidly between the hydration layers and the surrounding "free" water. Hydrogen nuclei will also exchange rapidly between water molecules. Over a brief interval of time, a hydrogen nucleus can therefore experience many different relaxation environments. This is known as the *rapid-exchange* case. The net relaxation rate we observe in this situation is a weighted average of the relaxation rates in each of the environments. The weighting for this average is the proportion of hydrogen nuclei in each environment, or equivalently, the proportion of time a hydrogen nucleus spends in each environment. For example, the transverse relaxation time, T_2, of the free water is estimated to be about 350 msec, and the T_2 of the water in a hydration sphere around macromolecules might be only 1 msec. At any given time, perhaps only 1% of the water is in the hydration layers around the macromolecules, and the remainder is in the free water. This of course depends on the concentration of the macromolecules. The resulting net transverse relaxation rate in this case is given by:

$$\frac{1}{T_2} = \frac{0.99}{0.350 \text{ sec}} + \frac{0.01}{0.001 \text{ sec}} = 12.83 \text{ sec}^{-1}$$

$$T_2 = 78 \text{ msec}$$

This example demonstrates the dramatic effect that a relatively small proportion of water in hydration layers can have on the relaxation times in tissues, and how increasing the concentration of macromolecules would decrease the value of T_2. The relaxation rate, $1/T_2$, also called R_2, has been shown to be linearly proportional to the concentration of macromolecules in solution (12).

The alternative to the rapid-exchange case is the *slow-exchange* case. This occurs when physical boundaries prevent water molecules or individual hydrogen atoms from moving freely between different relaxation environments. For example, the MR signal from neural tissues can arise from water in intracellular and extracellular spaces, blood plasma, and CSF. Within each of these spaces there are many different relaxation environments as well, and hydrogen nuclei can exchange rapidly between them. However, hydrogen nuclei cannot move rapidly, relative to the time scale of relaxation, across cellular membranes or across membranes into or out of the CSF or blood. The total MR signal we detect is a mix of the signals from these various environments, and each environment can have unique transverse and longitudinal relaxation times. In practice, the MR signal from neural tissues in the brain is typically composed of signals from three environments, which are distinguishable because of their different relaxation properties. These have been identified as CSF, intracellular and extracellular water, and water trapped in myelin sheaths (13). Within each of these three groups there are also different environments, but their relaxation times are too similar to be identified separately. In gray matter and white matter in the brain, the two main components are attributed to myelin water with a T_2 of around 15 msec, and the other is attributed to intracellular and extracellular water with a T_2 of approximately 80 msec. The difference is that gray matter has roughly 3% of the signal in the faster relaxing component ($T_2 \approx 15$ msec), whereas white matter has about 11% of its signal in this component (13,14). It is also worth noting, for comparisons of the signals, that the gray matter and white matter differ slightly in their proton densities, with 0.832 g of water per mL of tissue in gray matter, compared with 0.708 g/mL in white matter.

20. Both longitudinal and transverse relaxation times, T_1 and T_2, generally tend to be longer for more mobile water and shorter for water with more restricted mobility and for lipids.
21. In biological tissues there are many different relaxation environments, and we see the net effect of these environments on the MR signal.
22. In the *rapid-exchange* case, each hydrogen nucleus spends time in more than one relaxation environment. The MR signal has transverse and longitudinal relaxation rates that are, respectively, weighted averages of the transverse and longitudinal relaxation rates of each environment.
23. In the *slow-exchange* case, we observe the sum of MR signals from different environments, and we can observe components of the MR signal with different transverse and longitudinal relaxation rates.

3.7 Observing the Effects of Relaxation

Relaxation times are extremely useful because they can demonstrate differences in environments between tissues or changes between healthy and damaged or diseased tissues. As you may have noticed in the previous section, relaxation times are determined almost entirely by the mobility of the hydrogen nuclei and the materials they interact with. The only other factor is the magnetic field strength of the MRI system. This means that, for example, normal healthy gray matter in the motor cortex of the brain will have the same transverse and longitudinal relaxation times in any 3 tesla MRI system in the world. The difference in the relaxation times if instead we used a 7 tesla MRI system is also very predictable. If something was to cause local edema (15), for example, the accumulation of fluid would alter the chemical environment of the water in the tissues. From the explanations in the previous section of how relaxation occurs, we can predict that this would likely cause T_1- and T_2-values to increase because there would be more free water. Similarly, we could predict how relaxation times would be expected to change for almost any damage or disease process if we know how the tissue environment is altered. Conversely, from the changes in relaxation times we can interpret the likely changes in the tissues that have occurred, and the relaxation times therefore have diagnostic value (16–21).

We can observe the effects of the relaxation times in two different ways, corresponding to the two different modes of relaxation: transverse and longitudinal. If initially the magnetization is at equilibrium and we apply an RF pulse, then the magnetization will be tipped some angle away from B_0 toward the transverse plane. Now the magnetization will precess around B_0, and as described above, we could detect the transverse component of the magnetization. But, if we were to wait a bit of time before detecting the MR signal, then the transverse component will have decayed some amount due to transverse relaxation, and we will detect a weaker signal. If we wait too long, then the transverse magnetization will have decayed completely to zero, and we will not detect a signal, as illustrated in Figure 3.13.

The relaxation processes result in exponential return of the magnetization to its equilibrium values, and so the signal that would be measured at some time, t, after the excitation pulse is given by:

$$S(t) = S_0\, e^{-t/T_2}$$

The value of S_0 is related to the equilibrium magnetization, and so it is proportional to the proton density. S_0 is the signal magnitude that would be measured if there was no relaxation; or we could measure the signal instantaneously after the RF pulse, and this equation describes how

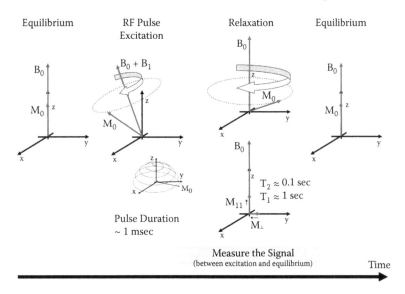

FIGURE 3.13 The timeline of disturbing, or *exciting*, the magnetization from its equilibrium position with an RF pulse, and then the return to equilibrium through the processes of relaxation. After the RF pulse, the resulting transverse component of the magnetization can be detected as it precesses around B_0. The rate of transverse relaxation, and the time allowed between applying the RF pulse and measuring the MR signal, determines how much the signal will have decayed due to transverse relaxation. The longitudinal relaxation rate determines how long it takes for the magnetization to recover to equilibrium, parallel to B_0.

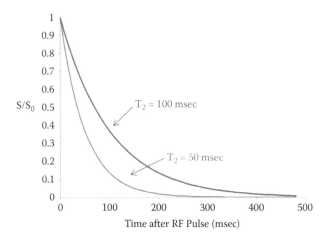

FIGURE 3.14 Examples of relaxation of transverse magnetization at two different values of T_2.

quickly the magnetization will decay, relative to T_2. For example, if $T_2 = 90$ msec, and we wait 90 msec after the RF pulse to measure the signal, then the signal will have decayed to $0.37\ S_0$. However, if we wait only 45 msec, then we have $0.61\ S_0$. In either case, the signal we measure would be weighted depending on the value of T_2 (Figure 3.14).

Another important factor is the recovery of the magnetization parallel to B_0. In practice we will need to apply RF pulses many times in order to measure the MR signal with different spatial encoding applied, as will be discussed in Chapter 5. Or, we may want to acquire images

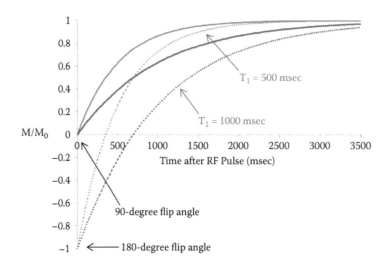

FIGURE 3.15 Examples of the recovery of longitudinal magnetization for two values of T_1 and two different flip angles.

repeatedly to monitor changes in time, such as for fMRI. After the RF pulse we could wait long enough for the magnetization to relax completely back to equilibrium. We could then apply an RF pulse again and measure the signal, and the signal strength would have no dependence at all on T_1. However, if we wait a shorter time after the RF pulse, then the magnetization may not have time to relax fully, depending on whether T_1 is long or short (Figure 3.15). The longitudinal magnetization at this point in time will depend on the value of T_1. If we were to apply another RF pulse at this instant, then the magnetization that is tipped toward the transverse plane would also depend on the value of T_1, and so would the signal we detect. As a result, we can make the signal strength depend on T_1. Another factor that plays a role is how far we tip the magnetization away from B_0 in the first place. If we tip it only a small angle, then the longitudinal magnetization does not have far to recover. Taking both of these factors into consideration, the longitudinal magnetization at a time, t, after an RF pulse that has tipped the magnetization an angle α from B_0 is given by:

$$M(t) = M_0[1-(1-\cos a)e^{-t/T_1}]$$

Looking at this equation we can see that if the time, t, is much longer than T_1, then $M(t) = M_0$, regardless of the flip angle. At a very short time after the RF pulse (i.e., t is very small), then $M(t) = 0$ for $\alpha = 90°$ and $M(t) = -M_0$ for $\alpha = 180°$.

We can choose the time between repeated RF pulses (called the *repetition time*, TR) and the flip angle to select the sensitivity of the MR signal to differences in T_1 between different tissues or fluids. We can also choose the time interval between the RF pulse and when we record the MR signal to select the sensitivity to differences in T_2 between different tissues or fluids. It is important to remember the distinction that with transverse relaxation the decay of the magnetization relates directly to the decay of the MR signal. But, with longitudinal relaxation, we are talking about the recovery of magnetization parallel to B_0, and we would have to apply another RF pulse to tip this magnetization toward the transverse plane to see the effects of the longitudinal relaxation.

One final factor to take into consideration is that the magnetic field inside any material tends to be different than outside the material. This effect can be very large for magnetic materials, but there are subtle but very important differences between air, tissues, bone, and blood. The relative change in the magnetic field between inside and outside a material is termed its *magnetic susceptibility* and in tissues is typically only around –9 ppm, meaning that the field inside is slightly

Table 3.2 Selected Values (in SI Units) of Volume Magnetic Susceptibilities of Various Fluids and Tissues	
Material	Volume Magnetic Susceptibility
Free space	0.00 ppm
Air	0.40 ppm
Water	−9.04 ppm
Fat	−7.79 ppm
Bone	−8.44 ppm
Blood	−8.47 ppm
Gray matter	−8.97 ppm
White matter	−8.80 ppm
Blood 98% O_2 saturation	−9.12 ppm
Blood 75% O_2 saturation	−8.79 ppm

Sources: Collins, C.M. et al., *Magn Reson Imaging* 20, 5, 413–424, 2002 (22); Spees, W.M. et al., *Magn Reson Med* 45, 4, 533–542, 2001 (23).

lower than outside. For the purposes of MR imaging of the brain or spinal cord, there may be a number of different materials within the region being imaged, with the magnetic susceptibilities listed in Table 3.2.

At the boundaries between materials the magnetic fields do not change abruptly but instead have smooth transitions. Observations from MR images and from numerical simulations (22) show that the shape of the head and the air spaces in the sinuses and ear canals produce magnetic field variations across the brain and brainstem. Similarly, imaging of the spinal cord and brainstem are affected by the bone in the spine and the proximity to the lungs.

The effect of magnetic susceptibility differences between the various tissues produces a magnetic field that is typically not perfectly uniform, even across a relatively small volume. We can think of the magnetic field as being $B_0 + \Delta B$, where the ΔB refers to the relatively tiny variation in the magnetic field across different positions. The Larmor frequency then also has tiny variations (again, $\omega = \gamma B$), and the hydrogen nuclei do not precess at exactly the same speed. Frequency differences between hydrogen nuclei in different materials and between different positions within molecules are called *chemical shift* effects and can be used to obtain chemical information about materials by means of spectroscopy or *chemical shift imaging*. The term *chemical shift* is often used to refer to frequency differences within materials, but does include the bulk magnetic susceptibility, such as between water and lipids. The influence these effects have on the appearance of an image is discussed in Chapter 5 (Section 5.6). Returning to magnetic susceptibility differences in particular, imagine a small cubic volume of tissue, just 1 or 2 mm across each side, which is near an air/tissue interface. The volume might have a magnetic field variation of 0.1 ppm across it, from one side to the other. This means that in a 3 tesla MRI system, having a Larmor frequency of 127.7 MHz, the frequency would vary by 12.77 Hz across the volume. In the time span of only 78 msec (in this specific example) the hydrogen nuclei on one side of the volume will have completed one more full rotation (precession) around B_0 than the hydrogen nuclei on the other side. The magnetic field is continuous and so all of the hydrogen nuclei across the volume will be spread out to point equally in all directions in the transverse plane, resulting in a net total transverse magnetization of zero. As this spreading out occurs over time we would therefore see the total transverse magnetization decaying more quickly than under the influence of transverse relaxation alone. It is necessary to characterize the effective transverse relaxation time, T_2^*. This reflects both the true transverse relaxation, characterized by T_2, and the effects of non-uniform magnetic fields:

$$\frac{1}{T_2^*} = \frac{1}{T_2} + \gamma \ B$$

The influence of inhomogeneous magnetic fields on the MR signal is of particular interest for fMRI, because the magnetic susceptibility of blood depends on its oxygen saturation, as indicated in Table 3.2. As will be discussed in Chapter 6, this is the basis of blood oxygenation–level dependent (BOLD) fMRI (24,25).

Key Points

24. Magnetization in tissues can be detected only after it has been tipped toward the transverse plane by an RF pulse.
25. After the magnetization has been tipped from equilibrium, the transverse component decays toward zero, and the longitudinal component grows to its equilibrium value.
26. At the same time, the magnetization precesses around B_0 and can be detected.
27. If we wait a time between the RF pulse and measuring the MR signal, the signal strength will depend on the decay rate, T_2.
28. If the signal is not allowed to recover fully to its equilibrium value parallel to B_0 before the next RF pulse is applied, then the signal after the next RF pulse will depend on T_1.
29. Magnetic susceptibility differences between tissues create spatial distortions in B_0, resulting in different rates of precession. This effect adds to the rate of transverse decay and has the characteristic decay time, T_2^*, which is shorter than T_2.
30. *Chemical shift* refers to frequency differences between nuclei due to adjacent nuclei within the molecular structure and can therefore depend on the position within a molecule.

References

1. Lauterbur PC. Image formation by induced local interactions—Examples employing nuclear magnetic-resonance. *Nature* 1973;242(5394):190–191.
2. Bonneville F, Moriarty DM, Li BSY, Babb JS, Grossman RI, Gonen O. Whole-brain N-acetylaspartate concentration: Correlation with T2-weighted lesion volume and expanded disability status scale score in cases of relapsing-remitting multiple sclerosis. *American J Neurorad* 2002;23(3):371–375.
3. Abragam A. *Principles of Nuclear Magnetism*. New York: Oxford University Press; 1961.
4. Hoult DI, Bhakar B. NMR signal reception: Virtual photons and coherent spontaneous emission. *Concepts in Magn Reson* 1997;9(5):277–297.
5. Bloombergen N, Purcell EM, Pound RV. Relaxation effects in nuclear magnetic resonance absorption. *Phys Rev* 1948;73:679–712.
6. Solomon I. Relaxation processes in a system of 2 spins. *Phys Rev* 1955;99(2):559–565.
7. Lu H, Nagae-Poetscher LM, Golay X, Lin D, Pomper M, van Zijl PC. Routine clinical brain MRI sequences for use at 3.0 tesla. *J Magn Reson Imaging* 2005;22(1):13–22.
8. Bottomley PA, Foster TH, Argersinger RE, Pfeifer LM. A review of normal tissue hydrogen NMR relaxation times and relaxation mechanisms from 1–100 MHz: Dependence on tissue type, NMR frequency, temperature, species, excision, and age. *Med Phys* 1984;11(4):425–448.

9. Donahue MJ, Lu H, Jones CK, Edden RA, Pekar JJ, van Zijl PC. Theoretical and experimental investigation of the VASO contrast mechanism. *Magn Reson Med* 2006;56(6):1261–1273.

10. Fullerton GD, Potter JL, Dornbluth NC. NMR relaxation of protons in tissues and other macromolecular water solutions. *Magn Reson Imaging* 1982;1(4):209–226.

11. Menon RS, Rusinko MS, Allen PS. Proton relaxation studies of water compartmentalization in a model neurological system. *Magn Reson Med* 1992;28(2):264–274.

12. Menon RS, Allen PS. Solvent proton relaxation of aqueous solutions of the serum proteins alpha 2-macroglobulin, fibrinogen, and albumin. *Biophys J* 1990;57(3):389–396.

13. Whittall KP, MacKay AL, Graeb DA, Nugent RA, Li DK, Paty DW. *In vivo* measurement of T2 distributions and water contents in normal human brain. *Magn Reson Med* 1997;37(1):34–43.

14. Jones CK, Xiang QS, Whittall KP, MacKay AL. Linear combination of multiecho data: short T2 component selection. *Magn Reson Med* 2004;51(3):495–502.

15. Kamman RL, Go KG, Berendsen HJ. Proton-nuclear magnetic resonance relaxation times in brain edema. *Adv Neurol* 1990;52:401–405.

16. Estilaei M, MacKay A, Whittall K, Mayo J. *In vitro* measurements of water content and T2 relaxation times in lung using a clinical MRI scanner. *J Magn Reson Imaging* 1999;9(5):699–703.

17. Bottomley PA. NMR in medicine. *Comput Radiol* 1984;8(2):57–77.

18. Hendee WR, Morgan CJ. Magnetic resonance imaging. Part II—Clinical applications. *West J Med* 1984;141(5):638–648.

19. Crooks LE, Hylton NM, Ortendahl DA, Posin JP, Kaufman L. The value of relaxation times and density measurements in clinical MRI. *Invest Radiol* 1987;22(2):158–169.

20. Poon CS, Henkelman RM. Practical T2 quantitation for clinical applications. *J Magn Reson Imaging* 1992;2(5):541–553.

21. Wansapura JP, Holland SK, Dunn RS, Ball WS, Jr. NMR relaxation times in the human brain at 3.0 tesla. *J Magn Reson Imaging* 1999;9(4):531–538.

22. Collins CM, Yang B, Yang QX, Smith MB. Numerical calculations of the static magnetic field in three-dimensional multi-tissue models of the human head. *Magn Reson Imaging* 2002;20(5):413–424.

23. Spees WM, Yablonskiy DA, Oswood MC, Ackerman JJ. Water proton MR properties of human blood at 1.5 tesla: Magnetic susceptibility, T(1), T(2), T*(2), and non-Lorentzian signal behavior. *Magn Reson Med* 2001;45(4):533–542.

24. Menon RS, Ogawa S, Kim SG, Ellermann JM, Merkle H, Tank DW, Ugurbil K. Functional brain mapping using magnetic resonance imaging. Signal changes accompanying visual stimulation. *Invest Radiol* 1992;27 Suppl 2:S47–S53.

25. Ogawa S, Tank DW, Menon R, Ellermann JM, Kim SG, Merkle H, Ugurbil K. Intrinsic signal changes accompanying sensory stimulation: Functional brain mapping with magnetic resonance imaging. *Proc Natl Acad Sci USA* 1992;89(13):5951–5955.

The Fundamental Building Blocks of MRI Methods
Spin Echoes and Gradient Echoes

To use the magnetic resonance (MR) signal that was described in Chapter 3, we need it to last long enough to be measured, and we need to be able to detect differences in relaxation times between different tissues or between different points in time. This is accomplished by creating an *echo* of the signal, by bringing the signal back briefly. There are two different techniques for creating an echo, as will be described in this chapter, and these are at the heart of every MR imaging method.

4.1 The Need for Echoes

In spite of the apparently wide variety of MR imaging methods and the abundance of jargon and acronyms used to name different methods, there are actually only two fundamental imaging methods: the *spin echo* (SE) and the *gradient echo* (GE). With very few exceptions, every MR imaging method is a simple variation of one of these two fundamental methods. Because it comes up often in relation to functional magnetic resonance imaging (fMRI), it is worth mentioning here that *echo-planar imaging* (EPI) is not a distinct imaging method but is one of the ways that spatial encoding is imposed onto the MR signal. EPI can be used with a spin echo or with a gradient echo and is discussed in detail in Chapter 5. The distinction between the imaging method and spatial encoding is important because the imaging method (GE or SE) determines the overall image appearance and determines how well different anatomical features, or neural functions, are contrasted from their surroundings. The spatial encoding method, on the other hand, affects the acquisition time and image quality, but not the contrast between features in the image.

In magnetic resonance, the term *echo* refers to the return of the MR signal after it has decayed. As the name implies, the idea is similar to echoes produced by sounds reflecting off walls or other large surfaces and that are probably more familiar. The MR signal decays quickly after a radio-frequency (RF) pulse, as described in Chapter 3, because all of the hydrogen nuclei in an object are not in exactly the same magnetic field (due to field imperfections, magnetic susceptibility differences, interactions between hydrogen nuclei, etc., which contribute to T_2^* as described in Chapter 3). As a result, the nuclei precess at different speeds, and the directions of the transverse magnetization at different points in the object spread out, or the magnetization *dephases*, in the transverse plane. An echo can be produced by reversing this dephasing.

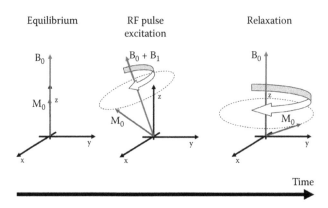

Equilibrium   RF pulse   Relaxation
excitation

FIGURE 4.1 Schematic of the initial steps needed to produce the MR signal by tipping the magnetization away from alignment with B_0.

The spin echo was first described by Erwin L. Hahn in 1950 (1,2), long before MR imaging was invented. The spin echo has been described as "a phenomenon of monumental significance" (3) because of its impact on a range of fields, in addition to MRI. The use of gradient-echoes did not appear until the development of MR imaging, more than 20 years later (4,5). The great advantage of producing echoes is that they enable measurement of much stronger signals and allow more time for the measurements than would be possible if we could only apply an RF pulse and then measure the signal before the transverse magnetization decayed to zero.

As described in Chapter 3 (and repeated in Figure 4.1), the magnetization in the object to be imaged (such as a head, arm, leg, etc.) must first be tipped away from alignment with B_0 by means of an RF pulse. Immediately after the RF pulse, the magnetization at all points is oriented in the same direction (i.e., it is *in phase*). This produces the greatest MR signal, and as the magnetization from different locations gets out of phase over time, the total net signal is reduced. In practice, steps must be taken to ensure that the magnetization is in phase as much as possible immediately after the RF pulse, because the B_1 field of the RF pulse is not always perfectly uniform, and magnetic field gradients and other field distortions cause the Larmor frequency to vary with position. Nonetheless, it is possible to compensate for the bulk of these effects, and for the purposes of this discussion we can assume that the magnetization is perfectly in phase immediately after the RF pulse.

It is helpful to clarify two points here to help with the descriptions that follow. First, we are concerned primarily with the transverse component of the magnetization, because this is the part of the magnetization that we can detect. To reiterate from Chapter 3, this is the component of the magnetization that is 90° to B_0. Second, it is easiest to think of the total net magnetization from tiny localized volumes (some texts will refer to these as *isochromats*). Each of these small volumes is small enough that the magnetic fields are perfectly uniform everywhere in the volume, and all of the hydrogen nuclei have exactly the same Larmor frequency. Even if these volumes are only 10 μm cubes, they may still contain roughly 10^{14} hydrogen nuclei, and we can consider the total net magnetization of the volume. To describe the total behavior of the magnetization within tissues, we can compare the Larmor frequencies and properties of small volumes at different locations.

Now, back to what happens to the magnetization immediately after the RF pulse. In a magnetic field, B_0, we know from the Larmor equation that the magnetization will precess at the frequency $\omega_0 = \gamma B_0$. But, imperfections in the static magnetic field, magnetic susceptibility differences between tissues, bone, air, and so forth, and also interactions between individual hydrogen nuclei all cause subtle variations in the frequency (this was discussed in detail in Chapter 3). As

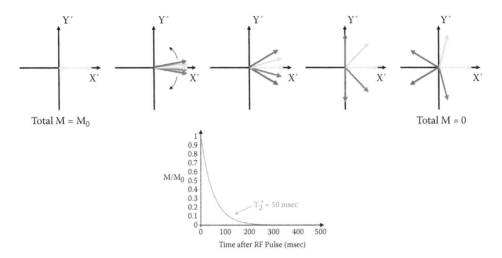

FIGURE 4.2 Schematic demonstration of the decay of the total transverse magnetization after an RF pulse. Each colored arrow represents the direction of the magnetization in a small volume, in the rotating frame of reference (rotating at the Larmor frequency), and shows how the magnetization spreads out in phase (*dephases*) over time. This results in the decay of the total net magnetization, and therefore the MR signal, with the characteristic decay time, T_2^*. In reality, the magnetization would be spread out in a continuum and not at a few discrete values as shown.

a result, the total magnetization summed across a large number of small volumes decays exponentially with the characteristic time T_2^*, as shown schematically in Figure 4.2. This is called *free induction decay*, and the curve shown in Figure 4.2 is therefore called the F. I. D. (pronounced by reading each letter, as opposed to reading it as a single word). Recall that the decay time, T_2^*, depends on the transverse relaxation time, T_2, and also the magnetic field variation across the entire volume of fluid or tissue that gives rise to the MR signal (6). The value of T_2^* can therefore depend on the proximity to bone/tissue or air/tissue interfaces because of magnetic susceptibility differences between the materials and can be changed by applying a magnetic field gradient.

$$\frac{1}{T_2^*} = \frac{1}{T_2} + \gamma \; B$$

The transverse relaxation characterized by the time T_2 occurs because of random thermal motion and interactions between individual hydrogen nuclei or with magnetic materials. The added effect of the magnetic field variations, on the other hand, is constant (unless we change the magnetic field). This difference will be shown in the following sections to be very important to the differences between a spin echo and a gradient echo.

The importance of creating signal echoes is demonstrated by the value of T_2^* in the brain, around 50–60 msec at 1.5 tesla, 30 msec at 3 tesla, and 15 msec at 7 tesla. If we also apply a magnetic field gradient, such as to encode spatial information into the MR signal, then the value of T_2^* is reduced significantly (depending on the gradient strength). Immediately after an RF pulse, we may have only a few milliseconds to measure the MR signal before it decays to zero. For example, if $T_2 = 90$ msec and $T_2^* = 30$ msec, then $\gamma\Delta B = 22.2$ sec^{-1}. With the addition of a typical gradient of 0.5 gauss/cm (equal to 2130 Hz/cm or 13380 sec^{-1} cm^{-1}), in the span of 1 mm then $\gamma\Delta B = 1338$ sec^{-1}, and the new effective value of T_2^* is 0.74 msec.

4.2 Spin Echo

A spin echo is formed by applying a second RF pulse some time, τ, after the initial excitation RF pulse, as shown in Figure 4.3. It is important to note that the value of τ is typically much shorter than the value of T_1. Each RF pulse is applied over a very short duration (typically a few milliseconds) and is essentially instantaneous compared with the relaxation of the magnetization. The second pulse is designed to rotate the magnetization 180° around any axis in the transverse plane. For example, if the initial excitation pulse rotates the magnetization 90° around the x-axis, it is designated as a 90_X pulse, and its effect is to rotate the magnetization onto the –y-axis. (The rotation is counterclockwise when looking along the direction of the B_1 field of the RF pulse.) The 180° pulse that follows can be a 180_X or 180_Y pulse (i.e., rotation around x or y) with equal effect. Immediately after the 180° pulse, both the transverse and longitudinal magnetization components at every point in the object being imaged are inverted compared with the axis they were rotated around. This process also inverts the phase distribution of the transverse magnetization relative to what it was before the 180° RF pulse. In order to avoid producing a large longitudinal magnetization component, the initial RF excitation pulse is typically set to rotate the magnetization 90°, so that the magnetization is entirely in the transverse plane both before and after the 180° RF pulse.

Now, the same magnetic field distribution that made the magnetization spread out in phase after the 90° RF pulse causes it to come back into phase after the 180° RF pulse, as shown in Figures 4.4 and 4.5. The field distribution could include the effects of magnetic field gradients we have applied, in addition to the inherent spatial magnetic field distribution that exists in the absence of any applied gradients. If we think of the magnetization in three small volumes,

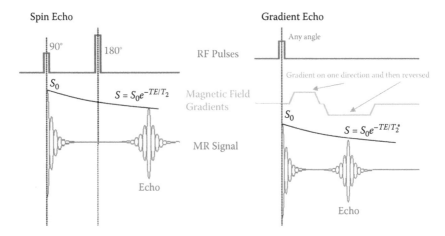

FIGURE 4.3 Schematic representations of the time sequences of RF pulses and gradients required to produce a spin echo and a gradient echo.

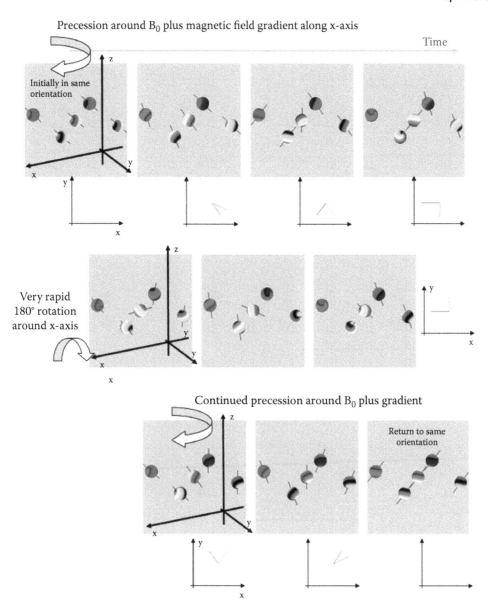

FIGURE 4.4 Graphical representation of the formation of a spin echo. Each sphere represents the magnetization in a very small volume of water. Under the influence of spatial variations in the static magnetic field (which may include a magnetic field gradient), the magnetization in the small volumes spreads out in phase (i.e., the direction they are pointing in the x–y plane). A 180° RF pulse inverts this phase distribution, and the same spatial magnetic field variations now cause the magnetization to come back into phase, briefly producing a net total larger magnetization and larger MR signal (the *echo*), until the magnetization from the small volumes goes back out of phase again.

chosen to be at locations with slightly different magnetic fields, then the magnetization is precessing at slightly different speeds after the initial excitation pulse. At some time, τ, later, the magnetization in the three volumes is at angles A, B, and C from the x-axis, for example (we could also have chosen some other axis). Now immediately after we apply the RF pulse to rotate the magnetization 180° around the x-axis, the magnetization will be at angles –A, –B, and –C,

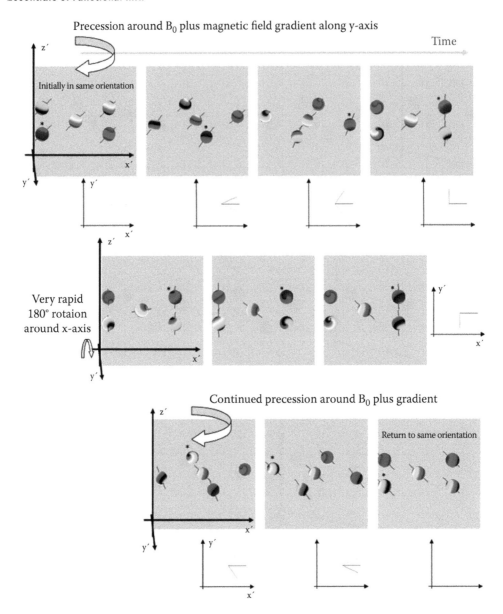

FIGURE 4.5 Graphical representation of the formation of a spin echo, as in Figure 4.3, except it is now shown in the rotating frame of reference. The sphere at the center is shown precessing at the same frequency as the rotating frame, and therefore appears to be stationary. An asterisk (*) marks one of the spheres so its position can be tracked from the point of view of the rotating frame of reference.

respectively, from the x-axis. The magnetization in the three small volumes continues to precess at the same speeds as before the 180° RF pulse, and after another time, τ, the magnetization directions in the three volumes are once again the same (i.e., the magnetization is back in phase). When the magnetization is back in phase, we can again detect an MR signal, and this is the spin echo.

However, it is important to note that the magnetization does not return to being perfectly in phase, and we do not recover the entire MR signal at the peak of the spin echo. The action of the 180° RF pulse cancels out the effects of any constant magnetic field variations. It does not alter the effects of transverse relaxation, characterized by the time, T_2, because the relaxation is caused by random interactions between hydrogen nuclei. At the peak of the echo (the point of the maximum MR signal), the signal magnitude, S, will have decayed from the value S_0 that would have been measured immediately after the RF pulse:

$$S = S_0\, e^{-TE/T_2}$$

The time, TE, is the *echo time* and is equal to 2τ, the time from the first RF excitation pulse to the center of the spin echo. The time τ is exactly the time between the RF excitation pulse and the 180° RF pulse, and so the value of TE is easily controlled by choosing the timing between the RF pulses.

After the peak of the spin echo, the magnetization in each small volume continues to precess at the same speed. After a relatively short time the magnetization becomes out of phase again and the total MR signal decays to zero. At this point we could apply another 180° RF pulse to invert the phase distribution again and form another spin echo. This would again cancel out the effects of the constant magnetic field variations but would not reverse the effects of transverse relaxation. The maximum signal at the center of the second echo would be reduced by the transverse relaxation that has occurred since the first RF excitation pulse.

As mentioned above, possible variations of the spin echo include rotating the magnetization around two different axes with the two RF pulses. For example, the initial excitation pulse may be a 90_X pulse to rotate the magnetization onto the –y-axis or a 90_Y pulse to rotate the magnetization onto the x-axis. The 180° pulse can be $180_{\pm X}$ or $180_{\pm Y}$. In every case, a spin echo is still formed exactly as described above. Two advantages of altering the direction of the 180° rotation are that (1) we can control the direction of the magnetization in the transverse plane, and (2) we can rotate the magnetization in one direction (such as 180_X) to produce the first spin echo, and in the opposite direction (such as 180_{-X}) to produce a second spin echo (2,7–9). By rotating the magnetization in two different directions, any imperfections in the RF pulse (such as rotating too much or too little in some places) are canceled out for every second echo that is formed, and the errors do not accumulate. Finally, it is also possible to form a spin echo by using RF pulses with lower flip angles than 90° and 180°. In this case the signal intensity is lower at the peak of the spin echo, but the amount of energy that is needed to produce the signal is much lower. This can be important when a large number of echoes must be formed in a short time for fast imaging methods, such as may be used for fMRI.

Key Points

4. Immediately after an RF excitation pulse, the MR signal decays exponentially due to transverse relaxation and dephasing caused by spatial variations in the static magnetic field and magnetic field gradients.
5. The dephasing caused by static spatial field variations, and by any gradients that are applied, can be canceled out after a 180° pulse.
6. The spin echo is T_2-weighted as a result.
7. Mnemonic for how a spin echo is created: SPICE
   Send it down 90°
   Precess around
   Invert the magnetization (180° pulse)
   Coherence returns
   Echo is formed

4.3 Gradient Echo

A gradient echo is formed after an excitation RF pulse by applying a magnetic field gradient for a brief time, and then applying a gradient in the opposite direction (Figure 4.3). Shortly after the gradient direction is reversed, the effects of the two gradients will cancel each other out, bringing the transverse magnetization back into phase as shown in Figures 4.6 and 4.7. When the magnetization is in phase, we can again detect a strong MR signal, and this is a gradient echo. The time between the RF pulse and the peak of the echo is called the echo time, TE, as it is with a spin echo.

Precession around B_0 plus magnetic field gradient along x-axis

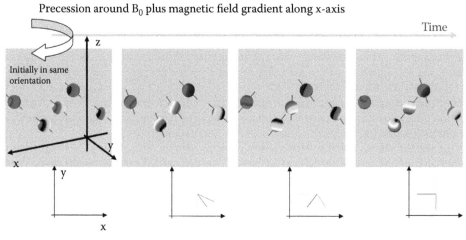

Continued precession around B_0 plus reversed field gradient

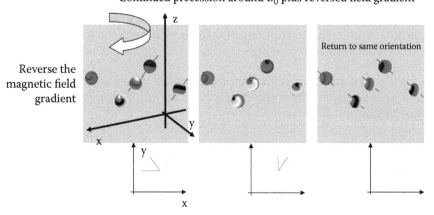

FIGURE 4.6 Graphical representation of the formation of a gradient echo. Each sphere represents the magnetization in a very small volume of water. Under the influence of a magnetic field gradient, the magnetization in the small volumes spreads out in phase (i.e., the direction they are pointing in the x–y plane). When the gradient direction is reversed, the magnetization comes back into phase, briefly producing a net total larger magnetization and larger MR signal (the *echo*), until the magnetization from the small volumes goes back out of phase again. With the gradient echo, only the effect of the gradient is canceled out, but the effects of constant spatial magnetic field variations are not.

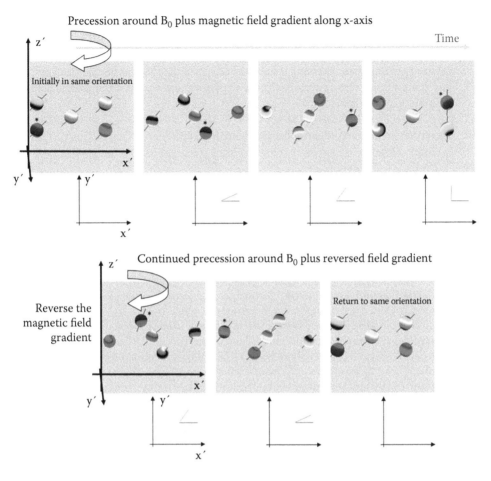

FIGURE 4.7 Graphical representation of the formation of a gradient echo, as in Figure 4.6, except it is now shown in the rotating frame of reference. The sphere at the center is shown precessing at the same frequency as the rotating frame and at the center of the MRI system so that the gradient has no effect, and therefore appears to be stationary.

The most important distinction between a gradient echo and a spin echo is that the inherent magnetic field variations are not canceled out with a gradient echo. Only the dephasing caused by the applied gradients is reversed. The peak signal therefore decays exponentially with the characteristic time, T_2^*, inherent to the tissues, in the absence of any applied gradient fields. At the peak of the gradient echo, the signal amplitude, S, will have decayed from the value S_0 that would have been measured immediately after the RF pulse:

$$S = S_0\, e^{-TE/T_2^*}$$

As with the spin echo, we can easily set the value of TE simply by choosing when to apply the gradient and when to reverse it. After the peak of the echo, the magnetization again dephases and the signal decays quickly, under the influence of the gradient and the inherent magnetic field variations. It is also possible to reverse the gradient a second time and form another gradient-echo. The peak of the next echo will again be reduced according to the decay with the characteristic time T_2^*, from the time of the RF excitation pulse.

Key Points

8. A gradient echo is formed after a single RF excitation pulse by applying a magnetic field gradient and then reversing it until the effects of the two gradients cancel out.
9. Only the gradients are canceled out; the dephasing caused by static spatial field variations are not affected.
10. A gradient echo is T_2^*-weighted as a result.
11. Mnemonic for how a gradient echo is created: SUGAR

  Single RF pulse
  Under 90° (typically)
  Gradient
  And
  Reverse

4.4 Steady-State Methods and Stimulated Echoes

Whenever RF excitation pulses are applied repeatedly at a rate that does not permit the magnetization to fully relax back to equilibrium between each pulse (such as $TR < T_1$), the signal intensity after each excitation reaches a new *steady state* that is different from that of a single excitation. The signal in this case depends on the repetition time, TR, and on the value of T_1, and so is T_1-weighted. A special case occurs when TR is very short (such as $TR < T_2$). This situation only occurs in practice when gradient-echo methods are used, with excitation flip angles that are reduced from 90°, because otherwise the longitudinal magnetization has insufficient time to recover and little to no signal is produced by each excitation. A very important factor contributing to the signal is whether or not the transverse magnetization just prior to each excitation is (1) completely dephased so that it cannot contribute to later echoes, (2) partially dephased, or in a regular pattern, so that it could be recovered later, or (3) brought completely back into phase. When the magnetization is completely dephased prior to each subsequent excitation, then the transverse magnetization prior to each RF excitation pulse is zero. In this case the echo signals are simply gradient echoes, possibly with T_1-weighting, as described above. If instead the transverse magnetization is in phase, or is in a pattern that can be reversed, then it could add to the signal in later echoes.

To describe the MR signal that would be produced by repeated RF pulses, we can picture a specific example of repeated RF excitation pulses that tip the magnetization by 30°, applied every 20 msec. Simulations of the echo signals produced by this sequence are shown in Figure 4.8. If the magnetization starts at equilibrium and we apply a 30° RF pulse, then a proportion of the magnetization is in the transverse component ($M_\perp = M_0 \sin \alpha = 0.50\,M_0$) and the longitudinal component is slightly reduced ($M_\parallel = M_0 \cos \alpha = 0.866\,M_0$). A gradient echo could be created by applying a gradient and reversing it. After the signal in the gradient echo has again decayed to zero, there is no transverse magnetization remaining.

In one situation we could simply apply the RF pulse and gradients again to produce another gradient echo. However, if we use a short repetition time between excitation pulses, the longitudinal magnetization does not fully relax to equilibrium. Assuming T_1 is around 1 sec, as with most tissues, with a very short repetition time, TR, of 20 msec as in this example, the longitudinal magnetization prior to the second RF pulse is:

$$M_\parallel = M_0[1-(1 - \cos \alpha)e^{-TR/T_1}] = 0.869\,M_0$$

The second RF pulse therefore does not act on the same magnetization that was affected by the first RF pulse. Another 30° RF pulse will again rotate a small amount of magnetization into the transverse plane ($M_\perp = 0.869\,M_0 \sin \alpha = 0.435\,M_0$) and further reduce the longitudinal magnetization ($M_\parallel = 0.869\,M_0 \cos \alpha = 0.753\,M_0$). As can be seen, each subsequent RF pulse rotates

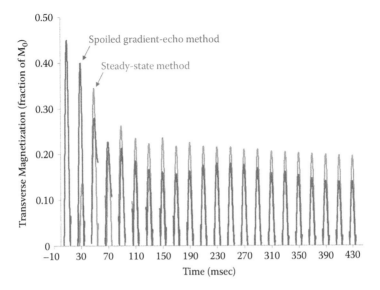

FIGURE 4.8 Simulations of the transverse magnetization that would be produced by 30° RF excitation pulses with a repetition time of 20 msec. The direction of the RF pulses are alternated, 30_x 30_{-x} 30_x 30_{-x} and so on, to avoid eliminating the longitudinal magnetization entirely, given the very short TR. After each RF pulse a gradient echo is formed by applying a magnetic field gradient and then reversing it. The red lines indicate the echoes that are produced when the transverse magnetization is completely dephased after each gradient echo. The blue lines indicate the echoes produced when the transverse magnetization is refocused (by making the total gradient effects equal to zero), after each gradient echo, thereby creating a new (larger) steady-state magnetization.

less magnetization into the transverse plane. However, the Bloch equations (10) demonstrate a key point with the portion of the equations that relates to longitudinal relaxation:

$$\frac{dM_{\parallel}}{dt} = -\frac{M_{\parallel} - M_0}{T_1}$$

The rate of change of the longitudinal magnetization ($dM_{\parallel}/dt$.) is greater the further $M_{\parallel}$. is from the equilibrium value of M_0. That is, the longitudinal magnetization changes faster when $M_{\parallel}$ is smaller (or even more negative). After several identical RF pulses have been applied, the transverse magnetization produced by each pulse is balanced by the recovery due to relaxation between pulses. As a result, the magnetization reaches a new steady state. The longitudinal magnetization prior to each RF pulse in the steady state is:

$$M_{\parallel} = M_0 \frac{\left(1 - e^{-TR/T_1}\right)}{\left(1 - \cos\alpha \, e^{-TR/T_1}\right)}$$

The signal at the center of the gradient echo that would be produced by the next RF pulse, including the effects of transverse relaxation, is (11–13):

$$S(GE)_\alpha = S_0 \sin\alpha \frac{\left(1 - e^{-TR/T_1}\right)}{\left(1 - \cos\alpha \, e^{-TR/T_1}\right)} e^{-TE/T_2^*}$$

This method is termed a *spoiled* or *incoherent* gradient echo, because the transverse magnetization is not brought back into phase prior to each subsequent RF pulse and does not contribute to later echoes. Although a new steady state is reached and the resulting echoes are T_1-weighted, this method does not fall into the category of what is generally termed a steady-state method, as described below.

4.4.1 Steady-State Methods

In a second situation that follows from the above example, if instead we refocused the magnetization after each gradient echo, just prior to each subsequent RF pulse, then the transverse magnetization can contribute to later echoes. Another important difference between the *steady-state* and the *spoiled* methods described in the previous section is that the steady-state methods require the RF pulses to be alternated in direction (14). That is, if the first RF pulse is 30_X, then the second one must be 30_{-X}, the third is again 30_X, and so on. After the first RF pulse, the transverse and longitudinal magnetization components are exactly as described above. However, it is now important to take into consideration the direction of the magnetization, being along the –y-axis after the first pulse (assuming a 30_X pulse in this example). Just prior to the second RF pulse, the longitudinal magnetization has recovered slightly (depending on TR and T_1), but the refocused transverse magnetization has an amplitude equal to M_0., sin $\alpha\, e^{-TR/T_2^*}$. The second RF pulse again does not act on the same magnetization that was affected by the first RF pulse, as in the example above, and now both the transverse and longitudinal components must be taken into consideration. The second RF pulse rotates the longitudinal magnetization toward the +y-axis and rotates the transverse magnetization 30° back toward the longitudinal axis. As a result, in the transverse plane these two components partially cancel each other out. But, the component that was rotated from the longitudinal axis is the larger of the two, primarily because the transverse component had undergone relaxation with the characteristic time T_2^*. The magnitude of the second gradient echo that can be produced is therefore much smaller than that of the first echo (Figure 4.8). But, the longitudinal magnetization that remains after the second RF pulse is larger than after the first pulse, because it also has a contribution from the magnetization that was rotated out of the transverse plane. Continuing this process, the transverse component of the magnetization after the third RF pulse is again along the –y-axis and is not as large as after the first RF pulse, but is larger than that after the second RF pulse, and will settle into a steady state after several more pulses. From this progression of the magnetization after each pulse, it can be seen that the net signal produced in the steady state will depend on the TR and flip angle, and on both T_1 and T_2^*. The signal amplitude at the center of each gradient echo will eventually settle into a new steady state, given by (14):

$$S = S_0 \sin\alpha \frac{\left(1 - e^{-TR/T_1}\right)}{1 - \left(e^{-TR/T_1} - e^{-TR/T_2^*}\right)\cos\alpha - e^{-TR/T_1}e^{-TR/T_2^*}} e^{-TE/T_2^*}$$

Note, however, that this equation has been modified slightly from the original source, by replacing T_2 with T_2^*, in the denominator. This change is needed because the original text (14) refers to transverse relaxation in general. In the case of RF pulses that are alternated in phase, the phase distribution of the magnetization is never inverted, and so the effects of the static spatial field variations are not canceled out and must be taken into consideration for MRI of the body. In addition, the term T_2^*, to differentiate the total observed transverse relaxation from the true T_2 relaxation, does not appear to have been defined until a few years after this equation was first published (6).

The magnitude of the gradient echoes that can be produced after each RF pulse with a steady-state method and a spoiled gradient-echo method are shown in Figure 4.8. The signals shown are simulations, based on a repetition time of 20 msec and a flip angle of 30°.

A key feature of steady-state methods is that the transverse magnetization needs to be brought back into phase prior to each RF excitation pulse for it to contribute to later echoes as much as possible. The effects of all of the magnetic field gradients must therefore total zero prior to each successive RF pulse. There are a number of jargon terms to describe this, for example, making the net gradient effect equal to zero is called *gradient moment nulling*, and methods of this type are sometimes called *balanced* acquisitions, meaning the gradients are balanced to sum to zero net effect. Several steady-state imaging methods use completely balanced gradients and have TE = TR/2, so that the RF excitation pulses and the signal echoes are evenly spaced in time. This particular method has several names, such as FISP (fast imaging with steady-state precession), SSFP (steady-state free precession) and True SSFP, balanced FFE (balanced fast field echo), and FIESTA (fast imaging employing steady-state acquisition).

The advantage of the steady-state method is that it enables very frequent or rapid sampling of the MR signal, with a larger signal than would be obtained with a spoiled (or *incoherent*) gradient echo, as shown in Figure 4.8. The trade-off is that the resulting signal strength depends on both T_1 and T_2. The details of the signal dependences with various methods are summarized in Table 4.1.

Table 4.1 Summary of the Signal Magnitudes Produced by Conventional Spin-Echo and Spoiled Gradient-Echo Methods, and with Methods Employing Repeated Identical Excitations with Short Repetition Times	
$S(SE)_{90-180} = S_0 e^{-TE/T_2}$	Conventional spin echo, ignoring T_1-weighting
$S(GE)_\alpha = S_0 \sin\alpha \, e^{-TE/T_2^*}$	Conventional spoiled gradient echo, ignoring T_1-weighting
$S(GE)_\alpha = S_0 \sin\alpha \dfrac{\left(1 - e^{-TR/T_1}\right)}{\left(1 - \cos\alpha \, e^{-TR/T_1}\right)} e^{-TE/T_2^*}$	Spoiled gradient echo with magnetization spoiled between pulses (i.e., incoherent gradient echo), accounting for the effects of T_1-weighting
$S(TF) = S_0 \sin\alpha \dfrac{\left(1 - e^{-TR/T_1}\right)}{1 - \left(e^{-TR/T_1} - e^{-TR/T_2^*}\right)\cos\alpha - e^{-TR/T_1} e^{-TR/T_2^*}} e^{-TE/T_2^*}$	FLASH (fast low-angle shot) coherent gradient echo, with very short TR, and TE, and the special case of TrueFISP with TE = TR/2
$S(STE)_{\alpha-\alpha-\alpha} = \dfrac{1}{2} S_0 \sin^3(\alpha) \, e^{-TM/T_1} e^{-2\tau/T_2}$	Stimulated echo produced by 3 identical RF pulses with flip angle α (as described in Section 4.4.2). This equation assumes TM $\geq \tau$

Sources: Hahn, E.L., *Phys Rev* 80, 4, 580–594, 1950 (1); Tkach, J.A., and Haacke, E.M., *Magn Reson Imaging* 6, 4, 373–389, 1988 (6); Ernst, R.R., and Anderson, W.A., *Rev Sci Instrum* 37, 93–102, 1966 (12); Edelstein, W.A. et al., *J Comp Assist Tomogr* 7, 3, 391–401, 1983 (13); Haase, A., *Magn Reson Med* 13, 1, 77–89, 1990 (15); Scheffler, K., and Hennig, J., *Magn Reson Med* 49, 2, 395–397, 2003 (16).

Key Points

12. Steady-state methods, in contrast to spoiled methods, return the transverse magnetization to being in phase prior to each subsequent RF pulse, although static field variations are not canceled out.

13. The refocused transverse magnetization contributes to the signal after subsequent RF pulses. The transverse component is T_2^*-weighted, and so the steady-state magnetization is both T_1- and T_2^*-weighted.

14. Steady-state methods produce transverse magnetization that can be used to create a gradient echo after each RF pulse but do not spontaneously produce an echo, such as a spin-echo method would.

15. Steady-state methods also require that the RF pulses alternate in direction between successive pulses.

4.4.2 Stimulated Echoes

A different situation from that described above arises when repeated RF pulses are applied that rotate the magnetization in the same direction. Steady-state methods described in the previous section use pulses that alternate in direction, in order to avoid reducing the longitudinal magnetization to zero as a result of repeated low-to-medium flip angle pulses and inadequate time for relaxation between pulses. However, with large flip angle pulses such as 90°, the cumulative effect of repeated pulses can rotate the magnetization through a large enough angle to return the magnetization again toward the transverse plane, thereby reversing the transverse phase distribution and producing an effect similar to a spin echo, called a *stimulated echo*. Recall that to produce a spin echo, an initial 90° pulse is applied, and after a brief time the resulting transverse magnetization spreads out in phase due to spatial field variations. A 180° pulse is then used to invert the magnetization and the phase distribution. The same spatial field variations then cause the magnetization to come back into phase, producing a spin echo.

To produce a stimulated echo we replace the 180° pulse with two identical 90° pulses, being now the second and third pulses in the sequence (Figure 4.9). If these two 90° pulses are in very rapid succession, they produce the same effect as a 180° pulse, and a spin echo is formed as usual. However, allowing time between the second and third 90° pulses changes the amplitude and weighting of the echo that is produced. The action of the second RF pulse rotates the transverse magnetization onto the longitudinal axis in the negative direction (i.e., it is inverted), as shown in Figure 4.10. More accurately though, it is necessary to consider that not all of the magnetization was along one axis prior to the second pulse, but was spread out in the transverse (x–y) plane. Assuming 90_X pulses are used, the magnetization distribution is rotated around the x-axis into the y–z plane (recall that the z-axis is the longitudinal axis). The magnetization undergoes longitudinal relaxation, and the transverse components of the magnetization spread out in the transverse plane, as a result of the static spatial field variations that are always present.

The third 90° RF pulse rotates the magnetization back into the transverse (x–y) plane, reversing the phase distribution that existed immediately prior to the second RF pulse. The longitudinal magnetization to be rotated into the transverse plane depends on the time between the second and third RF pulses (the interval allowed for longitudinal relaxation), and the value of T_1. If the time between the first and second RF pulses is τ, then at the time τ after the third RF pulse the static spatial field variations cause a portion of the magnetization to come back into phase, and a stimulated echo is formed.

A key feature of the stimulated echo that makes it distinct from steady-state methods or a spoiled gradient echo is that the magnetization comes back into phase spontaneously and an echo is formed. That is, gradients are not necessary to produce an echo; the inherent spatial

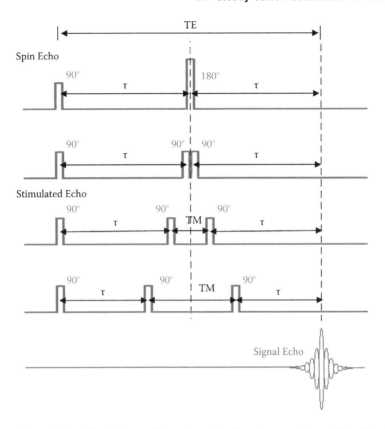

FIGURE 4.9 The relationship between the spin echo (top two rows) and the stimulated echo (bottom two rows) is shown by the continuity in the transition from one to the other. If the second and third 90° pulses rotate the magnetization in the same direction, then they sum to the effect of a 180° degree pulse when the *mixing time*, TM, is very close to zero. In this case, the spin echo and the stimulated echo are indistinguishable. As TM is increased, the differences become apparent, as described in the text.

magnetic field variations are sufficient. A stimulated echo can be produced with lower flip angle pulses as well, and the echo amplitude depends on the flip angle, α, the timing between the RF pulses, and as on T_1 and T_2:

$$S(STE)_{\alpha-\alpha-\alpha} = \frac{1}{2} S_0 \sin^3(\alpha)\, e^{-TM/T_1} e^{-2\tau/T_2}$$

For this equation, the times between the three pulses are as indicated in Figure 4.9, and the time between the second and third pulses, TM, is assumed to be at least as long as the time between the first and second pulses, indicated by τ (1). An assumption used for this equation is that the times between pulses (particularly TM) are long enough that the transverse magnetization becomes uniformly spread out in phase in the transverse plane between each pulse, due to the inherent spatial magnetic field variations. This is as shown in the lower row of Figure 4.10. The magnetization is therefore assumed to be uniformly distributed in phase in the y–z plane immediately after the second RF pulse, meaning that half of the magnetization components are along the y-axis and half along the z-axis. The transverse components (those parallel to the y-axis after the second pulse) then again become uniformly distributed in the transverse plane. After the third RF pulse, the magnetization components that were rotated onto the z-axis by the

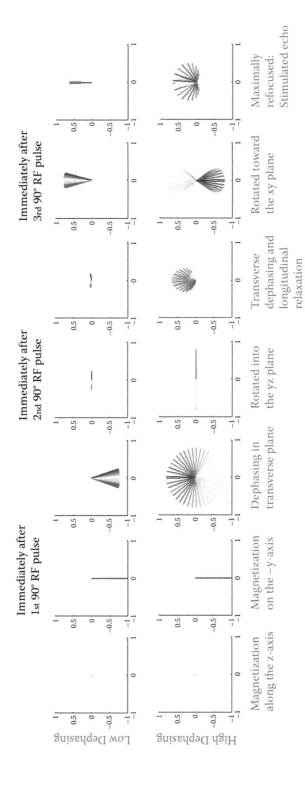

FIGURE 4.10 Simulations of the effects of three identical 90° RF pulses, resulting in a stimulated echo. The top row shows the case of short times between pulses relative to transverse dephasing due to static spatial field variations, resulting in little dephasing, whereas the bottom row shows the more general case of complete transverse dephasing between pulses. The bottom row also demonstrates the classic figure-eight pattern described by Hahn in 1950, just prior to, and after, the third RF pulse. (Hahn, E.L., *Phys Rev* 80, 4, 580–594, 1950 [1].)

second pulse are rotated back into the transverse plane, with the phase distribution reversed. The inherent spatial field variations that caused the dephasing now cause the magnetization to come back into phase, producing the stimulated echo, as described above. However, a key point is that only half of the magnetization is refocused, and this is the reason for the 1/2 factor in the equation above. The time interval between the first RF pulse and the second pulse is when the transverse dephasing occurs. The peak of the stimulated echo therefore occurs at this exact amount of the time again, after the third RF pulse. The echo time for the stimulated echo is therefore twice the time between the first and second RF pulses. The time interval between the second and third pulse, TM, is not included in the echo time. However, TM is a factor in the echo amplitude above, in relation to the longitudinal relaxation that occurs in this interval.

If between the RF pulses the magnetization does not become completely dephased (i.e., uniformly distributed in all directions in the transverse plane), then the situation is similar to that shown in the upper row of Figure 4.10. A stimulated echo is still produced and even has a higher peak amplitude. In the extreme case, when no time is allowed between the second and third pulses, then no transverse dephasing occurs in this interval, and the resulting echo formed after the third RF pulse is a spin echo.

In the more general case, similar to the steady-state methods described in the previous section, we can consider a sequence of RF pulses applied at equal intervals, with flip angles that may be 90° or less. The first three pulses in the sequence produce a stimulated echo, midway between the third and fourth pulses. However, the second, third, and fourth pulses do not produce another stimulated echo, because this set of three pulses does not begin in an initial state with magnetization along the longitudinal axis. Continuing the parallel to a spin-echo sequence, we can see that a spin echo requires a second 180° pulse to be applied to produce a second echo, and therefore two more 90° degree pulses would be expected to produce a second stimulated echo. A continuous sequence of RF pulses therefore does not produce a steady-state sequence of stimulated echoes. At best, it may be possible to produce repeated stimulated echoes after every second RF pulse in the sequence, similar to a train of spin echoes. A train of spin echoes is not a steady-state method either and cannot be sustained indefinitely. Each echo is attenuated by transverse relaxation, with the characteristic time, T_2. A repeated train of stimulated echoes would be similarly attenuated.

As described above for steady-state methods, however, a sequence of repeated RF pulses will eventually reach a new steady state, and a constant amount of transverse magnetization will be rotated into the transverse plane with each RF pulse. Simultaneously, the transverse magnetization that was produced by a prior RF pulse is rotated back into the transverse plane in a negative sense and can produce a stimulated echo. These two sources of transverse magnetization will be opposite in phase and will therefore tend to cancel each other out. The magnetization that has been rotated into the transverse plane by the most recent pulse may produce a larger amplitude gradient echo than the stimulated echo (depending on the echo time, repetition time, and flip angle), but in any case the net effect will be a reduced echo amplitude. The stimulated echo therefore can be a detriment to the total signal that is produced in this situation. But, it can be avoided simply by leaving the transverse magnetization out of phase in the transverse plane, or even applying strong *spoiling* gradients, prior to each subsequent RF pulse. This dephasing will eliminate the possibility of the formation of a stimulated echo by later RF pulses, and the net effect is identical to that of the spoiled gradient echo described above.

Key Points

16. Three RF pulses in a row can spontaneously produce a *stimulated echo* without the use of gradients.
17. The time, τ, between the first and second RF pulses is equal to the time between the third pulse and the center of the stimulated echo, and TE = 2τ.

18. The time between the second and third pulses is called the *mixing time*, TM, and is not included in the echo time.
19. A stimulated echo produced by three identical 90° RF pulses with very short TM is essentially identical to a spin echo produced by a 90° pulse and a 180° pulse.
20. Typically, the mixing time, TM, is sufficiently long to allow the transverse magnetization to become uniformly distributed in all directions in the transverse plane, resulting in a stimulated echo that is half as large as a comparable spin-echo.
21. A stimulated echo can be produced with flip angles of less than 90°, but repeated identical pulses at regular intervals will not produce repeated identical stimulated echoes similar to a steady-state method. A steady-state method based on identical RF pulses (i.e., not alternated in phase) may produce undesirable stimulated echoes that diminish the primary gradient-echo amplitude.

4.5 Signal Weighting and Contrast

The MR signal that is recorded with a spin echo or a gradient echo is weighted depending on the values of T_2 or T_2^*, respectively, of the tissues (cells, connective tissues, fluids, etc.) within the regions being imaged (Figure 4.11). The signal from each type of tissue also depends on the number of hydrogen nuclei that contribute to the signal (i.e., the *proton density*). The relaxation times and proton densities tell us something about the physical properties of the tissues. The magnitude of the MR signal also depends on the *flip angle*, which is the angle the magnetization is rotated away from B_0 by the RF excitation pulse. This dependence occurs for two reasons: the flip angle affects (1) the magnitude of the transverse magnetization that we detect to get the MR signal, and (2) the longitudinal magnetization that is available the next time the RF excitation pulse is applied. As described in Chapter 3, however, the longitudinal magnetization additionally depends on the time between successive applications of the RF excitation pulse (the repetition time, TR) and on the longitudinal relaxation times, T_1, of the tissues being imaged.

As a result, spin-echo and gradient-echo methods can both produce echo signals with peak amplitudes that are T_1-weighted, depending on the flip angle used for the RF excitation pulse, and on the repetition time, TR, as summarized in Table 4.2. A spin echo can also produce echo signals with peak amplitudes that are T_2-weighted, whereas a gradient echo produces echo signals that are T_2^*-weighted, and the amount of weighting in both cases is determined by the echo time, TE. For applications of MRI, critical imaging parameters to be selected are the choice of a spin-echo or a gradient-echo method, and the values of flip angle, TE, and TR. Although the MR signal can be made to be a combination of T_1- and either T_2- or T_2^*-weighting, typically parameters are selected so that one type of weighting is dominant. With the choice of a long TE, and long TR, the signal will be predominantly T_2- or T_2^*-weighted (depending on whether a

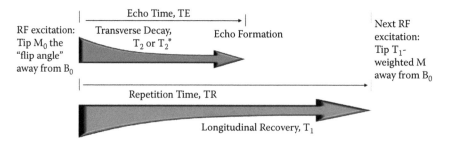

FIGURE 4.11 Schematic representation of how transverse (T_2 or T_2^*) and longitudinal (T_1) relaxation-time weighting can be imposed on the MR signal.

Table 4.2 Summary of Signal Weighting Factors for Spin Echoes and Gradient Echoes				
	TE	TR	Flip Angle, α	Resulting Weighting
Spin Echo				
Attenuation related to TE is: $T_2w = e^{-TE/T_2}$	Long TE ($>T_2$)	Long TR ($>2 \times T_1$)	Fixed at 90°	
Attenuation related to TR and flip angle, α, is (12,19): $T_1w = \sin\alpha \dfrac{\left(1 - e^{-TR/T_2}\right)}{\left(1 - \cos\alpha \, e^{-TR/T_1}\right)}$	$T_2w < 0.37$	$T_1w > 0.86$		T_2-weighted, little T_1-weighting
	Short TE ($< \frac{1}{2} T_2$) $T_2w > 0.72$	Short TR ($< \frac{1}{2} T_1$) $T_1w < 0.39$	90°	Little T_2-weighting, T_1-weighted
Gradient Echo	Long TE ($> T_2^*$)	Long TR ($> T_1$)	Examples: 68°	
Attenuation related to TE is: $T_2^*w = e^{-TE/T_2^*}$	$T_2^*w < 0.37$	$T_1w > 0.68$		T_2-weighted, little T_1-weighting
Attenuation related to TR and flip angle, α, is (12,20): $T_1w = \sin\alpha \dfrac{\left(1 - e^{-TR/T_1}\right)}{\left(1 - \cos\alpha \, e^{-TR/T_1}\right)}$		Long TR ($> 1.5 \times T_1$) $T_1w > 0.80$	76°	
	Short TE ($< \frac{1}{2} T_2^*$) $T_2^*w > 0.72$	Short TR ($< 10\% \ T_1$) $T_1w < 0.13$ $T_1w < 0.22$	Examples: 68° 25°	Little T_2-weighting, T_1-weighted

spin echo or a gradient echo is used), and not T_1-weighted. On the other hand, setting the TE as small as possible, and using a shorter TR and flip angle combination, will make the echo signal predominantly T_1-weighted, and not T_2- or T_2^*-weighted. Finally, if we instead choose a short TE value and a long TR value, then the signal is neither T_1- nor T_2- or T_2^*-weighted, and so is predominantly proton-density weighted.

The actual TE and TR values that are used in practice vary for different applications, and the need to balance signal strength, imaging speed, quality, and so forth will be discussed in Chapter 5. However, the following gives very rough ideas of the ranges of values that might be used (summarized in Table 4.2). A TE value could be considered long if it is greater than T_2 or T_2^* of the tissues being imaged, and short if it is less than 50% of T_2 or T_2^*, depending on if a spin echo or a gradient echo is used, respectively. Many tissues in the body, such as the brain, have T_2-values around 90 msec and T_2^*-values around 60 msec, 30 msec, and 15 msec at 1.5 T, 3 T, and 7 T, respectively. As a result, what would be considered a *long* or *short* value for TE is very different for a spin echo than for a gradient echo. Similarly, a TR value could be considered long if it is more than two times T_1, and short if it is half of T_1. In the body, many tissues have T_1-values around 1 sec, and these also depend on the magnetic field strength. The flip angle must be set depending on the repetition time, to get the degree of T_1-weighting that is desired. When excitation pulses are applied repeatedly at a regular rate, the flip angle that will give the maximum signal after each pulse is called the *Ernst angle*, α, and can be found with the equation $\cos\alpha = e^{-TR/T_1}$ (12). Higher flip angles will make the signal more T_1-weighted. Again, these values are by no means definitive; a range of values is used in practice, and what is considered a long or a short value can depend on the imaging application.

Another important consideration is the contrast (i.e., signal intensity difference) between different tissues. For example, when imaging the brain it is often desirable to distinguish gray matter and white matter, with T_1-values of approximately 1200 msec and 700 msec, respectively; T_2-values of around 99 msec and 79 msec, respectively; and T_2^*-values of around 30 msec for both. The maximum T_2- or T_2^*-weighted contrast will be obtained when the echo time, TE, is roughly midway between the T_2- or T_2^*-values of the tissues being contrasted. In the case of gray matter and white matter with the relaxation values assumed here, T_2^*-weighting will not provide any contrast, and T_2-weighting will provide some contrast at TE values around 80 to 90 msec, as shown in Figure 4.12. With T_1-weighting provided by repeated excitation, a spin echo will provide the greatest contrast at a repetition time, TR, which is approximately equal to the T_1-values of the tissues. A gradient echo can also be made to be T_1-weighted and provides the option of using lower flip angles and shorter repetition times. While the contrast is maximum

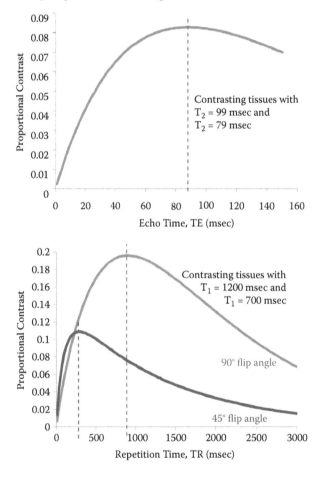

FIGURE 4.12 Examples of contrasts (i.e., signal differences) between tissues with different T_2- and T_1-values, depending on the echo time, repetition time, and flip angle. The signal differences are expressed as a proportion of the overall signal intensity from each tissue, assuming the tissues have the same proton density.

with a flip angle of 90° and a repetition time around 920 msec as with the spin echo, about half as much contrast remains with a flip angle of 45° and a repetition time of only 280 msec. As will be described in Chapter 5, this shorter repetition time translates to much faster imaging (taking 30% of the time). However, one very important factor that affects the overall contrast that we have not yet considered is the proton density of the tissues. The optimal contrast will be obtained when all of the differences in proton density and relaxation times are taken into consideration.

The method of T_1-weighting employed in the example above involves repeated stimulation without allowing time for complete recovery of the longitudinal magnetization to equilibrium, whereas the transverse magnetization is assumed to relax completely to zero between excitations. The transverse magnetization can be easily eliminated either by allowing enough time for transverse relaxation or by applying magnetic field gradients to dephase the magnetization. In the steady-state case described in Section 4.4, "Steady-State Methods and Stimulated Echoes," the signal instead tends to have a combination of T_1- and T_2^*-weighting, because the transverse magnetization is not eliminated between excitations (16).

Key Points

22. When then MR signal amplitude depends predominantly on one of the relaxation times, T_1, T_2, or T_2^*, then the signal is called T_1-weighted, T_2-weighted, or T_2^*-weighted, respectively; otherwise the signal is proton-density weighted.
23. The amount of T_1-weighting of the signal depends on the choice of TR and flip angle.
24. The amount of T_2- or T_2^*-weighting of the signal depends on the choice of TE.

4.6 Inversion-Recovery Methods

One common modification to imaging methods is the addition of a 180° RF pulse prior to the start of the conventional gradient-echo or spin-echo method. This is typically done only with the repetition time set long enough to allow full relaxation of the longitudinal magnetization before the sequence of RF pulses and gradients are applied. The action of the 180° RF pulse is to invert the longitudinal magnetization, and so it is called an *inversion pulse*. After the inversion pulse, the magnetization undergoes longitudinal relaxation:

$$M_\parallel = M_0 \left(1 - 2e^{-TI/T_1}\right)$$

Here $M_\parallel$ indicates the magnitude of the longitudinal magnetization.

Before the magnetization has time to fully relax back to equilibrium though, the conventional gradient echo or spin echo is applied. The time between the inversion pulse and the RF excitation pulse is called the *inversion time*, abbreviated TI. This imposes T_1-weighting on the echo signal, in addition to any T_2- or T_2^*-weighting it would normally have if the inversion pulse had not been applied. For this reason, the echo time, TE, is typically set to be as short as possible, so that the T_2- or T_2^*-weighting is minimal and can be ignored.

The resulting T_1-weighting is a bit different from the weighting obtained by setting the repetition time, TR, too short to allow full relaxation between repeated excitations. With an inversion pulse, tissues or fluids with very short or very long T_1-values can give higher signals, whereas those with intermediate T_1-values can have lower signal, as shown in Figure 4.13.

One specific application of the inversion-recovery method is to selectively eliminate the signal from cerebrospinal fluid (CSF) for images of the brain or spinal cord. CSF has a much longer T_1-value (4300 msec at 3 T; refer to Table 3.1) than gray matter or white matter (~1200 msec

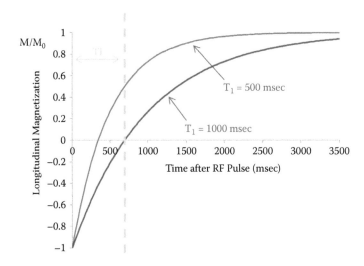

FIGURE 4.13 Inversion-recovery curves for tissues with two different T_1-values of 500 msec and 1000 msec. With an RF pulse applied at an inversion time, TI, as indicated, the tissues with T_1 = 1000 msec would have no longitudinal magnetization to be rotated into the transverse plane, and so this tissue would not produce any MR signal. The tissues with T_1 = 500 msec, however, would produce a signal that is roughly 50% of the signal magnitude it would have with no T_1-weighting at all. This difference could be reversed if the TI value was shortened to about 300 msec, to be at the point where the longitudinal magnetization from tissues with T_1 = 500 msec is passing through zero. At this point the longitudinal magnetization from the other tissues (with T_1 = 1000 msec) is at roughly −0.5, but only the magnitude of the longitudinal magnetization matters, and the signal it produces would still be roughly 50% of that with no relaxation weighting.

and ~700 msec at 3 T, respectively). At an inversion time of 2980 msec, the signal from CSF is essentially zero, while the signals from gray matter and white matter are at 83% and 97% of the amplitudes they would have in the absence of relaxation weighting, respectively. This method is termed *fluid-attenuating inversion recovery*, or FLAIR.

Key Points

25. An *inversion pulse* is a 180° RF pulse that is typically applied prior to a conventional imaging method, such as a gradient-echo or a spin-echo.
26. The time between the center of the inversion pulse and the center of the subsequent RF excitation pulse is called the *inversion time*, or TI.
27. The effect of the inversion pulse is to impose T_1-weighting on the longitudinal magnetization.
28. With the appropriate selection of inversion time, signals from tissues with a particular T_1-value can be selectively nulled at the time the imaging method is applied, as long as the T_1-values of the other tissues to be imaged are sufficiently different.

4.7 Magnetization Transfer Contrast

Contrast between different tissues can be produced by making use of the multiple relaxation environments that exchange quickly on the time scale of relaxation, as described in Chapter 3 (Section 3.6) (17,18). Even though hydrogen nuclei may move between two or more relaxation

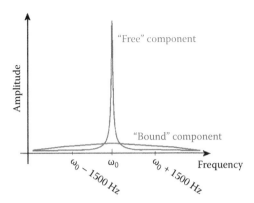

FIGURE 4.14 The estimated frequency spectrum of MR signal from tissues, with both bound and free signal components. This plot shows the relative amount of water at each precession frequency for the two water components.

environments and exchange energy between nuclei in different environments, each environment has specific properties that can be exploited. The water in hydration spheres around macromolecules and lipids relaxes very quickly because of strong interactions between adjacent nuclei and because of the altered movement of the water molecules. These interactions also create a broader spread of magnetic field intensities, and therefore a broader spread of precession frequencies, than in the "free" water. The distribution of precession frequencies for all of the hydrogen nuclei (i.e., the MR spectrum) has a Lorentzian line shape for exponentially decaying signals, and the peak width at half height is equal to $1/(2\,T_2)$. The free component has a relatively narrow peak because the T_2-value is estimated to be about 350 msec, whereas the "bound" component (the water in hydration spheres around macromolecules) has a very broad spectrum because the T_2 of this component is less than 1 msec (on the order of 10 μsec to 100 μsec) (17) (Figure 4.14). Both of these components will contribute to the total spectrum. As a result, it is possible to excite the entire system of hydrogen nuclei with an RF pulse that is at a frequency of 1000 Hz to 2000 Hz different from the Larmor frequency (ω_0), and it will affect only the bound component. The RF pulse is typically very long, or has a specially designed shape, so that it acts on only a very narrow frequency range. The pulse also produces a very high flip angle (several full rotations) so that it *saturates* the spins. This means the spins are evenly distributed along the ± z-axis and do not produce any net signal. Because hydrogen nuclei will exchange rapidly between different water components and will also exchange magnetic energy, the saturation becomes distributed throughout all of the water quite rapidly. The rate of exchange between the different water components is a key factor in determining the total effect of the RF pulse, but it also depends on the relative amounts of water and the relaxation times of each component. The amount of signal reduction that is produced is called the *magnetization transfer ratio*, or MTR, and it varies between different tissues and fluids. The effect of this frequency-selective excitation is termed *magnetization transfer contrast*, and it is particularly useful for contrasting tissues and fluids with different amounts of bound water and/or different exchange rates. For example, whereas gray matter and white matter have MTRs of 40% to 60%, in skeletal muscle it is 75%; and in blood, CSF, bile, and urine it is less than 5%.

4.8 Summary

Understanding the myriad of methods, or the key concepts behind a new method encountered in a scientific paper, is drastically simplified by first asking the question "Is it a spin echo, or is it a gradient echo?" Once you know the answer to this question, you know the key concepts that will influence the signal and tissue contrast that can be obtained with the imaging method. The MR signal is always inherently proton-density weighted, but relaxation-time weightings can be made to dominate. If the imaging method involves a 90° excitation pulse followed by one or more refocusing pulses (rotating the magnetization 180° or less), then it produces a spin echo, which can be T_1- and/or T_2-weighted. If the method instead involves a single excitation pulse, rotating the magnetization less than 90° in most cases, and a combination of gradients that are at some point in time reversed, then the method is a gradient echo. The echo signal that is produced can be T_1- and/or T_2^*-weighted. However, if the gradients used in this method are designed to refocus the transverse magnetization prior to the next RF excitation pulse, then it is likely a steady-state method. In this case, the signal that is produced is both T_1- and T_2-weighted. All other additions, such as an echo-planar imaging scheme for spatial encoding (discussed in detail in Chapter 5) or a preceding 180° inversion pulse, are simply variations superimposed on the two key underlying concepts, namely, the spin echo and gradient echo.

The acronyms used to describe the various MR imaging methods differ between MR system manufacturers, and a list of many of the common methods and their definitions is provided in Table 4.3. The classification of these methods in terms of spin echo or gradient echo is shown schematically in Figure 4.15.

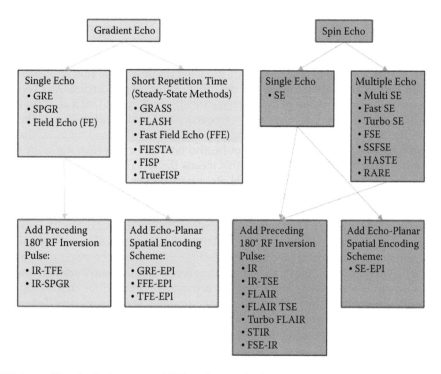

FIGURE 4.15 Flowchart of common MR imaging methods and the classification of each method. Gradient-echo methods are shown in green, spin-echo methods are shown in dark blue, and steady-state methods are shown in light blue.

Table 4.3 List of Acronyms Used to Name Common MR Imaging Methods and the Fundamental Type of Each Method

Acronym	Definition	Method Type
GRE	Gradient-recalled echo	Gradient echo
FFE	Fast-field echo	Gradient echo
TFE	Turbo-field echo	Gradient echo
GRASS	Gradient recall acquisition using steady states	Gradient echo
SPGR/Spoiled GRASS	Spoiled gradient recalled echo	Gradient echo
FLASH/Turbo FLASH	Fast imaging using low-angle shot	Gradient echo
FastSPGR	Fast SPGR	Gradient echo
FISP	Fast imaging with steady-state precession	Gradient echo*
PSIF	Reversed FISP	Gradient echo*
SSFP	Steady-state free precession	Gradient echo
TrueFISP	True FISP	Gradient echo*
FIESTA	Fast imaging employing steady-state acquisition	Gradient echo*
b-FFE	Balanced fast field echo	Gradient echo*
SE	Spin echo	Spin echo
FSE	Fast spin echo	Spin echo
Turbo-SE	Turbo spin echo	Spin echo
RARE	Rapid acquisition with relaxation enhancement	Spin echo
SSFSE	Single-shot fast spin echo	Spin echo
HASTE	Half-Fourier single-shot turbo-spin echo	Spin echo
IR	Inversion recovery	SE or GRE
STIR	Short tau (inversion time) inversion recovery	SE or GRE
FLAIR	Fluid attenuation inversion recovery	SE or GRE
EPI	Echo-planar imaging	SE or GRE

Source (for RARE): Hennig, J. et al., *Magn Reson Med* 3, 6, 823–833, 1986 (19).
*Indicates a steady-state free-precession method producing a mix of T_1- and T_2^*-weighting.

References

1. Hahn EL. Spin echoes. *Phys Rev* 1950;80(4):580–594.
2. Hahn EL. Spin echoes. *Phys Rev* 1950;77(5):746.
3. Pines A. Erwin L. Hahn—Scientist, mentor, friend. *J Magn Reson* 2006;179(1):5–7.
4. Edelstein WA, Hutchison JMS, Johnson G, Redpath T. Spin warp NMR imaging and applications to human whole-body imaging. *Phys Med Biol* 1980;25(4):751–756.
5. Mansfield P, Pykett IL. Biological and medical imaging by NMR. *J Magn Reson* 1978;29(2):355–373.
6. Tkach JA, Haacke EM. A comparison of fast spin echo and gradient field echo sequences. *Magn Reson Imaging* 1988;6(4):373–389.
7. Carr HY, Purcell EM. Effects of diffusion on free precession in nuclear magnetic resonance experiments. *Phys Rev* 1954;94(3):630–638.
8. Bloombergen N, Purcell EM, Pound RV. Relaxation effects in nuclear magnetic resonance absorption. *Phys Rev* 1948;73:679–712.
9. Meiboom S, Gill D. Modified spin-echo method for measuring nuclear relaxation times. *Rev Sci Inst* 1958;29(8):688–691.
10. Bloch F. Nuclear induction. *Phys Rev* 1946;70(7–8):460–474.
11. Pelc NJ. Optimization of flip angle for T1 dependent contrast in MRI. *Magn Reson Med* 1993; 29(5):695–699.
12. Ernst RR, Anderson WA. Application of Fourier transform spectroscopy to magnetic resonance. *Rev Sci Instrum* 1966;37:93–102.
13. Edelstein WA, Bottomley PA, Hart HR, Smith LS. Signal, noise, and contrast in nuclear magnetic-resonance (NMR) imaging. *J Comp Assist Tomogr* 1983;7(3):391–401.
14. Mansfield P, Morris PG. *NMR Imaging in Biomedicine*. New York: Academic Press Inc.; 1982.
15. Haase A. Snapshot FLASH MRI. Applications to T1, T2, and chemical-shift imaging. *Magn Reson Med* 1990;13(1):77–89.
16. Scheffler K, Hennig J. Is TrueFISP a gradient-echo or a spin-echo sequence? *Magn Reson Med* 2003;49(2):395–397.
17. Mehta RC, Pike GB, Enzmann DR. Magnetization transfer magnetic resonance imaging: a clinical review. *Top Magn Reson Imaging* 1996;8(4):214–230.
18. Wolff SD, Balaban RS. Magnetization transfer contrast (MTC) and tissue water proton relaxation *in vivo*. *Magn Reson Med* 1989;10(1):135–144.
19. Hennig J, Nauerth A, Friedburg H. RARE imaging: A fast imaging method for clinical MR. *Magn Reson Med* 1986;3(6):823–833.

5

Creating an Image from the Magnetic Resonance Signal

The ideas presented in the previous chapters have covered how the magnetic resonance (MR) signal is produced and how we can get physiologically relevant information from it, and now it is necessary to introduce spatial information. By making the signal depend on the position of the tissues and the like from where it originates, we can map out the spatial distribution of the signal and create an image. The spatial information that is encoded into the signal does not interfere with the physiological information that we need, and so we can create images that reflect both anatomical (spatial) and physiological information. In this chapter the basic methods for creating MR images are described, as well as the choices that are available for the imaging parameters, such as the time it takes to acquire an image, spatial resolution, and relaxation-time weighting.

An image is a representation of the spatial distribution of objects, structures, materials, and so forth within a region. This might seem somewhat obvious, but it is worthwhile to begin by recalling exactly what it is we are trying to accomplish. In Chapter 3 it was shown how different materials, or more specifically, tissues or fluids for MR imaging (MRI) of the body, have different proton densities and relaxation times and can be distinguished by their different MR signal intensities as a result. To create an image it is therefore necessary to put (or *encode*) spatial information into the MR signal as well. That is, the signal we detect must somehow reflect where it originated; otherwise, we will not be able to create an image.

The easiest way to make the MR signal depend on the position in space where it originates is to apply magnetic field gradients. Linear magnetic field gradients, and how they are created, are described in Chapter 2. A gradient can be produced in any direction and is defined in terms of how much the magnetic field changes between two positions. For example, a typical gradient strength is around 1 gauss/cm, meaning that the field changes by 1 gauss for every difference in position of 1 cm. Gradient fields are designed to have zero effect at the exact center of the MRI system, and so the position within the MRI system is defined in terms of displacement from the center. It is also important to recall that the magnetic field is always parallel to the static magnetic field of the MRI system, B_0, so the gradient field adds to or subtracts from B_0. The *gradient direction*, however, refers to the direction in which the magnetic field changes. As a result, the total magnetic field now depends on the position when a gradient is applied:

$$B = B_0 + G_X X$$

for a gradient in the X-direction, and

$$B = B_0 + G_X X + G_Y Y + G_Z Z$$

for the general case of a net gradient in any direction.

In the first example above, the gradient is applied in the x direction, and the value of x is the distance (in cm) from the center of the MRI system. Standard definitions for positive and negative have to be used for consistency, and these are usually specified. For example, the RAS convention means the positive directions are to the right (R), in the anterior direction (A), and in the superior direction (S) (i.e., toward the head). Negative values would therefore mean toward the left, posterior, or inferior, directions. The second equation above is for the more general case of a gradient in any direction in three dimensions. To produce a gradient 45° from the x-axis in the x–y plane, for example, we could apply equal gradients in the x and y directions at the same time, and no gradient in the z direction. In the general case, the total gradient magnitude, G_{mag}, is:

$$G_{mag} = \sqrt{G_X^2 + G_Y^2 + G_Z^2}$$

The effect of the gradient is to change the Larmor frequency of the precessing nuclei, depending on their position. Again we rely on the Larmor equation:

$$\omega = \gamma B$$

When we explicitly write out the fact that the total field, B, is the sum of B_0 and the gradient field, then:

$$\omega = \gamma(B_0 + G_X X)$$

$$\omega = \omega_0 + \gamma G_X X$$

Now, we need to apply this spatial encoding in three dimensions, in a way that we can use to create the desired image.

Key Points

1. Spatial information can be encoded into the MR signal by measuring the signal while a magnetic field gradient is applied.
2. The gradient causes the frequency of the signal, or the phase of the signal at any fixed point in time, to depend on the position where the signal originates.
3. The gradient can be in any direction by applying simultaneous x, y, and/or z gradients.

5.1 Spatially Selective Radio-Frequency Pulses

One of the ways we can encode spatial information into the MR signal in one dimension, is by acquiring the signal from only a thin *slice*—meaning a narrow range of space in one direction, but spanning all space in the other two directions. A slice can also be thought of as a plane, except that it has some small thickness. The slice can be in any orientation, but the primary anatomical slice orientations are as follows (Figure 5.1):

Axial (or Transverse)	Slice plane is right–left (R/L) × anterior–posterior (A/P)
Sagittal	Slice plane is superior–inferior (S/I) × anterior–posterior (A/P).
Coronal	Slice plane is right–left (R/L) × superior–inferior (S/I).

Note: The superior–inferior direction is also known as the head–foot direction.

To acquire the signal from only a thin slice, we need to make the radio-frequency (RF) excitation pulse affect only the nuclei within a narrow range of a space in one direction. In brief

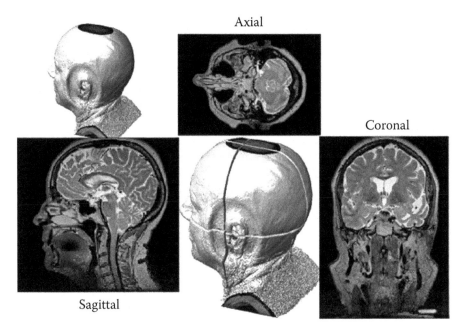

FIGURE 5.1 Definition of *slices* with respect to MRI. In this example, each 2D slice represents a 3-mm section through the head, as shown by the cutaway sections of the 3D head. One slice is shown for each of the three principal anatomical directions: Sagittal—an A/P × S/I plane; Axial—an R/L × A/P plane; and Coronal—an S/I × R/L plane.

terms, this selective excitation is accomplished by using two facts: (1) an RF pulse affects only the nuclei within a narrow range of Larmor frequencies, and (2) when a gradient is applied, the nuclei have a range of Larmor frequencies related to their positions along the gradient direction. We can therefore apply an RF pulse while a gradient is applied and excite only the nuclei within a selected range of frequencies. Since the frequency depends on the position, this is the same as exciting the nuclei only within a selected range of positions.

The frequency range that is affected by an RF pulse can be tailored to our needs (with some limitations) with the selection of the duration and amplitude envelope of the pulse. Think of the RF pulse as having two components that are multiplied together (Figure 5.2): (1) a magnetic field of constant amplitude that oscillates or rotates at the frequency we choose (at the Larmor frequency), and (2) the *amplitude envelope* that specifies the amplitude and phase of the pulse at each instant in time. The frequency of oscillation of the first of these two components determines the center of the narrow range of Larmor frequencies that will be affected. The shape and duration of the second component (the amplitude envelope) determines the frequency span,

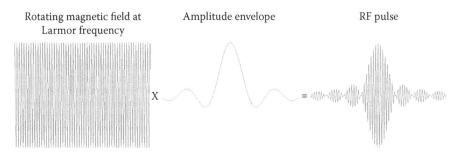

FIGURE 5.2 Components of an RF pulse.

and how the effect of the pulse varies across this span. Ideally, for exciting only a thin slice, the RF pulse would affect all nuclei equally (i.e., tip them through the same flip angle) across some frequency span and have no effect at all on nuclei with Larmor frequencies outside this span. In practice, this ideal case is impractical to achieve, and we have to be satisfied with getting as close as is reasonably possible, for reasons described below.

In Chapter 3 (Section 3.4) it was mentioned that the frequency of the RF pulse (i.e., the frequency of the rotating magnetic field) had to be very close to the Larmor frequency for the pulse to have any effect. Here we will have to be much more specific because we need to know exactly what range of frequencies is affected and whether or not nuclei at all frequencies within this range will be affected exactly the same. It turns out that the frequency span that is affected by an RF pulse can be approximated by the Fourier transform (FT) of the amplitude envelope. This approximation is most accurate for lower flip angle pulses, is fairly good for pulses of 90° or less, and is fairly poor for higher flip angle pulses such as 180°. Nonetheless, the effects of higher flip angle pulses can be accurately modeled by other means. The FT approximation also helps to demonstrate some consistent features for all pulses, such as the fact that narrower pulses affect a wider frequency span. For now we can use the FT approximation for the general discussion of how RF pulses behave. Using this estimate, we can also see that to get a perfectly rectangular frequency response (i.e., uniform over some range and zero everywhere else), we should use a pulse with an amplitude envelope that is the inverse FT of a rectangular function.

Figure 5.3 shows that the FT of the rectangular function is a *sinc* function (i.e., has the form of $\sin(x)/x$). Two key problems with the sinc function as an amplitude envelope are (1) it extends infinitely in both positive and negative directions from the center, and (2) most of the rotation of nuclei from alignment with B_0 is accomplished within a very narrow central lobe. The nuclei will be rotated at the frequency given by $\omega_1 = \gamma B_1$ as described in Chapter 3, and the total rotation angle (the flip angle) depends on how long the pulse is applied. If almost all of the rotation is accomplished in the narrow central lobe, then this lobe needs to have a very high peak B_1 value, or else the RF pulse must have a very long duration. These two problems make the sinc function impractical for use as an RF pulse amplitude envelope.

However, a practical solution is to multiply the ideal sinc function with a Gaussian function. The FT of these two functions multiplied together is equal to the convolution of the FT of the sinc function with the FT of the Gaussian function (this is the *Fourier convolution theorem* described in Chapter 2). We know from above that the FT of the sinc function is the ideal rectangular response that we want, and the FT of a Gaussian function is itself a Gaussian function. The convolution results in an effective smoothing of the rectangular function (Figure 5.4). The example in Figure 5.4 (bottom row) shows an amplitude envelope that is practical and is commonly used in practice. By multiplying the sinc function with a Gaussian function, we can use only the

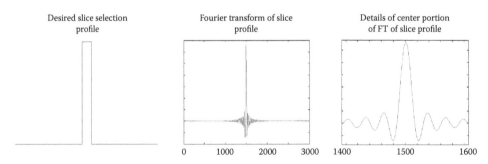

FIGURE 5.3 The ideal RF pulse frequency response profile (left), and its Fourier transform, which approximates the amplitude envelope needed to obtain this frequency response (middle) for a pulse with a duration of 3 msec. The panel on the right shows the detail of the center portion of the Fourier transform.

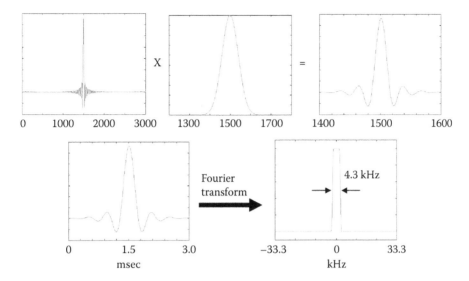

FIGURE 5.4 The top row shows the effect of multiplying a sinc function (left) with a Gaussian function (center). Note the different horizontal axis scales on the three panels. The resulting sinc × Gaussian function is zero outside the range displayed. In the bottom row this shape is shown as an amplitude envelope for an RF pulse with a duration of 3 msec. The Fourier transform of this shape is a slightly smoothed rectangular function (the convolution of a narrow Gaussian function and the original ideal rectangular function). In this example, since 200 points are used to describe a pulse with a duration of 3 msec, each point represents 15 µsec, and so the frequency range spanned after the FT is 1/15 µsec = 66.7 kHz (as shown in Chapter 2).

nonzero portions of the result as our amplitude envelope and ignore the rest. This new amplitude envelope can be scaled to span a practical duration for an RF pulse, such as 3 msec. The central lobe, where most of the work of rotating the magnetization is done, is now a larger proportion of the amplitude envelope, and the peak B_1 amplitude that is needed to rotate the magnetization has a realistic value. The resulting frequency response (i.e., the FT of the amplitude envelope) is also a reasonably good approximation of the ideal rectangular response shape.

The slice profile shown in this specific example will excite a frequency range of approximately 4.3 kHz, which is relatively large. In most cases the RF pulses used on current MRI systems will have narrower frequency responses, such as 1.5 to 2 kHz. In the example above we could apply this RF pulse while a magnetic field gradient is also applied, and only those nuclei with Larmor frequencies within this 4.3 kHz range would be rotated away from alignment with B_0. The Larmor frequency at the center of the slice is equal to the frequency of the rotating magnetic field, B_1, whereas the frequency spread that is affected is determined by the amplitude envelope shape. We could therefore choose to apply a gradient that makes the Larmor frequency change by 5 kHz/cm. In other words, a gradient of 1.17 gauss/cm (recall $\omega = \gamma B$ and ω equals 4257 Hz/gauss). We could select the center frequency of the RF pulse to be at the Larmor frequency at the center of the MRI system, ω_0 (equal to γB_0), and the pulse would affect a slice through the center of the MRI system. Alternatively, increasing the frequency to $\omega_0 + 10$ kHz would shift the slice by 2 cm to one side. If the gradient was in the right–left direction relative to a person inside the MRI system, then this would be a shift of 2 cm to the right. Because we have chosen a gradient of 5 kHz/cm to use with a pulse that affects a span of 4.3 kHz, the pulse will excite a slice with a thickness of 0.86 cm (equal to 4.3 4.3 kHz/(5 kHz/cm)). Clearly any direction and strength of the gradient field, and center frequency of the RF pulse, can be used (within the limits of the MRI system). The key point here is that we are quite easily able to control the exact orientation, slice thickness, and location of the slice we wish to excite.

In practice, on most current MRI systems, you will rarely need to think about the shape of the RF pulse amplitude envelope, the frequency response, or the direction and amplitude of the gradients that are applied. These will be computed automatically by the MRI system depending on your specified slice position and so forth. However, you will need to consider related effects of the characteristic features of the RF pulse. The frequency response of the RF pulse relates directly to the thickness of the slice that is excited and to the consistency of the flip angles across this slice thickness. For example, if the peak of the response function varies in amplitude by 5%, then the flip angles across the slice thickness vary by approximately 5% as well. Recall that depending on the imaging method being used, the signal intensity and the degree of T_1-weighting can both depend on the flip angle. As a result, some portions of the slice thickness (such as the center) could be more represented in the total signal than the portions of the slice near the edges. Ideally, the variation in flip angles across the slice will be very small to negligible.

A greater concern for functional MRI (fMRI) and anatomical imaging that is unavoidable is that the transition at the edge of the slice has some width that is greater than zero. In the example shown in Figure 5.4, the transition at the edge of the pulse spans 1340 Hz. This is roughly 30% of the width of the profile, and would be rather poor performance for a RF pulse. The width of the resulting slice is typically defined as the span that receives at least 50% of the full effect of the RF pulse, as shown in Figure 5.5. Modern MRI systems typically provide RF pulse shapes that have only 5% to 10% transition widths or less. Nonetheless, at the edge of every slice there is a very narrow region with a range of flip angles. If contiguous slices are desired, then there will inevitably be a region between any two slices that is affected, at least partially, by the RF pulses for both slices. The transition edge can be a concern for multi-slice imaging and time-series imaging, such as for fMRI, if there is no gap allowed between the slices. In most cases, images of multiple slices will be obtained in order to span the three-dimensional (3D) extent of a region of the body, and a gap may not be desired in case details of the anatomy are missed. If the slices are acquired sequentially, from one side of the body to the other, then the edge of the second slice that is excited may be affected by the preceding excitation of the first slice (and the third slice will be affected by the second slice, etc.). This will generally cause the lower signal to be detected from the zone of overlap between the slices, unless there is sufficient time allowed between the excitations for the longitudinal magnetization to relax to equilibrium. In practice, it is assumed that there is no cross-excitation between slices (or that it is small enough that it can be ignored),

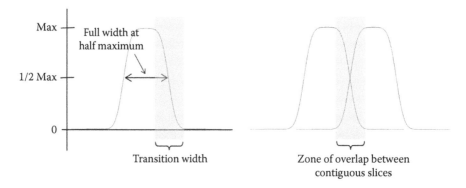

FIGURE 5.5 Schematic showing the definition of the pulse width and transition width, given that the transition between "inside" the slice to "outside" the slice has some spatial span that is greater than zero. The slice width is therefore defined as the full width at half-maximum (FWHM), or the span that has at least 50% of the full effect of the RF pulse. Two contiguous slices (i.e., with no gap between them) would therefore produce a region that is affected by both of the RF pulses that excited the two slices. Note that the transition width and overlap between slices are grossly exaggerated for demonstration purposes. In practice, the transition width is typically less than 10% of the slice width.

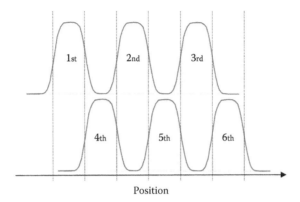

Position

FIGURE 5.6 Schematic of multiple slice profiles in an interleaved acquisition order, to reduce cross-excitation effects between slices. Again, the slice profiles are shown with exaggerated widths of the transition edges, to demonstrate the problem of cross-excitation between adjacent slices.

and multiple slices are imaged in each repetition time, TR, period. This approach is to make efficient use of time and enable large volumes to be imaged at a reasonable rate. An alternative approach that is commonly used is to image the slices in an interleaved order (Figure 5.6). For example, every other slice will be excited, and the signal from each will be detected; and then the remaining slices will be excited, and the signal detected.

Interleaved acquisitions reduce the cross-excitation effects in the zone of overlap between slices because one-half of a TR period can pass between excitation of adjacent slices, allowing time for relaxation. For example, the excitation of the even-numbered slices can be distributed in time to span the first half of the TR period, and then the excitation of the odd-numbered slices can span the second half of the TR period. Because the region in the transition width does not receive the full effect of the RF pulse, the nuclei are not tipped as far from the longitudinal axis as at the center of the slice, and therefore the transition zone requires less time to relax toward equilibrium. The extra time allowed for relaxation by the interleaved acquisition can therefore reduce the cross-excitation effects, although perhaps not entirely, depending on the repetition time and the flip angle of the RF pulses. In practice, the effect of cross-excitation between slices can be observed, even with interleaved slices, because the slices that are excited after their immediate neighbors may have slightly lower signal intensity.

For repeated multiple-slice acquisitions, such as to acquire a time series for fMRI, it is also necessary to consider the previous excitations of the adjacent slices (in the preceding TR period). The second and subsequent excitations of the slices can reach a steady state, with the zone of overlap between slices potentially having slightly more T_1-weighting than the center portions of the slices. However, the outer edges of the outermost two slices are not affected by cross-excitation (because these edges have no adjacent slices being excited), and so only these outer slices may have slightly higher signal intensity. Another way to avoid the cross-excitation is to leave a gap between the slices. In practice this gap would only need to be 5%–10% of the slice thickness (depending on the capabilities of the MRI system). However, this gap can present other problems for time-series acquisitions if the person being imaged moves slightly between TR periods, or with blood flow, because some nuclei may move between the slices and the gap. This effect is discussed in detail in Chapter 6 in relation to imaging methods specifically for fMRI.

Some MRI systems also provide options for the type of slice-selective RF pulses to use. These are typically specified as being *normal* or *sharp*, and sometimes there are options for *quiet* or *whisper* pulses. A *sharp* pulse means that it has a narrower transition width, so the slice profile is more rectangular. This is achieved usually with a longer RF pulse (e.g., perhaps 5 or 6 msec instead of 3 msec) and/or a broader frequency span being affected by the pulse. As a result, the RF pulse may take a bit more time to apply, and stronger magnetic field gradients may be required

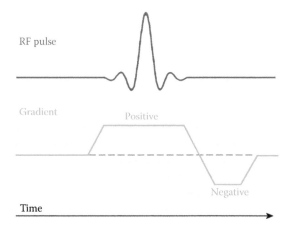

FIGURE 5.7 Schematic representation of a spatially selective RF pulse with a shaped amplitude envelope to produce the desired slice profile, the slice gradient that creates a frequency span across the object being imaged, and the subsequent slice rewind gradient to bring the magnetization in the slice into phase. The height of the green line indicates the strength of the gradient, as a function of time, and the height of the blue line indicates the amplitude of the B_1 field of the RF pulse.

to achieve the same slice thickness as with a *normal* pulse. Stronger gradients produce louder sounds in the MRI system when they are applied, and the minimum slice thickness that can be achieved may be increased. Whisper pulses, on the other hand, are designed to be quieter, meaning that they use weaker magnetic field gradients. The trade-off is that the whisper RF pulses may take longer to apply than the normal pulses, and/or they may have a wider transition width.

One last point to mention is that after the RF pulse has been completed and the gradient has been turned off, there is one very important job left to do before a signal can be detected. The gradient provided a frequency span across the object so that a specific frequency range, and therefore a specific position range, could be selectively excited. As a consequence, the frequency varied across the thickness of the slice, leaving the magnetization spread out in phase across the slice thickness after the RF pulse has been completed. After the RF pulse it is therefore necessary to apply a gradient again in the slice direction, but reversed from that used to select the slice. When this *slice rewind* gradient has been applied for the correct duration, it brings the magnetization in the slice into phase, and then a strong MR signal can be detected. As an interesting historical note, the need for the subsequent gradient to bring the magnetization back into phase was the subject of strong debate (1–3), but the necessity of this gradient is now well known. A schematic representation of the spatially selective RF pulse and gradient combination is shown in Figure 5.7.

Key Points

4. One dimension of spatial encoding can be applied with the use of RF pulses that affect only the magnetization within a thin range in one direction, or a *slice*.
5. A *spatially selective* RF pulse is created by applying the RF pulse while a gradient is also applied.
6. An RF pulse only affects the magnetization precessing at frequencies within a narrow range around the frequency of rotation of the B_1 field.
7. The frequency range that is affected is determined by the shape of the RF pulse amplitude envelope, meaning the pattern in time of the amplitude of the B_1 field.
8. At the edges of the slice region that is affected by the RF pulse is a narrow transition zone between the full effect of the RF pulse and zero effect outside of the slice.

5.2 Encoding Spatial Information into the MR Signal to Create an Image

Spatial encoding within the plane of a slice is also accomplished with the use of magnetic field gradients. For this discussion we consider the signal that we can detect after a slice-selective RF pulse has been applied. Now, we also apply a magnetic field gradient in one direction in the plane of the slice. Recall that the effect of a gradient in the x direction is to change the net magnetic field, and so it also changes the Larmor frequency:

$$\omega = \omega_0 + \gamma G_X X$$

The signal will therefore have a range of frequencies, depending on how the nuclei are distributed in the x direction. If we record the MR signal for a period of time, then we can apply the Fourier transform to determine the magnitude of the signal at each frequency. Since the frequency and the position are directly related (because of the linear gradient) this process is equivalent to determining the magnitude of the signal at each position. This is part of the information that we need to create an image. If we knew the magnitude of the signal at every position in two dimensions within the plane of the slice, we could represent the magnitude with a gray scale and display it as an MR image. However, we cannot simply apply two gradients at once to get two-dimensional (2D) spatial information; we have to resort to other methods.

First, it is useful to consider again how we can get spatial information in one direction at a time. The effect of a gradient depends on both the gradient strength and the duration it is applied. Ignoring the constant component of the frequency, ω_0, and focusing on the change in frequency $\Delta\omega$ that is caused by the gradient (remember that Δ is used to indicate a change in a value), we can see that $\Delta\omega = \gamma G_X X$. After the gradient has been applied for a time, t, the transverse component of the magnetization has changed in phase, ϕ, by the amount $\Delta\phi = \gamma G_X X t$. To put this into more familiar terms, imagine you are traveling some distance (walking, cycling, driving, etc.) and you want to keep track of your progress to your destination. The frequency is analogous to your speed of travel, and the change in phase is analogous to how far you have traveled since you began. The total phase depends on the initial phase, just as your position at any point in time would depend on where you started from.

The effect of a gradient on the magnetization in the transverse plane is shown in Figures 5.8 and 5.9, for a sample object. In the first of these two examples, two different gradients are applied, and in the other, the same gradient is applied for two different durations, producing the same net effects in the two cases. The total signal that would be detected is shown to decrease with a stronger gradient, or when applied for a longer duration, in this example.

However, these examples also demonstrate that there may be certain gradient strengths and durations that can return the magnetization in the transverse plane to having similar phases, thereby producing a stronger signal. This effect can only occur when the spacing of the features of the object being imaged matches the distance between full rotations of the magnetization. For example, if two features in an object are 5 cm apart, and the gradient and duration cause a phase change of 360° (or 2π radians) over a 5 cm span, then the magnetization in these two features is in phase. Figure 5.10 shows the same object as in the previous example, and the evolution of the transverse magnetization while a gradient is applied for increasingly long durations. Initially, the magnetization is in phase and the total magnetization (and therefore the MR signal) is very large. After the gradient is applied, the magnetization spreads out in phase, but at one specific duration the magnetization returns to being somewhat in phase because of the spacing of the features in the object. At this duration the total magnetization is larger than at shorter or longer durations, although not as large as when the magnetization was completely in phase, because the features are not point objects but have some width as well. The variation in phase across the width of the object also demonstrates that the oscillating pattern that is identified by the total signal is a sine or cosine function. Figure 5.11 illustrates this point, since a model function composed of a single sine wave oscillation has nonzero total signal at only one

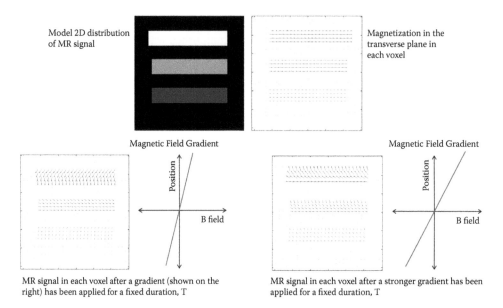

FIGURE 5.8 Simulations of the effects of gradients with two different strengths, applied for a fixed duration, T. The effect of the gradient is shown on the simulated MR signal in the transverse plane for a model object, as shown in the upper left of the figure. The grayscale image represents a simple set of three rectangular objects, each with different proton density (and therefore MR signal strength). The arrows in the illustrations of the MR signal at each voxel indicate the signal strength (corresponding to the arrow length) and the direction. The total signal measured at any instant in time is the sum over all voxels, and therefore depends on the gradient that is applied.

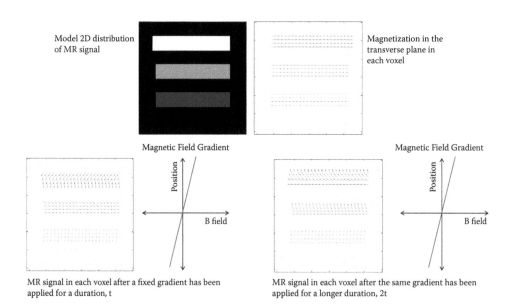

FIGURE 5.9 Simulations of the effects of one gradient applied for two different durations, t and 2t. The effect of the gradient is shown on the simulated MR signal in the transverse plane for a model object, as shown in the upper left of the figure. As in Figure 5.8, the arrows represent the magnitude and direction of the signal in each voxel.

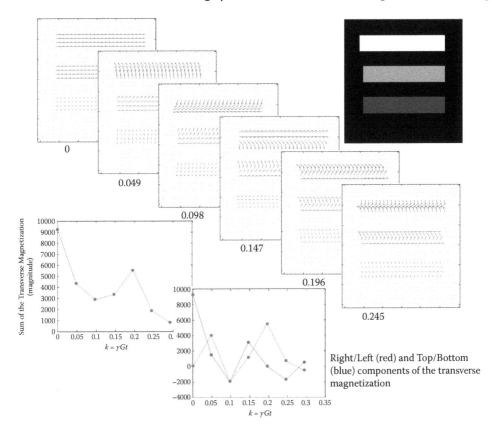

FIGURE 5.10 Simulations of the transverse magnetization in a sample object (shown in upper right corner), with a fixed gradient G, applied for a range of durations, t. The net effect of the gradient is quantified with $k = \gamma Gt$ and is indicated below each frame (arranged diagonally across the center of the figure) in units of radians/pixel. The total magnitude of the magnetization at each value of k, and the horizontal and vertical components (equivalent to the real and imaginary components), are shown in the bottom left part of the figure.

specific combination of gradient strength and duration, and zero for other combinations. The total signal at each duration and gradient strength therefore gives information about the spacing of the features of the object. In fact, this duration and gradient strength gives information about only a *specific* spacing of features. That is, it demonstrates whether or not the features in the object have a spatial component that repeats at a specific spacing.

The total effect of the gradient and duration is quantified by $k_X = \gamma G_X t$, or in the second direction we would instead use $k_Y = \gamma G_Y t$, where now X and Y can be any two orthogonal directions, such as R/L and A/P for a transverse slice, or R/L and S/I for a coronal slice. When we measure the signal over a period of time, at many different values of t, we are therefore recording at many different values of k_x or k_y. Alternatively, we could get the exact same effect by measuring the signal at one fixed duration, after applying different gradient strengths. As long as the product of γGt has the same value, we get the same sensitivity to spatial components of the object, regardless of how the gradients and timing are applied. Spatial information can therefore be encoded into the signal in two directions, by first applying a gradient in the y direction for a fixed duration, T, before removing the gradient, and then applying a gradient in an orthogonal direction, x, while we record the MR signal across a range of times, t. The resulting data would be at one value of $k_Y = \gamma G_Y T$ and over a range of values of $k_X = \gamma G_X t$. Since the values of k_X are measured by making the frequency of the precession depend on the position, k_X is called the

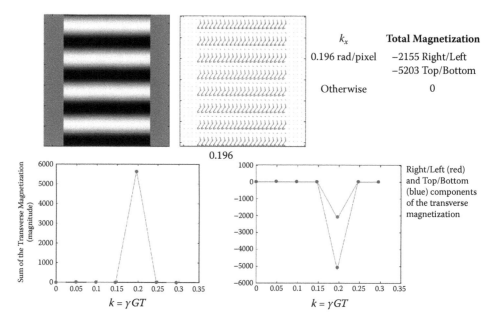

	k_x	Total Magnetization
	0.196 rad/pixel	−2155 Right/Left
		−5203 Top/Bottom
	Otherwise	0

Right/Left (red) and Top/Bottom (blue) components of the transverse magnetization

$k = \gamma GT$ $k = \gamma GT$

FIGURE 5.11 Simulations of the transverse magnetization in an object composed of a single sinusoidal oscillation in one direction (upper left corner) under the influence of a fixed gradient, G, applied for a range of durations, t. The total magnetization is shown to be zero, except at one particular value of $k = \gamma Gt$. This result demonstrates that the total magnetization at each value of k reflects the component of the features in the object that varies sinusoidally at one specific frequency. Moreover, the total magnetization at this value of k reveals both the magnitude and the phase of the sinusoidal variation.

frequency-encoding direction, or the *readout* direction. The values of k_Y are imposed by making the phase of the signal depend on the position, and so this is called the *phase-encoding* direction. Finally, recall that the Fourier transform demonstrates the magnitude and phase of each frequency component in an image, and therefore k_x and k_y are exactly the elements of the FT of the image we are trying to construct.

The units of k_x and k_y help to demonstrate what is physically represented by these values. The product γGt includes γ with units of [(radians/sec)/gauss], G with units of [gauss/cm], and t with units in [seconds]. Therefore, the result (k_x or k_y) has units of [radians/cm]. This indicates an angle of phase change across a given distance, exactly as described above. The magnitude and phase of the MR signal at each value of k_x and k_y therefore show the contribution from a pattern that varies with position as a sine or cosine function with a specific spacing between the peaks (i.e., a specific *spatial frequency*). Again, k_x and k_y are exactly the elements of the FT of the image that we are trying to construct.

Key Points

9. Spatial information in one direction is encoded by measuring the signal at the same time that a magnetic field gradient is applied—called the *frequency-encoding* direction.

10. Spatial information in a second direction, orthogonal to the first direction, is encoded by applying a magnetic field gradient for a fixed duration, T, and then turning off the gradient before the signal is measured—called the *phase-encoding* direction.

11. The effect of the gradient is determined by the gradient strength (i.e., how much the field changes over a given distance), and how long the gradient is applied before the signal is measured.

12. The effect of the spatial encoding gradients in two directions, x and y, is quantified with the values k_X and k_Y, respectively, where $k_X = \gamma G_X t$ and $k_Y = \gamma G_Y T$.

13. The MR signal is sampled at different values of k_X by using a fixed value of G_X and measuring at different points in time, t. The signal is sampled at different values of k_Y by using different values of G_Y, and keeping time T fixed.

14. The two-dimensional space spanned by values of k_X and k_Y is called k-space, and the MR signal measured at all points on a two-dimensional grid spanning k-space is equal to the Fourier transform of the MR image we wish to construct.

5.3 Constructing an Image from k-Space

All of the preceding discussion shows that if we measure the total transverse magnetization at enough values of k_x and k_y, then we can apply the inverse FT and get the image. The big question is, then, how many different values of k_x and k_y need to be sampled? The answer is provided by the general explanation of the Fourier transform in Chapter 2. First, we need to recognize that we record the MR signal at discrete points in time, and therefore at discrete values of k_x and k_y. Naturally this means we will be using the discrete form of the Fourier transform and using the key underlying assumption of the discrete FT, as described in Chapter 2. One key feature of the discrete FT is that to create a 2D image represented with N_x pixels by N_y pixels, then we must have data recorded at the same number of k_x and k_y values, respectively. For example, if we want a 256 × 256 image then we need to sample 256 k_X values at each of 256 different values of k_Y. The distances spanned by the image in each direction, called the *field of view* (FOV), are given by $2\pi/\Delta k_X$ and $2\pi/\Delta k_Y$, where Δk_X and Δk_Y are the spacing between the sampled k_X and k_Y values. For example, in the previous examples, if the MR signal was recorded at k_X values 0, 0.049, 0.098, 0.147 … radians/mm, then the spacing is 0.049 radians/mm and the FOV in the x direction would be 128 mm. The distance represented by each pixel is the size of the field of view, divided by the number of pixels, N_X and N_Y, in each direction. If we had 128 pixels in each direction, then continuing the previous example, 1 pixel would span 1 mm in each direction. The size of each pixel in the resulting image can therefore also be calculated with $2\pi/(N_X\Delta k_X)$ and $2\pi/(N_Y\Delta k_Y)$. This assumes that k_x and k_y values are sampled symmetrically from $-k_{xmax}$ to $+k_{xmax}$, and from $-k_{ymax}$ to $+k_{ymax}$, where k_{xmax} and k_{ymax} are the highest values of k_x and k_y that are sampled, at values of $\Delta k_X(N_X/2)$ and $\Delta k_Y(N_Y/2)$. The value of $N_X\Delta k_X$ is the total span of measured k_X values, and $N_Y\Delta k_Y$ is the span of k_Y values. Putting this together, we can now see that if we want to image a region with (for example) a 20 cm × 20 cm field of view, with 1 mm × 2 mm per pixel, then we need to sample 200 × 100 values of k_X and k_Y, respectively. The spacing between k-points needs to be $2\pi/200$ mm in each direction, so Δk_X and Δk_Y are both 0.0314 radians/mm. The total range of k-values is 6.28 radians/mm for k_X and 3.14 radians/mm for k_Y. Choosing 2 msec for the duration of the gradient for k_Y, then G_Y needs to range from −0.0294 gauss/mm to 0.0294 gauss/mm, in steps of 0.000587 gauss/mm, since $\Delta k_Y = \gamma \Delta G_Y T$. We can only apply one value of k_Y each time we record the MR signal, so we will have to record the signal 100 times to get 100 different values of k_Y. In each acquisition, after G_Y has been applied for the selected duration, it is then turned off, G_X is turned on, and we sample the MR signal at 200 values of k_X. We have to choose the sampling rate, so we will use 20 µsec/point for this example, and the total sampling time is 4 msec for 200 points. Because $\Delta k_X = \gamma G_X \Delta t$, the value of G_X needs to be 0.0587 gauss/mm. As a check on these values, we can use the fact that the frequency range, BW (i.e., the *bandwidth*), that we are measuring is related to the sampling rate, DW (the *dwell time*), with BW = $2\pi/$DW. (Refer to Chapter 2 for an explanation of the frequency range we can accurately measure.) At a sampling

Table 5.1 Summary of Gradients and Durations Calculated from Selected Image Properties

Choose parameters:

1. Field of view (FOV)

2. Desired pixel size, ΔX by ΔY

3. Frequency range to sample (bandwidth, BW)

4. Value of T (duration of G_Y) is typically fixed in the MRI pulse program at a suitable value

To determine:

$N_X = FOV_X/\Delta X$

$N_Y = FOV_Y/\Delta Y$

$DW = 2\pi/BW$ for BW in units of rad/sec

$DW = 1/BW$ for BW in units of Hz

DW is the time between measured points

$T_{read} = DW\ N_X$ total time to read N_X values

$\Delta k_X = 2\pi FOV_X$

$\Delta k_Y = 2\pi FOV_Y$

$\Delta k_X = \gamma G_X\ DW$

$\Delta k_Y = \gamma \Delta G_Y T$

$$G_X = 2\pi/(\gamma\ DW\ FOV_X)$$

$$\Delta G_Y = 2\pi/(\gamma\ T\ FOV_X)$$

Note: Use $\gamma = 4257$ Hz/gauss, gradients in units of gauss/cm, FOV and pixel size values in cm, and seconds for the times DW and T, in order to have consistent units.

rate of 20 μsec, we are then sensitive to a frequency span of 2π radians/20 μsec or approximately 314,000 rad/sec (equivalent to 50 kHz given that 1 Hz is 2π rad/sec). Instead of choosing the sampling rate above, we could have chosen the bandwidth if it was more convenient, to get the same result. With a gradient of 0.0587 gauss/mm for G_X (equivalent to 1570 rad/sec/mm), then the frequency range we create with this gradient over a 200-mm span is 314,000 rad/sec. This confirms that we have calculated the correct value for G_X because the frequency range created by this gradient matches the frequency range we can measure at our chosen sampling rate, as expected.

This example demonstrates that we can choose the physical parameters of the image we want to construct (field of view, pixel size, and bandwidth) and from these values determine the gradient strengths needed and the durations they must be applied, as summarized in Table 5.1. The order the points of k-space are sampled is irrelevant, making a wide range of k-space sampling schemes possible.

The zero point of k-space ($k_X = 0$ and $k_Y = 0$) corresponds to the center of the FT of the image we are trying to construct, and is at the center of k-space, with points measured on all sides. In the preceding description, it is obvious that we can sample negative and positive values of k_Y simply by making G_Y positive or negative, since $k_Y = \gamma G_Y T$. It is not obvious, though, how we can sample negative and positive values of k_X, since $k_X = \gamma G_X t$, and t in this equation is time. The solution is to first apply a negative G_X gradient to set k_X to a negative value, and then switch to a positive G_X gradient while the MR signal is recorded. While we are recording the MR signal with a gradient on, the points in k-space that we are sampling are changing in time, so we can think of this as traveling through k-space as we record the MR signal. The value of γG is then the speed and direction of the travel in k-space. We can use this idea as a tool to graphically represent the method and order we use for sampling points in k-space (Figure 5.12).

Using this form of graphical representation of k-space, we can now see how data points can be sampled everywhere within a 2D grid spanning N_X by N_Y points to complete the data needed to construct an image. It also shows how the formation of spin echoes and gradient echoes (described in Chapter 4) are essential to sampling the data at all values of k-space. The example

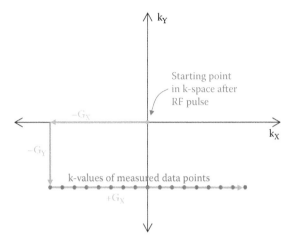

FIGURE 5.12 Graphical representation of how the application of gradients changes the value of k-space as a function of time. Immediately after the RF pulse, both k_X and k_Y are equal to zero. When a gradient is applied, the coordinates of the point that could be sampled in k-space change at a rate that depends on the gradient strength and direction, analogous to "traveling" through k-space. The sequence of gradients therefore can be displayed in a plot of k-space, along with the k-space values for each of the data points when the MR signal is recorded.

in Figure 5.12 actually shows the production of a gradient echo, since a gradient is applied, and then reversed, bringing the magnetization briefly back into phase, as it crosses $k_X = 0$. However, the magnetization in this example would not be completely in phase at the center of the echo because of the G_Y gradient (causing a nonzero k_Y value). The resulting gradient echo would therefore have a lower peak amplitude than one that passes through the very center of k-space, but is nonetheless a gradient echo.

A more general k-space depiction of a gradient-echo imaging method is shown in Figure 5.13, along with a schematic of the sequence of the RF pulse and gradients (called the *pulse sequence*). Immediately after the RF pulse, no spatial encoding has been applied, and therefore k_X and k_Y are both zero, and we are at the origin in k-space. When the negative G_X gradient is applied, we travel in the negative direction along the k_X axis. The gradient is then reversed, and we travel in

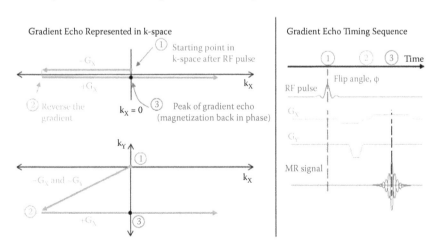

FIGURE 5.13 Representation of gradient-echo acquisitions in k-space (left), and the corresponding gradient-echo sequence of RF pulse and gradients (right).

the positive direction along the k_X axis. At some point $k_X = 0$ again, meaning that any dephasing caused by the negative gradient has been canceled out by the positive gradient, and a gradient echo is formed. It is not necessary for the speed of travel in the negative direction (the magnitude of the $-G_X$ value) to have been the same as the speed of travel when moving in the positive direction (the magnitude of the $+G_X$ value). For example, we could apply a strong negative gradient for a brief duration, followed by a weaker positive gradient for a much longer duration, to get the same effect. That is, at some point $k_X = 0$ and a gradient echo is formed. In the left panel of Figure 5.13 the effects of two different G_Y gradient amplitudes are shown, one zero and one negative. From these examples it can be seen how k_Y can be set at any one value after each RF excitation pulse, whereas k_X is spanned over a range of values while the MR signal is sampled. Repeated acquisitions of the MR data are required so that different k_Y values can be set with each value of k_X, and all of the 2D range of k-space can be spanned. At the very center of k-space, when the value of $k_Y = 0$ and the gradient echo is formed when the k_X value returns to zero, the signal peak is T_2^*-weighted.

The k-space representation of a spin echo is shown in Figure 5.14, along with the corresponding pulse sequence. The key difference from the gradient echo is the addition of the 180° pulse that inverts the values of both k_X and k_Y. Immediately after the 90° excitation pulse, k_X and k_Y are equal to zero. With the first G_X gradient we travel in the positive direction along the k_X axis to $k_X = k_{X\max}$, and the 180° pulse flips the position in k-space onto the $-k_X$ axis at $k_X = -k_{X\max}$. The G_Y gradient is typically applied after the 180° pulse so that imperfections in the 180° pulse do not cause errors in the value of k_Y. In Figure 5.14 two examples are shown with different values of G_Y; one is equal to zero and the other is negative, to set two different values of k_Y. The G_X gradient after the 180° pulse causes the k_X value to increase once again, and at some point $k_X = 0$. If the timing of the gradients is set specifically so that $k_X = 0$ (i.e., the gradients are canceled out) exactly at the point in time when the spin echo forms, then both the gradients and the static spatial magnetic field variations are canceled out. In this case the MR signal measured at the center of k-space is T_2-weighted. As with the gradient echo, k_Y can be set at any one value after each RF excitation pulse, whereas k_X is spanned over a range of values while the MR signal is sampled. Repeated acquisitions of the MR data are required so that different k_Y values can be set with each value of k_X, and all of the 2D range of k-space can be spanned.

The final outcome of the measurements described above is a complete two-dimensional grid of the MR signal magnitude and phase, at each (k_X, k_Y) coordinate (Figure 5.15). The inverse Fourier transform can now be applied to this data to construct the image of the slice that was excited with the spatially selective RF pulse. The resulting image has a field of view that is

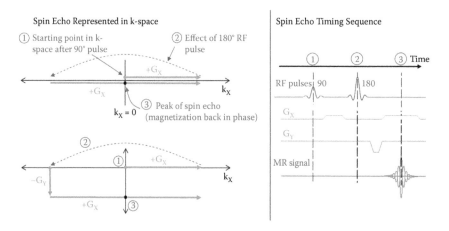

FIGURE 5.14 Representation of spin-echo acquisitions in k-space (left), and the corresponding spin-echo sequence of RF pulses and gradients (right).

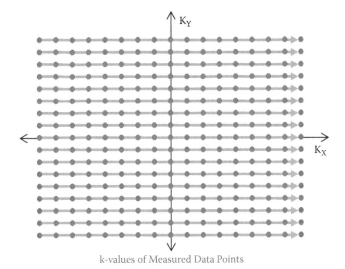

k-values of Measured Data Points

FIGURE 5.15 Two-dimensional grid of regularly spaced points sampled in k-space, with the center at $k_X = 0$ and $k_Y = 0$.

determined by the spacing between the points in k-space, and the size of each point (pixel) in the image is determined by the highest magnitude of k_X and k_Y values that are measured. The image weighting (i.e., T_1, T_2, T_2^*, or proton density) is determined by the weighting of the point at the center of k-space (k_X and k_Y both equal to zero), whereas the other points in k-space provide spatial information.

The spatial resolution of the resulting image is defined as the minimum separation required between two objects for them to be distinguishable in the image. The image resolution is therefore typically the size of 1 pixel. A common exception to this rule occurs when images are interpolated to change the size for display purposes, or to make two images the same size, and so forth. In this case the actual image resolution is still determined by the highest magnitude of k_X and k_Y values that were sampled to create the image, even though the end result might have smaller pixels. Interpolating images is fairly common, partly because it is easily done in k-space before the FT is applied. It is useful to discuss here because it helps to demonstrate how the sampling space and range in k-space affects the resulting image. If we were to sample a regular 2D grid of points in k-space and then fill in a larger range of k-space with zeros (called *zero filling*), then the resulting image would be constructed with smaller pixels. Because the measured MR signal at higher absolute values (meaning large positive or large negative values) of k_X and k_Y is quite small, the approximation made by replacing these values with zero is only a small deviation from the actual values. However, replacing the values with zero instead of measuring them still means we lose information about the image. Similarly, filling in more k-space than we have measured does not add more information about the image. The effects of zero filling are demonstrated in Figure 5.16.

The opposite action to zero filling is to crop the k-space image data by replacing the outer regions, furthest from the center of k-space, with zeros. This has the same effect as measuring the MR signal over a small range of k-space values and zero filling to a larger range, as shown in Figure 5.17. In the examples shown in this figure it appears that a relatively small portion of k-space is really needed to construct a rather detailed image. However, the ability to resolve fine details can be noticeable with even a partial loss of k-space information. For example, if an image has a resolution of 1 mm × 1 mm, then cropping the k-space data by 50% will reduce this resolution to 2 mm × 2 mm, and small features such as blood vessels, small tumors, small lesions from multiple sclerosis, and the like may be obscured. Similarly, for the purposes of fMRI, sampling

 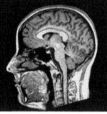 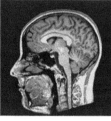 

| 900 × 900 point of k-space | 900 × 900 pixel image | 1800 × 1800 pixel interpolated image | 1800 × 1800 points of k-space, created with zero-filling |

FIGURE 5.16 k-space data for a sample 900 × 900 pixel MR image, and the resulting interpolated image after zero-filling k-space to 1800 × 1800 data points. The resulting interpolated image is essentially identical to the original image, except that the dimensions of each pixel are half as large.

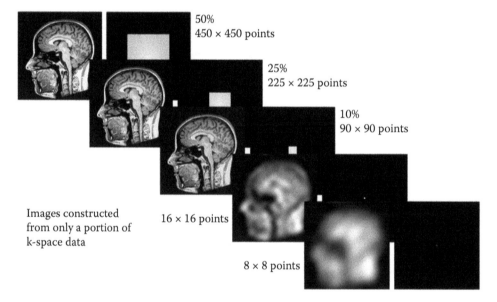

50%
450 × 450 points

25%
225 × 225 points

10%
90 × 90 points

Images constructed from only a portion of k-space data

16 × 16 points

8 × 8 points

FIGURE 5.17 Examples of the effects of cropping the k-space data (i.e., replacing outer values with zeros). The resulting images are identical to what would be obtained by only sampling a small portion of k-space data and zero filling to a larger size of k-space before applying the FT to construct the image.

a smaller portion of k-space can make it possible to acquire images faster, as will be discussed in more detail in later sections, but will be at the expense of spatial resolution.

The same methods that are used to produce 2D images can also be extended to three dimensions. A very thick slice, or *slab*, can be excited with a spatially selective RF pulse, and then spatial encoding can be applied in three dimensions within this slab. Frequency encoding is applied in the x direction exactly as with 2D images, and two phase-encoding gradients are applied in y and z, to sample all k_X values at all combinations of k_Y and k_Z values. Again, x, y, and z can be any three orthogonal axes as needed for the desired image orientation. If we want to image a 3D volume that is 20 cm × 15 cm × 10 cm, for example, with pixels that are 1 mm × 1 mm × 1 mm, then we need to sample a 3D k-space matrix that is 200 × 150 × 100 points. It makes the most sense to sample 200 k_X values each time the data are sampled (each TR period), and this must be repeated 150 times for each k_Y value and 100 times for each k_Z value, meaning that 15,000 TR periods will be needed to sample all of the necessary data. If a gradient echo method is used with a very short TR value and low flip angle, then the data can be acquired in a reasonable amount time. For example, a 30-msec repetition time can be used with a 13° flip angle, to produce a total image acquisition time of 7.5 minutes.

Key Points

15. The spacing of the sampled points in k-space, Δk_X and Δk_Y, determines the size of the field of view of the resulting image in the x and y directions, respectively.
16. The upper limits of the ranges of k_X and k_Y values that are sampled determine the dimensions of the pixels, ΔX and ΔY, in the resulting image.
17. The center of k-space, where k_X and k_Y are both equal to zero, determines the relaxation-time weighting and contrast for the entire resulting image.

5.4 Signal Strength, Imaging Speed, and Spatial Resolution—You Cannot Have It All

The limit of the spatial resolution that can be achieved is an important issue for all applications of MRI, including fMRI. One might ask, why don't we image at high enough resolution to see individual neurons for fMRI? There are several reasons why this is not currently possible, and may never be, even with advances in technology. One key concept that has not yet been introduced is that of *noise* in the MR data that are recorded. The standard definition of noise, with respect to electronic equipment, is an unwanted random component of a signal. For MRI, the term *noise* is often extended to include both the random fluctuations in the measured signal as well as unwanted signal components that arise due to physiological motion, such as breathing and heartbeat, and are therefore not entirely random. The effects of physiological motion can be avoided or reduced somewhat, and will be discussed more in detail in relation to fMRI in the chapters that follow. Here we will focus on the random noise that cannot be avoided.

The cause of the random fluctuations (i.e., noise) in the MR signal is thermally driven movement (Brownian motion) of electrons (4) in the body being imaged and in the receiver electronics of the MR equipment. Electrical noise can be generated at all stages of the receiver chain from the MR receiver coil, the preamplifier to the analog-to-digital (A/D) converter (5–8). For this reason, modern MRI systems are designed to convert the signal from analog to digital as close as possible to the receiver coil. Independent of the distance between the receiver coil and the A/D converter, the noise generated by the body is the dominant noise source for human MRI. Efforts to further reduce noise generated by the receiver electronics therefore yield only minor improvements in the resulting image quality (owing to the very high quality of the electronics in modern MRI systems). The noise generated by a live human body depends on several factors, including the frequency range that is measured (i.e., the bandwidth, BW), the B_0 field strength, and factors relating to the shape and sensitivity of the MR coil (4,6). The coil(s) used to detect the MR signal should be fit close to the area being imaged to be sensitive to the least possible volume of tissue and, as a result, detect less noise from the body. For example, when imaging a person's head, using a coil designed to fit around the head will produce a signal with less noise than would a large body coil. In this example the head coil would also produce a stronger signal because the elements of the head coil are closer to the precessing magnetization in the head. In addition to factors of the coil design and shape, the strength of the MR signal that is measured immediately after a 90° RF pulse depends on B_0 and the number of protons contributing to the signal. Note that here we are assuming the signal is from tissues at body temperature, and we are ignoring relaxation effects. Because the MR signal, S_{signal}, increases approximately as a function of B_0^2, and the noise, S_{noise}, increases approximately with

B_0, the signal-to-noise ratio, S_{signal}/S_{noise} (also called SNR) increases approximately linearly with the field strength, B_0 (4). Another factor that must be considered, though, is how many times the signal is measured to contribute to the result. For example, we could measure the same signal several times, N_{sample}, and sum the results. The random noise increases with the square root of the number of measurements ($S_{noise} \times \sqrt{N_{sample}}$), but the constant signal increases the same amount with each measurement, ($S_{signal} \times N_{sample}$). Therefore, S_{signal}/S_{noise} increases with $\sqrt{N_{sample}}$. If we measure the same signal four times and sum the measurements, we increase the SNR by a factor of 2. The strength of the MR signal that we measure is more easily manipulated than the noise because it depends on the number of hydrogen nuclei that are contributing, and therefore on the volume of tissue we are measuring from. This point brings us to how the SNR and the image resolution are intertwined.

When measuring the MR signal to construct an image, we measure the signal from the entire slice or volume that was excited by the RF pulse. Each measured point in k-space contributes to the entire image, but the signal is effectively distributed among all of the pixels in the image. The physical volume that is represented by one pixel in an MR image is called a *voxel* and is determined by the pixel dimensions and the slice thickness, and directly affects the number of hydrogen nuclei that are contributing to the signal for that pixel. The SNR of the resulting image therefore depends on the pixel dimensions and slice thickness, and on the number of k-space points that are measured. The number of k-space points that are measured has a direct effect on how long it takes to acquire all of the data to construct an image. The time to acquire the data of course also depends on the sampling rate, and the sampling rate determines the frequency range that is measured (the bandwidth). As mentioned above, the noise in the measured signal also increases as a function of a square root of the bandwidth. Now, this description might seem to be going in circles, but the key point here is that the image resolution, the time to acquire the image data, and the SNR are all inextricably linked. This interaction will be shown in Chapter 7 to be extremely important for fMRI, when we want to obtain images with very high SNR and high resolution in a very short time.

Fortunately, there is a way to summarize these interactions and estimate how a change in the resolution, the imaging speed, or the SNR will require (or cause) a change in the other two values. The SNR is proportional to the square root of the number of measurements that contribute, $\sqrt{(N_X N_Y)}$, because the MRI signal measured at each point in k-space contributes to the entire image. The noise in each pixel of the resulting image is therefore the average of the noise across all of the points in k-space. The noise itself is proportional to the square root of the bandwidth, $\sqrt{BW}$, and so the SNR is proportional to $\sqrt{(N_X N_Y/BW)}$. This equation can be simplified slightly because the sampling rate, DW (recall it is also called the dwell time) is related to the bandwidth: $DW = 2\pi/BW$. The SNR is therefore proportional to $\sqrt{(N_X N_Y DW)}$, which is equal to $\sqrt{(total\ sampling\ time)}$. Now we also need to include the dependence on the voxel volume, V, and so:

$$SNR \propto V \sqrt{(total\ sampling\ time)}$$

In some cases, it is useful (but perhaps less intuitive) to expand this expression into all of the contributing terms:

$$SNR \propto \frac{FOV_X}{N_X} \times \frac{FOV_Y}{N_Y} \times slice\ thickness\ \sqrt{(N_X N_Y/BW)}$$

$$SNR \propto \frac{(FOV_X \times FOV_Y \times slice\ thickness)}{\sqrt{(N_X N_Y BW)}}$$

$$SNR \propto \frac{Total\ imaging\ volume}{\sqrt{(N_X N_Y BW)}}$$

While this latter equation is simply another way of expressing the same dependence of the SNR on the imaging parameters, it can be helpful when choosing the imaging parameters in order to be able to see exactly how they will affect the SNR. For example, the total time required to sample the k-space data can be reduced by increasing BW or by decreasing the total number of sampled k-space points (i.e., $N_X \times N_Y$). This reduction in total sampling time could be achieved without also suffering a reduction in SNR, by balancing the increase in BW with a decrease in $N_X \times N_Y$ or a reduction of the total imaging volume. In any case, the equations above show that the total sampling time cannot be reduced without decreasing the SNR or increasing the voxel volume (reducing the image resolution).

There is an important distinction between the *total sampling time* and the total time it takes to acquire an image, which we will call the *acquisition time*. The total sampling time is only the time spent actually recording the k-space data and does not include the time required to apply RF pulses and gradients and to allow for relaxation between repeated RF excitation pulses. As discussed in preceding sections, the k-space sampling is centered on the formation of the echo, at the time TE from the center of the RF excitation pulse. A longer total sampling time therefore means the minimum TE that can be used is also longer. Subsequently, longer times for TE and the sampling time cause a longer minimum repetition time, TR. The total acquisition time depends on the repetition time, TR, and the number of TR periods that will be needed to collect all N_Y lines of data spanning N_X points in k-space. The total acquisition time may also depend on the number of slices being imaged.

With conventional imaging methods, as opposed to fast imaging methods, which will be described in the next section, one line of data consisting of multiple k_X values at a single value of k_Y is sampled in each TR period. The time needed to excite one slice and sample the data is approximately equal to the echo time, TE, plus one half of the sampling time, as shown in Figure 5.18. If the TR is several times longer than this minimum time needed for one slice, then other slices can be excited and sampled during the same TR period. For example, if the echo time is 75 msec and the sampling time is 8 msec, then at least 79 msec are needed for each slice, and in reality a few more msec are needed. If the TR is set at 1 sec, then it may be possible to acquire data from 12 slices. If the data are to be sampled over a k-space grid, $k_X \times k_Y$, that is 256×128 points, then 128 TR periods are needed to sample all of the data, for a total acquisition time of 128 sec. However, if 13 to 24 slices are to be imaged in this example, then the slices must be imaged in two separate series of images and concatenated to produce one series. The total acquisition time would therefore be twice as long. Most MRI systems will automatically divide the acquisition into the necessary number of concatenations and will acquire adjacent slices as far apart in time as possible in order to avoid cross-excitation effects. For example, if two sets of images are acquired, the odd-numbered slices would be imaged in the first set of images, taking $N_Y \times$ TR seconds, and the even-numbered slices would be imaged in the second set, taking another $N_Y \times$ TR seconds. In practice, in some cases the total acquisition time can be reduced by

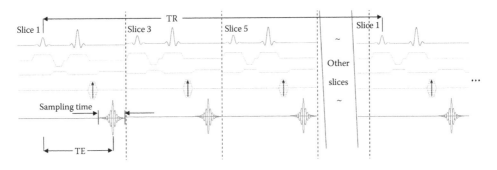

FIGURE 5.18 Schematic of the sampling scheme for multiple slices.

increasing the TR to reduce the number of concatenations that are needed. Of course, this is only possible if the increase in TR still meets the desired conditions for T_1-weighting the images.

Technical limitations of the image resolution and field of view are imposed because there are upper limits to the gradient strengths that can be produced and the bandwidth that can be sampled. Many of the current commercial MRI systems produce maximum gradients of 4.5 gauss/cm or 19.2 kHz/cm. With a slice-selective RF pulse affecting a frequency range of 1500 Hz, then the absolute minimum slice thickness that can be excited is 0.8 mm. In theory, with a 19.2 kHz bandwidth for the measured signal, then a 1-cm field of view could be imaged. However, this would mean that the data points are measured every 8.3 μsec (1/(2π 19,200 cycles/sec)), taking only 0.53 msec to measure 64 points of k-space in the frequency-encoding direction. Each voxel would span only 0.156 mm in the x direction. In the phase-encoding direction, the gradient steps between k_Y lines is 0.141 gauss/cm in order to span the maximum range from 4.5 gauss/cm to –4.5 gauss/cm in 64 steps. The phase-encoding gradient would have to be applied for 10.5 msec to achieve a field of view of 1 cm (recall that $\Delta G_Y = 2\pi/\gamma\ T\ FOV_Y$). (Limiting the gradient duration to a more reasonable value of 2 msec, then the minimum field of view that can be achieved is 5.23 cm.) Thus the upper limits of the gradients create lower limits on the slice thickness and image resolution that can be achieved, independent of the limits imposed by the need for an adequate SNR.

In order to estimate the SNR we can compare this image acquisition with 64 × 64 points, 0.8-mm slices, 1 cm × 1 cm FOV, and 19.2 kHz bandwidth, to a more conventional image acquisition with 256 × 256 points, 2-mm slices, 22 cm × 22 cm FOV, and a 19.2 kHz bandwidth, and we must assume that the TE, TR, and flip angles are identical for the two acquisitions. While the more conventional acquisition might be expected to have an SNR in the vicinity of 100, the high-resolution acquisition we are considering would have an SNR of roughly 0.33. (Recall that $\propto$ *Total imaging volume*/$\sqrt{(N_X\ N_Y\ BW)}$). To produce images with an adequate SNR to reliably detect features in the anatomy, we would therefore have to acquire the same data multiple times. As discussed above, summing (or averaging) the data from repeated acquisitions will increase the SNR as a function of the square root of the number of acquisitions. Therefore, if we want to increase the SNR to the modest value of 33, we would have to sum the data from 10,000 identical acquisitions of all N_X by N_Y points, and this is not practical.

As a bit of an aside, it is interesting to see how the SNR is affected by 3D acquisitions. We can look again at the previous 3D example with a 20 cm × 15 cm × 10 cm field of view, with 1-mm cubic voxels (recall that a *voxel* is the volume represented by a pixel in an image). In this example we set the TR to 30 msec and the flip angle to 13° to attain a reasonable acquisition time of 7.5 minutes. For comparison, we can look at a conventional 2D acquisition with the same TR, flip angle, and bandwidth, but with a 1-mm slice and 20 cm × 15 cm field of view, with a 200 × 150 point sampling matrix in k_X and k_Y. The image resolution is therefore identical in the two cases, and the time to image of one slice in two dimensions will be 4.5 sec. The SNR of the 3D image data in this example is therefore 10 times higher than the SNR of the 2D image, because of the factor of 100 in the total imaging volume and in the number of k_Z points, N_Z. (Extending the expression for the SNR to 3D: $\propto$ *Total imaging volume*/$\sqrt{(N_X\ N_Y\ N_Z\ BW)}$).

Returning to the small field of view imaging example, another problem with trying to image only a 1-cm field of view within part of the body, as in the previous example, is that the body will surely extend beyond this field of view. When the k_X gradient is applied, the parts of the body that are outside of the desired field of view will be precessing at frequencies that are outside the frequency range that can be measured with a sampling interval of 8.3 μsec. This problem is avoided by filtering the signal to exclude any frequency components outside of the desired range. However, in the phase-encoding direction a filter cannot be applied, since only one value of k_Y is sampled at a time. The signal from hydrogen nuclei that are outside of the field of view in the Y direction will be changed in phase by more than 180° (or π radians) between each successive value of k_Y, and is therefore indistinguishable from a smaller phase change in the opposite

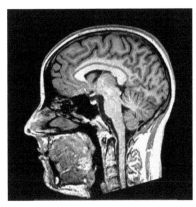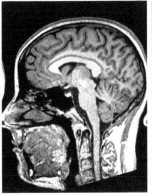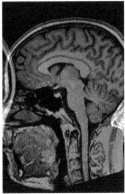

FIGURE 5.19 Example of aliasing (wrap-around) caused by selecting a field of view that is smaller in the phase-encoding direction than the anatomy being imaged. The unaliased image is shown on the left, followed by two examples of different amounts of aliasing in the middle and right frames.

direction. For example, a phase change of 185° is identical to one of −175°. In successive steps of k_Y the phase change would be indistinguishable from a rotation at a lower frequency in the opposite direction. After the FT is applied to the k-space data, the signal from outside of the field of view is misrepresented at a location within the field of view. This type of error is called *aliasing* and is described in detail in Chapter 2 (Section 2.3.3) and shown in the example in Figure 5.19. To avoid aliasing in the phase-encoding direction, we need to eliminate the signal from outside of the field of view by applying spatially selective RF pulses to tip the magnetization in these regions away from equilibrium prior to each RF excitation in the imaging sequence. These pulses are called *suppression* pulses or *saturation* pulses, depending on how they are designed to eliminate the signal. The *saturation* method is to use a long-duration, low-power pulse to rotate the magnetization through a very large flip angle (several times 360°) while relaxation processes also occur, in order to produce zero net transverse and longitudinal magnetization. The other method, the *suppression* method, is to rotate the magnetization through a flip angle such that the added effect of the subsequent RF excitation pulse in the imaging sequence will rotate the magnetization onto the longitudinal axis (for a total flip angle of either 0° or 180°).

We could use a different approach to obtain very high resolution images and also use a large field of view of 22 cm to span the entire brain (for example), and thereby avoid the problem of aliasing. If we use the full capabilities of the gradients, assuming a limit of 4.5 gauss/cm, then we must sample the data points every 14.9 μsec (this is the value of DW). (Recall that $G_X = 2\pi/$ $\gamma DW\ FOV_X$)). The resulting bandwidth is 67.1 kHz. To achieve a pixel size of 0.150 mm in the x direction, then we will have to sample 1467 points to span the 22-cm field of view, and this will take 21.9 msec. To achieve the same resolution and field of view in the phase-encoding direction, the phase gradient will have to span from −4.5 gauss/cm to 4.5 gauss/cm in 1467 steps and will have to be applied for 10.9 msec. As discussed above, the minimum slice thickness is expected to be around 0.8 mm, and so the resolution of 0.15 mm will not be achieved in all three directions. The minimum echo time possible with a gradient-echo sequence would be approximately 23.4 msec (1/2 of the RF pulse + phase gradient duration + ½ of data sampling time), and the time to acquire data from each slice (RF pulse + phase gradient duration + data sampling time) would be at least 35.8 msec. If we use a long repetition time of 5 sec, we will have time to sample data from 139 slices in each TR period to span 11.1 cm. This would provide quite good coverage of the 3D volume of the brain but would require a total imaging time of 7335 sec (5 sec × 1467 TR periods), or just over 2 hours. To estimate the SNR, again we can compare this image acquisition with 1467 × 1467 points, 0.8-mm slices, and 67.1 kHz bandwidth with a more conventional image acquisition with 256×256 points, 2 mm thick slices, and a 20 kHz bandwidth, and we must assume that the TE, TR, and flip angles are identical for the two acquisitions. While the more conventional acquisition might be expected to have an SNR in the vicinity of 100, the high-resolution acquisition we are considering would have an SNR of roughly 3.8. The end result is therefore that we would spend two hours to acquire an image with poor SNR. We have not even yet considered the challenge of asking a person to lie motionless inside the MRI system for this length of time, tolerating the considerable acoustic noise that would be produced by the rapid switching of the gradients at the upper limits of their design capabilities. Rapid gradient switching can also induce peripheral nerve stimulation, resulting in a slight feeling of tingling or pressure in the brief time that the gradients are turned on or off (9). Stimulation occurs more often when the person's body forms a closed conducting loop such as when their hands are clasped together or their ankles are crossed and the skin makes contact. As an example of the gradient strength needed to produce this effect, the stimulation threshold has been estimated to be approximately 2.2 gauss/cm when the time to switch on the gradient was 300 μsec (9). At extremely high rates of changing magnetic fields, the sensations may be uncomfortable or painful, and cardiac stimulation is also possible (10). Safety guidelines are set by the Food and Drug Administration (FDA) and International Electrotechnical Commission (IEC) to limit the gradient switching to approximately 20 tesla/second, although this is scaled for different modes of operation and with the duration of the gradient. The upper limit of magnetic field gradients that can be applied is therefore not strictly a limitation of the MRI hardware design.

In summary, the answer to the question why don't we image at high enough resolution to see individual neurons for functional MRI is that (1) it is not practical to try to acquire the amount of data that would be needed while a person lies motionless inside an MRI system, (2) the body cannot comfortably tolerate the changing magnetic fields that would be needed to acquire this data faster, and (3) the MR signal is too small compared with the electrical noise that is generated by the body.

Key Points

18. *Noise* is random fluctuations in the MR signal, and it originates primarily from motion of electrons in the body being imaged.

19. The impact of the noise, N, depends on its magnitude relative to the strength of the MR signal, S, and is quantified by the ratio S/N. This value is called the *signal-to-noise ratio*, or SNR.

20. The signal strength changes with the voxel volume, and the SNR increases when more data points are sampled. As a result, SNR $\propto V \sqrt{\textit{total sampling time}}$, where V is the voxel volume and the *total sampling time* refers to the time spent recording the data.

21. Since the time spent recording the data is determined by the number of points sampled ($N_X \times N_Y$ for the two-dimensional grid of points), and the frequency range (i.e., bandwidth) that is sampled, BW, the SNR can also be estimated with SNR $\propto$ (*total imaging volume*)/($\sqrt{N_X N_Y \text{BW}}$).

5.5 Fast Imaging Methods

A limitation of the methods discussed above is that one line of k-space data, at a single k_Y value, is measured after each RF excitation pulse. It is possible to acquire the k-space data faster by (1) using a shorter repetition time, TR, (2) sampling data at fewer k_Y values, or (3) sampling data at more than one k_Y value in each TR period. Typically, such "fast" imaging methods gain imaging speed at the expense of image quality, such as susceptibility to image artifacts (to be discussed in the following section), lower spatial resolution, and/or lower SNR.

Speeding up the image acquisition by reducing the TR is quite easily done but changes the T_1-weighting of the resulting images. Spin-echo methods have a fixed excitation pulse of 90°, and at TR values of 2 sec, 0.5 sec, and 0.1 sec will result in relative signal intensities of 100%, 44%, and 10% (assuming T_1 = 1100 msec), respectively, as discussed in Chapter 4. Gradient echoes with the optimal flip angle (the Ernst angle), however, will have relative signal intensities of 100%, 56%, and 25% at the same respective values of TR (with flip angles of 81°, 51°, and 24°, respectively). This approach is therefore much better suited for use with gradient echoes and can provide a substantial reduction in imaging time, but at a considerable cost in signal intensity and therefore in the SNR.

Sampling fewer lines of k-space data can be accomplished by decreasing the field of view in the phase-encoding direction, by increasing the pixel size in the phase-encoding direction, or by simply not sampling data at some k_Y values and filling in the data values with zeros (i.e., zero filling). It is often the case when imaging some regions of human anatomy that one dimension is smaller than another, and a rectangular field of view is a better fit than a square one. In this situation the frequency-encoding (x) direction should be along the larger dimension, and the phase-encoding (y) direction be along the smaller dimension. For example, a transverse section of the brain will often fit into a 22 cm × 18 cm field of view, and then k-space can be sampled with a 256 × 209 matrix (k_X × k_Y) so that the pixel size is 0.86 mm × 0.86 mm. Compared with a square field of view of 22 cm × 22 cm sampled with a 256 × 256 matrix, this approach will reduce the total acquisition time by almost 20% with no loss of image resolution. However, it will also reduce the SNR by about 10%, because fewer data points are measured and used to construct the image (recall that $\propto$ *Total imaging volume*/$\sqrt{(N_X N_Y BW)}$). Alternatively, if we decrease N_Y by 20% (such as decreasing from 256 to 205) without changing the field of view, then the SNR will be increased by 12%. These two approaches therefore achieve almost the same effect of reducing the imaging time, while one is at the expense of a loss of field-of-view size and decrease in SNR, and the other is at the expense of image resolution with a gain in SNR.

The third option, omitting some data and replacing it with zeros, will also result in a reduction in SNR but without a loss of field-of-view size or image resolution. Data must still be sampled at k_Y values up to $\pi/\Delta Y$ on one side of the center of k-space (where Δy is the pixel size in the y dimension), or the image resolution will be decreased, as with zero filling described earlier. However, because of the symmetry of k-space the corresponding values on the other side of center (such as the value at $k_Y = -\pi/\Delta y$) can be replaced by zeroes, up to a limit. Figure 5.20 shows examples of the effects of replacing portions of k-space with zeros (based on the same original image as in the examples in Figures 5.16 and 5.17). It can be seen in the example shown that up to 45% of the k-space data on one side of k-space center can be replaced with zeroes with no apparent change in image quality. In practice, MR images are commonly acquired with sampling matrices of 256 × 256 (i.e., k_X points × k_Y points), and up to 116 lines can be replaced by zeros, with good resulting image quality. The number of lines that are sampled symmetrically on both side of k-space center is called the *oversampling*. With 12 lines of oversampling to fill out a 256 line matrix of points, then 116 out of 256, or 45% are filled with zeros, in agreement with the example shown in Figure 5.20. Nonetheless, with less data sampled, the SNR is reduced accordingly. The method of increasing the imaging speed is termed *partial-Fourier*, or *half-Fourier* when almost half (but never fully half) of k-space is zero-filled.

The third option mentioned above, sampling data at more than one k_Y value in each TR period, can be achieved by producing more than one echo after each RF excitation pulse. In the case of a spin echo, 180° refocusing pulses can be applied repeatedly to create multiple spin echoes sequentially, as shown in Figure 5.21. Each echo must be produced while the same x gradient is applied so that the effects of the x gradients are canceled out exactly at the center of each echo. In effect, the k_X value passes through zero exactly at the center of each spin echo, as

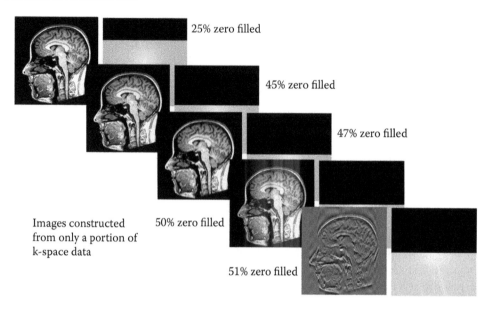

FIGURE 5.20 Examples of the effects of sampling only a portion of k-space and replacing the missing portions with zeros (i.e., zero filling). The original image is identical to that shown in Figures 5.16 and 5.17. Each row shows the k-space data on the right, and the resulting image created by the Fourier transform of the k-space data on the left.

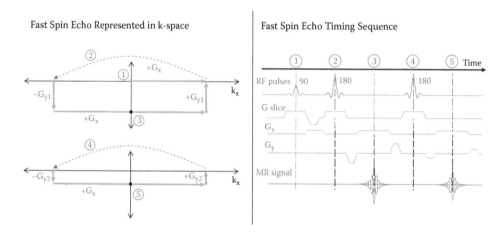

FIGURE 5.21 Schematic representations of the fast spin-echo imaging method in k-space (left) and with the timing of the RF pulses and gradients (right). The reduction of imaging time, making the method fast, is achieved by sampling the full range of k_X values at a different value of k_Y for each echo, thus providing data for more than one line of k-space in each TR period.

with a conventional spin-echo method. In addition, a y gradient must be applied prior to each echo to set a value of k_Y for each line of k_X values to be sampled, and then after the echo the opposite y gradient is applied to set the value of k_Y back to zero. A new value of k_Y is then set for each subsequent echo. In this manner, almost any number of lines of data at different values of k_Y can be acquired after each RF pulse, and the k_Y values need not be sequential but can be in any order desired. However, each echo is T_2-weighted from the time of the initial RF excitation pulse, and so later echoes are increasingly reduced in amplitude due to relaxation effects. The effects of this variable T_2-weighting across lines of k-space is discussed in detail in Section 5.6 on image artifacts and distortions. This method is typically termed a *fast spin echo* or a *turbo spin echo*, and the number of lines of k-space, and therefore the number of echoes, that are sampled after each RF excitation pulse is called the *echo train-length* or the *turbo factor*. A special case of this approach is when all of the k-space data for a complete image is acquired after a single RF excitation pulse, and this is called a *single-shot* fast spin echo. However, the single-shot method would require an exceedingly large number of echoes to be produced if every k_Y line of k-space was to be sampled. This method is therefore typically combined with the partial-Fourier method to reduce the number of echoes that are needed. For example, the first echo could be sampled at $k_Y = -12\ \Delta k_Y$, the second echo at $k_Y = -11\ \Delta k_Y$, and so on, so that the 13th echo is at k-space center, $k_Y = 0$. This sequence continues until the 141st echo, when $k_Y = 128\ \Delta k_Y$. This approach would give a half-Fourier acquisition with 12 lines of over-sampling as described above. It is called a *half-Fourier single-shot turbo spin echo*, or HASTE on some MRI systems, or SSFSE for *single-shot fast spin echo*.

The corresponding approach with a gradient-echo sequence is achieved by repeatedly reversing the x gradient (i.e., frequency-encoding gradient) to produce repeated gradient echoes. Recall that to produce one gradient echo we apply a gradient, and then reverse it to bring the transverse magnetization back into phase briefly while we record the MR signal. Here we simply reverse the gradient again to produce another echo, and again if we wish, and so on. Between the formation of each echo, while the x gradient is being reversed, an additional y gradient (i.e., phase-encoding gradient) is applied to change the k_Y value. However, because there is little time available, the k_Y value cannot be changed very much and so small incremental shifts in k_Y are applied. Typically, a large initial y gradient is applied before the formation of the first gradient echo to set k_Y at a large negative value, $-k_{Ymax}$. The incremental shifts between each successive echo then increase k_Y sequentially through the desired values to span k-space, until it arrives at $+k_{Ymax}$. Returning to the idea of traveling through k-space as we apply the gradients (as shown in Figure 5.22), we can see that this method of applying the gradients will cause the k-space values to sweep back and forth across k-space in the k_X direction, and we slowly also move across k-space in the k_Y direction, to span every point in the 2D plane. As a

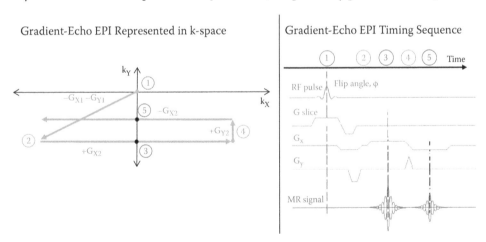

FIGURE 5.22 Schematic representations of the echo-planar spatial encoding scheme with a gradient-echo imaging method, shown in k-space (left) and with the timing of the RF pulse and gradients (right). Again, the reduction of imaging time is achieved by sampling the full range of k_X values at a different value of k_Y for each echo, thus providing data for more than one line of k-space in each TR period. In practice it is common to produce a string of 64 echoes to sample a complete 64 × 64 2D grid of points in k-space, sufficient to construct a complete image, after a single RF excitation pulse. For clarity, only two echoes are shown in these schematics, and so a complete EPI acquisition is not shown.

result, this method is called an *echo-planar* sampling scheme to produce *echo-planar images* or EPI. However, as with the fast spin echo described above, each successive echo is increasingly diminished in amplitude due to relaxation effects. In this case the echo amplitudes decay exponentially with the characteristic time T_2^* from the time of the RF excitation pulse. As a result, the MR signal at all points of k-space must be measured very quickly before the transverse magnetization decays to zero, and so smaller sampling matrices are typically used, such as 64×64, as opposed to the more common 256×256 for conventional spatial-encoding methods. The smaller matrix size generally means that a larger pixel size (lower spatial resolution) is obtained, but this is offset by the gain in SNR and the fact that images can be obtained very quickly, often in well under 1 sec per imaging slice. An alternative to the single-shot approach in which all of k-space is sampled after one RF excitation pulse is to *segment* the acquisition and only sample a portion of k-space (such as one-fourth) after each excitation, with several excitations (such as four) then required to complete the sampling. However, the single-shot and segmented approaches are each susceptible to image artifacts to different degrees, as will be discussed in the next section.

The same approach used for gradient-echo imaging with an EPI scheme for spatial encoding can be used with spin-echo imaging. This method is quite different from the fast spin echo described above because only one spin echo is produced. But, just like with the gradient echo EPI, the x gradient is applied in a positive direction, then negative, then positive, and so forth, to sweep back and forth across k-space in the k_X direction (Figure 5.23). At the same time, the value of k_Y is stepped across k-space with a small y gradient that is applied each time the x gradient is reversed. As a result, the data sampling sweeps across all values of k_X and k_Y in a 2D plane, to produce the same echo-planar sampling as described above for the gradient echo. A key point, however, is that the gradients must be applied so that the center point of k-space, $k_X = 0$ and $k_Y = 0$, is reached exactly at the center of the spin echo. In this way, the center point of k-space, and therefore the entire resulting image, is T_2-weighted instead of T_2^*-weighted. Some sources

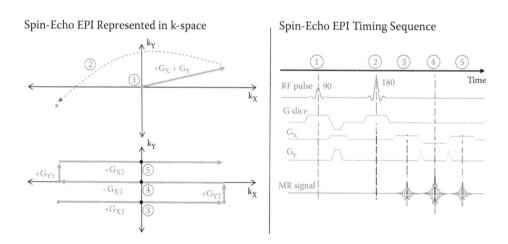

FIGURE 5.23 Schematic representations of the echo-planar spatial encoding scheme applied with a spin-echo imaging method, shown in k-space (left) and with the timing of the RF pulse and gradients (right). Again, the reduction of imaging time is achieved by sampling the full range of k_X values at a different value of k_Y for each echo, thus providing data for more than one line of k-space in each TR period. In contrast with the fast spin echo, only one spin echo is produced and the gradients are applied so that the center of k-space is sampled at the center of the spin echo. Again, for clarity, only three lines of space are shown being sampled; a complete EPI acquisition is not shown. In this example the second line of k-space that is sampled passes through $k_Y = 0$ and is aligned with the center of the spin echo.

describe the resulting images as having a combination of T_2- and T_2^*-weighting, but this is a misconception that is explained in Section 5.7.

Key Points

22. *Fast* imaging methods are designed to shorten the time needed to acquire an image, usually at the expense of image quality (spatial resolution, signal-to-noise ratio, sensitivity to image artifacts).
23. One approach is to use a gradient-echo method with a small flip angle and very short repetition time, TR, in order to achieve a short total image acquisition time.
24. Another approach is to sample multiple lines of k-space data at different values of k_Y, where each line spans the full range of k_X values, after each RF excitation pulse.
25. With the fast spin-echo method, multiple echoes are produced and one line of k-space data (spanning a range of k_X values) is sampled at a different value of k_Y for each echo.
26. With *echo-planar imaging* (EPI), gradients are applied to sweep the k_X values back and forth across k-space while also stepping across k_Y values.
27. The EPI spatial-encoding method can be used with a gradient-echo or a spin-echo.
28. *Single-shot* methods sample all values of k-space after a single RF excitation pulse and can therefore produce an image with a single excitation.

5.6 Parallel Imaging

The term *parallel imaging* refers to acquiring different parts of k-space data at the same time to reduce the time taken to acquire an image. This approach is very different from the fast imaging methods described in the previous section, and it can be used in combination with most other imaging methods, whether or not they are considered fast. The fundamental idea behind parallel imaging methods is to combine the signals from multiple receiver coils, making use of the fact that the receivers have different spatial distributions of sensitivity, to obtain more image data in a given amount of time. There are generally two different approaches to take, one in which reconstruction takes place in image space (SENSE, PILS), and the other in which reconstruction is done in k-space (SMASH, GRAPPA), with the acronyms for each method defined as follows (11):

SENSE   Sensitivity encoding

PILS   Partially parallel imaging with localized sensitivities

SMASH   Simultaneous acquisition of spatial harmonics

GRAPPA Generalized autocalibrating partially parallel acquisitions

The underlying idea for the PILS method can be illustrated by looking at a typical image (Figure 5.24a, left) and the corresponding k-space data (Figure 5.24a, right). If the k-space was instead sampled only in every second phase-encoding line to reduce the acquisition time, then the effect would be a field of view that is half as large in the phase-encoding direction, as shown in Figure 5.24b. Because the field of view is reduced, the resulting image suffers severe aliasing (wrap-around).

Two receiver coils can be used instead, each of which is sensitive to a smaller portion of space, as shown in Figure 5.24c. If the coils could be designed to be sensitive to precisely half of the field of view, then the two images could be combined to form a complete image. In practice, however, the coil sensitivity drops off gradually, and some aliasing remains that must be corrected. Figure 5.24d shows the k-space data sampled with each coil with the reduced acquisition time, and the resulting half field-of-view image from each coil. If the area of sensitivity of each coil

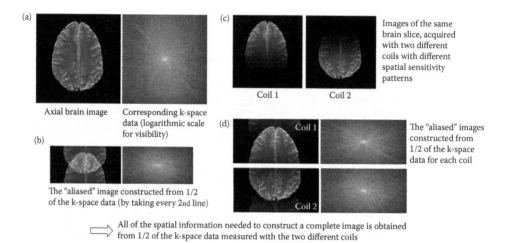

(a)

Axial brain image | Corresponding k-space data (logarithmic scale for visibility)

(b)

The "aliased" image constructed from 1/2 of the k-space data (by taking every 2nd line)

(c) Coil 1 | Coil 2

Images of the same brain slice, acquired with two different coils with different spatial sensitivity patterns

(d) Coil 1 | Coil 2

The "aliased" images constructed from 1/2 of the k-space data for each coil

All of the spatial information needed to construct a complete image is obtained from 1/2 of the k-space data measured with the two different coils

FIGURE 5.24 The concepts underlying parallel imaging methods. If the number of phase-encoded lines that are acquired is reduced by a factor of 2 (panel a compared with b) then the field of view in the phase-encoding direction is reduced by 2, and the image is likely to be aliased (wrapped around). If the image data are acquired with coils that are more sensitive to one-half of the field of view (panel c), then the appearance of the aliasing is reduced (panel d). It can be seen that by combining the k-space data that is under-sampled by a factor of 2, from two coils with sensitivities to different halves of the field of view, then all of the information that is needed to construct the complete image has been obtained. The image data can be acquired with the two coils simultaneously (i.e., in parallel), thereby reducing the total image acquisition time.

is precisely mapped, the aliasing can be corrected and the images combined to create a single image with the original larger field of view, in half the total acquisition time. The PILS method is therefore restricted to combinations of coils with very localized sensitivities.

The SENSE method is less restricted in its use because every coil can provide signal from every point in the resulting image. However, the contribution of each coil to a given pixel depends on its sensitivity at that location. Also, the reduced field of view that results from sampling fewer k-space lines produces aliasing (wrap-around). The signal in the resulting reduced field-of-view image therefore contains signal from a number, R, of equally spaced pixels in the full field-of-view image. If the field of view is reduced by a factor of 4, for example, then the signal can be a combination of 4 pixels in the full field-of-view image (i.e., R = 4)

$$I_k(x, y) = \sum_{l=1}^{R} C_k(x, y_l) \rho(x, y_l)$$

where I_k is the image intensity from coil k, C_k is the spatial distribution of the sensitivity of coil k, and ρ is the image pixel intensity that we wish to determine. The signal from all coils must be combined to get the full image information desired, and so the value of k must be spanned from 1 to the number of coils (N_c). The combined set of linear equations to solve can be expressed as a matrix equation:

$$I = C\rho$$

The solution is $\rho = (C^T C)^{-1} C^T I$ as this demonstrates the signal from each pixel in the full field-of-view image. To solve for ρ requires that the field-of-view reduction, R, is less than or equal to N_c.

The SMASH method also employs the combination of signals from a number of coils and makes use of the varying sensitivities across space of the coils. However, with SMASH the signals

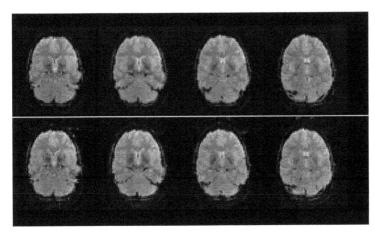

GE-EPI, TR = 2 sec

GE-EPI with
parallel imaging
TR = 1.76 sec

FIGURE 5.25 Comparison of axial brain images acquired with single-shot EPI, with and without SENSE parallel imaging. Images were acquired spanning the brain (only four slices are shown) with a typical acquisition method as used for fMRI at 3 tesla (gradient-echo EPI, TE = 30 msec, 3.3-mm cubic voxels). With the application of SENSE with an acceleration factor of 2, the image acquisition time decreased from 2 sec to 1.76 sec, but the SNR also dropped by 58% from approximately 95 to 55.

from the multiple receiver coils are combined in k-space to generate the missing phase-encoding steps. Extensions of the theory for SMASH include Auto-SMASH and VD-Auto-SMASH, and these are extended further to develop the theory for GRAPPA. As these techniques are quite complex, the math will not be described here. The important points to understand, to use these methods, are that the reduction factor, R, in the acquisition time is invariably less than or equal to the number of coils used, and that SMASH is applied only in the phase-encoding direction whereas GRAPPA can be applied in both phase- and frequency-encoding directions. GRAPPA is also more robust in areas of low field homogeneity or with fast methods such as EPI.

It is important to understand, however, that while parallel imaging methods can reduce the data acquisition time, and therefore reduce artifacts that may be caused by movement and non-uniform magnetic fields, there is also a cost. Because less data are acquired (as discussed in Section 5.4) the signal-to-noise ratio (SNR) of the resulting image is reduced. With the PILS method, the SNR is reduced by at least $1/\sqrt{R}$, meaning that if we reduce the field of view by a factor of 2 and correspondingly accelerate the data acquisition by a factor of 2, then the SNR would be reduced by $1/\sqrt{2}$, or 0.71 (11). The SNR is further reduced with the SENSE method by a factor that depends on the geometry of the coil configuration, g, resulting in a SNR reduction of $1/g\sqrt{R}$. Examples of images acquired with a single-shot EPI method, with and without SENSE with an acceleration factor of 2, are shown in Figure 5.25. In this specific case, images were acquired of axial slices spanning the entire brain in 2 sec, and this time was reduced to 1.76 sec with parallel imaging. However, the SNR decreased by a factor of 0.58 when parallel imaging was applied, from approximately 95 to 55.

5.7 Causes of Image Artifacts and Distortion

An MR image *artifact* is an error resulting in altered signal intensity (higher or lower) that does not correspond with the anatomy that is supposed to be depicted. Distortions are a particular form of artifact in which the signal in a voxel is displaced from its correct location, resulting in a change in the apparent shape of anatomical structures. The most common sources of artifacts and distortions in fMRI data are movements of any kind (blood flow, breathing, body movement, etc.), nonuniform magnetic fields, and inherent imperfections in imaging methods.

The effect on image quality and spatial resolution that is most often encountered, and that is most often misunderstood, is the one caused by measuring the MR signal for different points in k-space at different times after the RF excitation pulse. As a result, every data point in k-space does not have the same T_2- or T_2^*-weighting. This difference in weighting across k-space occurs with both conventional spatial encoding schemes when one line of k-space is acquired after each RF excitation pulse, and with fast imaging schemes when multiple lines of k-space are sampled. However, it is most often discussed in relation to fast imaging. As points of k_X are sampled, transverse relaxation processes continue, as always, and so the first points sampled are less T_2- or T_2^*-weighted than the later points simply because they are sampled at different times. With fast imaging methods each line of k_Y can also be differently T_2- or T_2^*-weighted. To see how this might affect the resulting image, we must first consider the ideal image, I_{ideal}, that would be constructed from ideal k-space data in which every point is sampled with the exact same weighting. This could only be achieved by sampling a single point of k-space (i.e., at one k_X and k_Y value combination) at a single fixed time interval after each RF pulse, to obtain ideal k-space data, S_{ideal}. We create the ideal image by applying the Fourier transform to the ideal k-space data: $I_{ideal} = FT(S_{ideal})$. In practice, the actual k-space data that we acquire, S_{actual}, is scaled according to the time, t, that has lapsed since the RF excitation pulse, due to transverse relaxation. The scale factor is therefore equal to e^{-t/T_2} or e^{-t/T_2^*}, depending on the imaging method. Regardless of the relaxation time that causes the weighting difference and whether the points of k-space are differently weighted in the k_X or k_Y direction or both, in the general case we can describe the weighting at each point with a 2D array, W. The actual k-space data is therefore $S_{actual} = S_{ideal} \times W$, and the actual image, I_{actual}, is equal to $FT[S_{ideal} \times W]$. Now, recall from Chapter 2 (Section 2.3.4), the Fourier convolution theorem:

$$FT[A \times B] = FT[A] \otimes FT[B]$$

In the present example then: $\quad FT[S_{ideal} \times W] = FT[S_{ideal}] \otimes FT[W]$

$$FT[S_{actual}] = FT[S_{ideal}] \otimes FT[W]$$

$$I_{actual} = I_{ideal} \otimes FT[W]$$

(Recall that $\times$ indicates point-by-point multiplication, whereas $\otimes$ indicates the convolution operation.) This equation therefore shows that the image we obtain in practice, I_{actual}, with variable weighting of the MR signal across points of k-space, is the convolution of the ideal image and the FT of the weighting pattern. Convolving an image with any function affects spatial features in the image. For example, an image can be convolved with a certain 2D pattern to smooth or blur edges between features, or a different pattern can be used to sharpen or enhance edges between features. Spatial smoothing is typically achieved in practice by convolving an image with a function with a narrow central peak, such as with the example shown in Figure 5.26. This is similar to the effects of variable relaxation-time weighting, as simulated in Figure 5.27, showing the effects of an exponentially decaying signal amplitude as one moves across one direction of k-space. Since the FT of the exponentially decaying function is a Lorentzian function, the net effect is spatial smoothing across a width that is determined by the exponential decay rate. The smoothing effect will be slightly greater (larger smoothing span) for tissues with short T_2- or T_2^*-values than for those that relax more slowly. Nonetheless, the effect is that of spatial smoothing; the contrast between features is not changed otherwise, and the relaxation-time weighting is not altered. The relaxation weighting of the entire image is determined by the weighting of the center point of k-space. This fact is demonstrated in Figure 5.28 with actual images of a human brain, comparing a conventional spin echo with a spin-echo EPI acquisition.

$$\begin{array}{cccccccccc}
0&0&0&0&0&0&0&0&0&0\\
0&0&0&0&0&0&0&0&0&0\\
0&0&0&0&0&0&0&0&0&0\\
0&0&0&1&1&1&0&0&0&0\\
0&0&0&1&1&1&0&0&0&0\\
0&0&0&0&0&0&0&0&0&0\\
0&0&0&0&0&0&0&0&0&0\\
0&0&0&0&0&0&0&0&0&0\\
0&0&0&0&0&0&0&0&0&0\\
0&0&0&0&0&0&0&0&0&0
\end{array}
\otimes
\begin{array}{ccc}
1/8&1/4&1/8\\
1/8&1/4&1/8\\
1/8&1/4&1/8
\end{array}
=
\begin{array}{cccccccccc}
0&0&0&0&0&0&0&0&0&0\\
0&0&0&0&0&0&0&0&0&0\\
0&0&1/8&3/8&1/2&1/2&3/8&1/8&0&0\\
0&0&1/4&3/4&1&1&3/4&1/4&0&0\\
0&0&1/4&6&1&1&3/4&1/4&0&0\\
0&0&1/8&3/8&1/2&1/2&3/8&1/8&0&0\\
0&0&0&0&0&0&0&0&0&0\\
0&0&0&0&0&0&0&0&0&0\\
0&0&0&0&0&0&0&0&0&0\\
0&0&0&0&0&0&0&0&0&0
\end{array}$$

$$\begin{aligned}
&0 \times 1/8 + 0 \times 1/4 + 0 \times 1/8\\
&+ 0 \times 1/8 + 1 \times 1/4 + 1 \times 1/8\\
&+ 0 \times 1/8 + 1 \times 1/4 + 1 \times 1/8 = 3/4
\end{aligned}$$

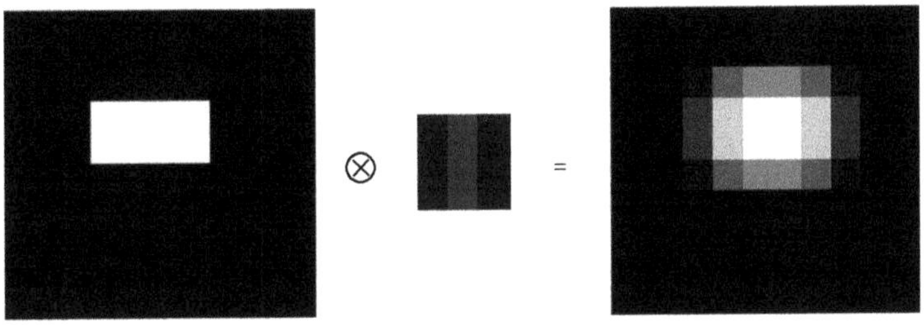

FIGURE 5.26 An example of the convolution operation between two 2-D grids of numbers (top), or alternatively, the same example is shown as images (bottom). The center portion of the figure shows the mathematical operation of the 2D convolution, and the result shows how: in this example, the result is a spatial smoothing or blurring of the original image.

Several common image artifacts can also be explained by how the causes of these artifacts affect the k-space data. For example, blood flow can cause the phase of the signal to change, thereby causing errors in the phase encoding, or causing imperfect formation of echoes and resulting in lower signal. The effect depends on the imaging method being used, and the speed and pulsatility of the blood flow, as shown in Figure 5.29. If the flow causes the signal to be very low at all points of k-space, then the blood vessel will simply appear dark in the image. In the case of random phase errors, then the signal from the blood will be spread uniformly across the phase-encoding direction, resulting in a streak across the image. The streak results from the signal not being correctly localized to one location in the phase-encoding direction. In the case of pulsatile blood flow in arteries, the flow may cause low signal in only some lines of k-space, resulting in variable k-space weighting as discussed above. However, in this case, the effect causes *ghosts* or replications of the signal from the blood at several points across the image in the k-space direction.

Similar artifacts are caused by any moving tissues or fluids, such as the heart, lungs, and cerebrospinal fluid. Since the appearance of the artifact depends on how the k-space signal is affected, it depends entirely on the type of imaging method and the timing of the signal acquisition relative to the periodic motion of breathing or the heartbeat. Nonperiodic movement, such as movement of the eyes or swallowing, can also cause similar artifacts. Again, the appearance of the artifacts depends on how the movement changes the MR signal, and whether or not it affects all lines of k-space, regularly spaced lines, or a single line. The formation of artifacts is of great importance for fMRI because the image can clearly be affected far from the area of movement or flow. The fMRI time-series in a region may be affected at only discrete points in time due to intermittent movement or periodic movement

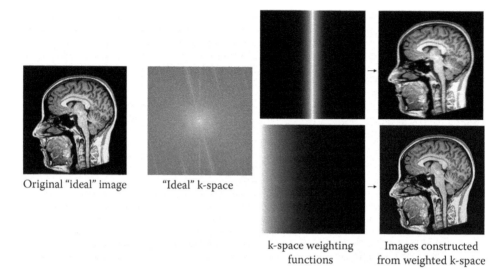

Original "ideal" image   "Ideal" k-space

k-space weighting   Images constructed
functions   from weighted k-space

FIGURE 5.27 Demonstration of the effects of variable weighting across lines of k-space on the resulting image. On the left, the so-called ideal image is shown, constructed from k-space data with perfectly uniform weighting of sampled points. On the right, two examples of the effects of weighting the lines of k-space are shown. In the top example k-space is weighted with a function that decays exponentially from center in the horizontal direction, and in the bottom example the exponential decay is from one side. The images that would be constructed from this weighted k-space are shown, and in comparison with the "ideal" image, can be seen to be affected only by spatial smoothing.

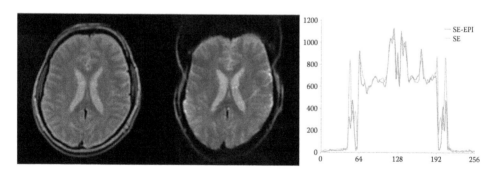

FIGURE 5.28 Actual examples of the effects of variable k-space weighting on MR images. The left-most image was acquired with a 3 tesla MRI system from a healthy person, using a conventional spin-echo (SE) imaging sequence. The image in the center was acquired from the same person, at the exact same slice location, field of view, and sampling matrix, using a spin echo with an EPI spatial-encoding method (SE-EPI). The EPI method results in variable weighting across lines of k-space. The plot on the right shows a comparison of the signal intensities along a horizontal line through both images, after scaling to have the same median intensity for both images. Except for the areas of skin and bone around the skull, this plot shows that the image contrast is essentially identical for the two methods, only the spatial resolution is different, and the distortion due to EPI sampling is apparent.

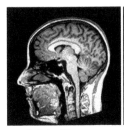 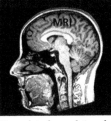 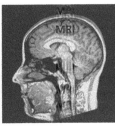 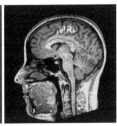

Original image | Complete loss of signal due to flow | Random phase error in signal from flow region | Signal loss in regularly spaced lines of k-space due to pulsatile flow

FIGURE 5.29 Simulations of artifacts caused by blood flow in a region shaped like the letters MRI. The left-most frame shows the original image, whereas the second frame from the left shows the case of flow causing complete dephasing of the magnetization and loss of signal at all points of k-space. The third frame shows the case in which flow causes random phase errors resulting in random spread of signal in the phase-encoding direction. The rightmost frame simulates the case of pulsatile blood flow, in which the magnetization is completely dephased (resulting in zero signal) only in regularly spaced lines of k-space, resulting in repeated ghosts or replications of the signal from the flow area in the phase-encoding direction.

due to breathing and heartbeat. It is therefore very important that the artifacts extend only in the phase-encoding direction, and so where the signal may be affected by movement, artifacts can be predicted. The phase-encoding direction can also be selected so that movement artifacts do not extend across important areas of interest in the brain or spinal cord for the purposes of fMRI.

Spatial distortions are another common form of image artifacts (as seen in Figure 5.28) and can typically be explained as incorrect sampling of k-space. That is, the point in k-space that is believed to be sampled is not the actual location that is sampled. Returning to the definition of k_X and k_Y:

$$k_X = \gamma G_X t$$

$$k_Y = \gamma G_Y T$$

It is clear that if there are errors in the gradients, or as is more often the case, there are other sources of magnetic field gradients, which we will call G_{Xerr} and G_{Yerr} (such as magnetic susceptibility differences between tissues or magnetic materials in the body), then k_X and k_Y may be altered:

$$k_{Xactual} = \gamma (G_X + G_{Xerr})t$$

$$k_{Yactual} = \gamma (G_Y + G_{Yerr})T$$

The effect of the magnetic field distortions may be very localized and affect only a small portion of the image. Nonetheless, these effects can be seen in practice, in areas of the brain such as the frontal lobe and temporal lobes, and in the brainstem, where there are nearby air/tissue interfaces, and can have a significant effect on fMRI. In the spinal cord the adjacent bone/tissue interfaces also cause image distortion. The magnitudes of the error terms, G_{Xerr} and G_{Yerr}, that are needed to cause significant distortions of the image are relative to the magnitudes of G_X and G_Y and depend on the magnitudes of t and T as well. For this reason, imaging methods that use higher values of G_X and G_Y tend to have less distortion, and methods with longer data acquisition times, such as EPI methods, suffer greater distortion as shown in Figure 5.28. This effect causes an important trade-off to be considered for fMRI, because fast imaging methods, which are desirable for fMRI, typically have long data acquisition times and therefore suffer greater spatial distortions, which are undesirable for fMRI. As with all imaging methods, it is therefore necessary to balance the speed with the image quality to obtain suitable levels of both of these features.

Key Points

29. Sampling lines of k-space at different times after the excitation pulse, or during the formation of different spin-echoes, affects the spatial resolution of the resulting image, and not the image contrast of effective relaxation-time weighting.
30. Variable weighting (i.e., scaling of the magnitude) across points of k-space causes the same effect as convolving the image with the Fourier transform of the weighting function, thereby affecting the appearance of spatial features.
31. Image artifacts can be caused by errors, such as periodic or intermittent movement, that produce loss of signal at only certain points in k-space, (again, causing variable weighting across points of k-space), resulting in blurring or replications of features (i.e., *ghosting*) in the resulting image.
32. Another form of image artifacts is caused by errors in the expected k-space values when the data points are sampled. Typically, this occurs due to distortions in the static magnetic field, B_0, as a result of magnetic susceptibility differences or magnetic materials in or around the body.

References

1. Mansfield P, Maudsley AA, Baines T. Fast scan proton density imaging by NMR. *J Physics E-Sci Instr* 1976;9(4):271–278.
2. Hoult DI. Zeugmatography: Criticism of concept of a selective pulse in presence of a field gradient. *J Magn Reson* 1977;26(1):165–167.
3. Mansfield P, Maudsley AA, Morris PG, Pykett IL. Selective pulses in NMR imaging: Reply to criticism. *J Magn Reson* 1979;33(2):261–274.
4. Redpath TW. Signal-to-noise ratio in MRI. *Br J Radiol* 1998;71(847):704–707.
5. Hoult DI, Lauterbur PC. Sensitivity of the zeugmatographic experiment involving human samples. *J Magn Reson* 1979;34(2):425–433.
6. Hoult DI, Chen CN, Sank VJ. The field dependence of NMR imaging. II. Arguments concerning an optimal field strength. *Magn Reson Med* 1986;3(5):730–746.
7. Chen CN, Sank VJ, Cohen SM, Hoult DI. The field-dependence of NMR imaging. 1. Laboratory assessment of signal-to-noise ratio and power deposition. *Magn Reson Med* 1986;3(5):722–729.
8. Hoult DI, Richards RE. Signal-to-noise ratio of nuclear magnetic-resonance experiment. *J Magn Reson* 1976;24(1):71–85.
9. Abart J, Eberhardt K, Fischer H, Huk W, Richter E, Schmitt F, Storch T, Tomandl B, Zeitler E. Peripheral nerve stimulation by time-varying magnetic fields. *J Comput Assist Tomogr* 1997;21(4):532–538.
10. Shellock FG. *Reference Manual for Magnetic Resonance Safety, Implants, and Devices.* 2009 ed. Biomedical Research Publishing Group; 2009.
11. Blaimer M, Breuer F, Mueller M, Heidemann RM, Griswold MA, Jakob PM. SMASH, SENSE, PILS, GRAPPA: How to choose the optimal method. *Top Magn Reson Imaging* 2004;15(4):223–236.

6

Principles and Practice of Functional MRI

The essential steps needed to create maps of neural function out of the magnetic resonance imaging (MRI) methods described in the preceding chapters are (1) to repeatedly acquire images over time in order to describe a time series, and (2) to make the MRI signal intensity depend in some way on neural activity. This chapter describes how the MRI signal can be made to vary with changes in neural activity because of changes in tissue properties, particularly by means of blood oxygenation–level dependent (BOLD) contrast. The physiological changes that accompany changes in neural activity, that underlie changes in MRI signal properties, are at the core of all functional MRI methods and have a strong influence over the timing of changes that can be detected, the spatial specificity, and in how data should be acquired for optimal sensitivity. The optimal parameters and how they are determined are described in this chapter, as well as how they must be adapted for different applications. Finally, although BOLD contrast is used for the vast majority of all functional imaging, it is useful to know that there are other contrast mechanisms that can be used for functional MRI (fMRI) as well, and these are described at the end of the chapter.

6.1 How MRI Becomes Functional MRI

A number of applications of MRI are referred to as being *functional* imaging, such as cardiac function, joint function (i.e., mechanical function), and so forth, but the term *functional MRI* is reserved to mean specifically neural function. All of these methods are based on acquiring MR images at different states of the function in question, and the changes in image signal intensity are related to the function. For example, for cardiac function the heart can be imaged at several phases of the cardiac cycle (by means of cardiac-gated acquisition of k-space data), and the ventricular and atrial volumes can be measured as a function of cardiac phase. Functional MRI to characterize neural function is no less straightforward, and we simply image the brain, brainstem, and/or spinal cord at different states of neural function. We infer that any regions of the central nervous system (CNS) that changed in the images between the two states must be somehow involved with the function that was varied.

For fMRI, images are acquired multiple times to describe a time series, while the person being studied performs tasks (cognitive, sensory, motor, etc.) to systematically vary their neuronal activity so that the differences in the images corresponding to different neural states can be reliably detected. The great challenge for fMRI is to determine which image changes were due to the neural function and which may have been caused by random noise, physiological motion, movement of the person being studied, or subtle changes in the MRI system itself (such as due to heating, vibration, electrical power supply fluctuations, etc.). The basic approach underlying the fMRI method results in both its effectiveness and its key limitations and technical challenges. The primary limitation of fMRI is that it can only show differences in neural function between states. The design of the reference or baseline state is therefore just as important as the design of the stimulus state. One of the key technical challenges is to understand the relationship between neural activity and the MR image intensity. A second key challenge is to determine which signal changes are truly related to neural functions and which are not. The most important developments with fMRI have been to respond to these challenges, and these are discussed in the sections that follow.

6.2 Contrast Mechanisms: Linking the MR Signal and Neural Function

The original development of fMRI began with the discovery that veins appear dark in T_2^*-weighted MR images because of the deoxygenated hemoglobin in the blood (1). This finding built upon the 1936 discovery by Drs. Pauling and Coryell that the magnetic properties of hemoglobin are different between the oxygenated and the deoxygenated states (2). Because the MR signal can be made to depend on the blood oxygenation level as it changes in response to neural function, images can therefore be acquired with BOLD contrast to reveal information about neural activity (3–6). While BOLD contrast is at the origin of fMRI and essentially all of the key developments in the field, and is used for the vast majority of fMRI studies, it is important to know that other fMRI contrast mechanisms exist as well, and these are described at the end of this chapter in Section 6.6. This chapter will focus on BOLD contrast, and the principles involved can generally be extended to other contrast mechanisms as well. Regardless of the mechanism used to detect changes in neural function, no fMRI methods (to date) can reveal neuronal activity directly, but instead we must rely on indirect observations to infer changes in activity.

6.2.1 BOLD Contrast

As mentioned in Chapter 3, *contrast* refers to differences in MR image intensity between features (objects) in the image or in one image feature between two different points in time. Contrast arising from the BOLD effect occurs due to a combination of three key effects; changes in neural activity are accompanied by local changes in blood oxygenation, relaxation times are influenced by the level of blood oxygenation, and the MRI signal can be made to depend on the relaxation times. Oxygenated hemoglobin contains no unpaired electrons and is therefore diamagnetic, which means that it has very little effect on the magnetic field around it (2). More specifically, when placed inside an MRI system, the magnetic field in the oxygenated hemoglobin is lower than that outside, although the effect is very weak, similar to the effect in water. On the other hand, when hemoglobin gives up all of its oxygen, it contains four unpaired electrons per heme unit, and as a result is paramagnetic (2). This means that the magnetic field is higher in the hemoglobin than it is outside, and the field change is linearly proportional to the strength of the external magnetic field. Accordingly, partially oxygenated hemoglobin (since there are four heme units per hemoglobin molecule) will also contain fewer unpaired electrons and will be less paramagnetic (i.e., lower magnetic susceptibility). The net effect is that the magnetic field around individual red blood cells, and in total inside blood vessels, can be altered depending on the blood oxygenation level. This is the first key element of BOLD contrast for fMRI.

To have an idea of the magnitude of the effect of the blood oxygen level on the magnetic field within the blood and tissues, we can look again at the volume magnetic susceptibility, as introduced in Chapter 3. The susceptibilities of water and of 98% oxygenated blood are similar, estimated to be –9.04 parts per million (ppm) (7) and –9.12 ppm (8), respectively. In comparison, blood that is 75% oxygenated has a relatively higher susceptibility of –8.79 ppm. In the extreme (nonphysiological) case of completely deoxygenated blood, the susceptibility is expected to be –7.69 ppm (8). While in this latter case the total susceptibility is still negative (it is diamagnetic), it is higher than that of water or oxygenated blood because of the paramagnetic contribution from the hemoglobin that is not bound to oxygen.

When a person is lying inside an MRI system, the magnetic field is therefore slightly different between inside and outside the blood vessels. (We could equally consider individual red blood cells, but it is easier to consider entire blood vessels for now.) The difference depends significantly on the oxygen saturation (SO_2) and on the proportion of volume taken up by red blood cells (i.e., the hematocrit) (8). The three-dimensional (3D) distribution and magnitude of the magnetic field distortion also depends on the diameter and orientation of the blood vessel (maximum field shift outside the blood vessel when it is oriented 90° to B_0, and minimum (zero) when it is parallel to B_0) (9). In general, though, we are most interested in the capillaries that directly supply oxygen to the tissues, and the capillary vessels have a fairly constant size (5–10 μm in diameter) and are effectively randomly oriented. As described in Chapter 3, the effect of a spatial magnetic field distortion, such as that caused by a magnetic field gradient between inside and outside a blood vessel, is to alter magnetization relaxation times, particularly T_2 and T_2^*. This is the second key element of BOLD contrast (although this prioritization of first, second, etc., key elements simply refers to the order they are being listed here).

The link to neural activity is created by the fact that an increase in neuronal firing rate involves a local increase in oxygen consumption ($CMRO_2$, the cerebral metabolic rate of oxygen consumption), and a change in the local blood oxygenation (10). However, the total effect is somewhat counterintuitive because the increase in oxygen consumption occurs simultaneously (this will be discussed in more detail in Section 6.2.2) with a local increase in blood flow. The increase in oxygen delivery exceeds the increase in oxygen demand, and the net change is an *increase* in blood oxygen level at sites of increased neural activity (11). This is the third key element of BOLD contrast.

6.2.2 Physiological Origins of BOLD

It is necessary to look at the physiology underlying BOLD signal changes in greater detail because there has been considerable research and debate devoted to how and why this sequence of changes occurs, the rates and relative timings of the different events, and how the physiological changes alter magnetization relaxation times for MRI. In spite of almost two decades of intensive research, the biological and physical origins of BOLD contrast are still not completely understood, and new and important findings could yet develop.

Early characterizations of the BOLD response (6,12) demonstrated that when a stimulus is applied (to evoke a change in neural activity), the MR signal change lags the stimulus onset and spans several seconds, as shown in Figure 6.1a. Similarly, the return to the prestimulus, or *baseline*, value after the stimulus is stopped takes several seconds to occur. A good understanding of this spread of the response in time is critical if we are to reliably detect it after a change in neural activity (12). To predict the time course of signal changes that would occur in general for any stimulation pattern, the response to an extremely brief stimulus (called a *delta* function) was determined (Figure 6.1b). It was found that after the onset of a very brief stimulus, it takes roughly 2 sec for MR signal changes to begin to occur, and then the signal rises and reaches a peak after 4 to 6 sec. After the cessation of the stimulus, the MR signal drops and reaches a minimum in about 12 sec, at a value that is below the baseline value (13,14). The signal then recovers fully back to its baseline value by about 20 sec after the stimulus ended (Figure 6.1b). Because the primary factors determining the shape of the response are the changes in blood flow and oxygen consumption, the response function is called the *hemodynamic response function*, or *HRF*. The response pattern described here is the one that has been adopted for widely used analysis programs such as Statistical Parametric Mapping, or SPM (15), and because it is widely accepted it is often referred to as the *canonical* HRF. The observed signal change time-course, $I(t)$, can

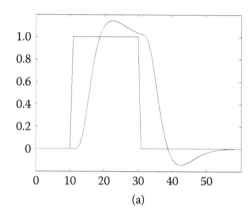 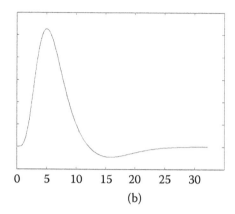

(a)   (b)

FIGURE 6.1 (a) BOLD response to a block stimulus, and (b) the hemodynamic response to a brief stimulus (i.e., the *hemodynamic response function*). The blue line in (a) is the convolution of the stimulus function (red line) with the hemodynamic response function shown in (b).

therefore be expressed as $I(t) \propto P(t) \otimes \mathrm{HRF}(t)$. where $P(t)$ is the stimulation paradigm (i.e., pattern in time, t), and $\otimes$ is the convolution operation (12). An example is shown in Figure 6.1a, where the stimulus pattern in time, $P(t)$, is indicated with the blue line and the predicted MR signal response, $I(t)$, is indicated with the red line. The important result is that now we have a good idea of the temporal pattern of MRI signal changes that we need to look for, to find the regions of the CNS that responded to the stimulus.

The speed of the response demonstrated by the HRF shows that the rate of BOLD signal change is very slow compared with the spiking rate of neurons, and the net effect observed is the cumulative response to the total change in blood oxygenation in each voxel. We can only observe the net effect of a great many neural signaling events, and each event consists of many excitatory and inhibitory inputs to dendrites and neuronal cell bodies, new action potentials being generated, neurotransmitters being recycled by astrocytes, membrane potentials being restored, and so forth. We cannot assume that the neuronal spiking rate is the key factor that determines the BOLD response, so it is best for now to consider all possible elements of neural signaling. Regardless, it is clear that the responses to the individual neural signaling events merge, and the magnitude of net BOLD signal change is therefore expected to reflect the total change in neuronal activity. More accurately, the magnitude of the BOLD response is expected to reflect the total change in energy metabolism.

The net change in the BOLD signal in response to many neural signaling events is demonstrated by considering the stimulation block as if it was a set of very brief events in rapid succession. Figure 6.2 shows how the responses are similar between that expected for a continuous

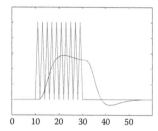 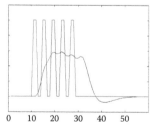 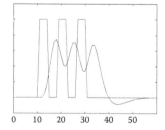

FIGURE 6.2 Examples of the predicted BOLD signal change response, based on the convolution of the stimulation paradigm and the hemodynamic response function. The response to a string of brief stimuli in rapid succession (left) is essentially identical to the response to a continuous stimulus.

stimulus block and one composed of a number of brief successive blocks. When the brief stimulus blocks are closely spaced in time (relative to the speed of the BOLD response), the expected time-course shape of the BOLD response is unaltered from that of a single continuous block, but the amplitude (the peak height) of the response is reduced by the intermittent periods without stimulation. The upper limit of this progression is to model the stimulus as a brief spike for each neural signaling event, so that we can observe the total cumulative effect. From the examples in Figure 6.2 we can see that this would look exactly like the response to the continuous block stimulus, and the magnitude of the response would depend on the rate of neural signaling events. Now it is necessary to verify that this model of the BOLD response is accurate and to determine how the individual responses sum. For example, if we compare two different stimuli, and one produces twice the change in neuronal spiking rate as the other, then is the BOLD signal change also twice as large? The short answer is "probably not," but to properly answer this question we still need to take a closer look at the physiology underlying the BOLD signal changes.

Models of the physiology underlying the BOLD effect have been developed and refined since fMRI was first developed (16). As described above (i.e., the three key elements discussed earlier) when some stimulus such as a visual cue, motor task, cognitive process, and so on causes the neural activity to increase, the neurons are more metabolically active and consume more oxygen (10). Vasodilators are released as a consequence of, or in relation to, the increase in neuronal firing rate, and local blood vessels dilate, thereby providing more oxygenated blood and increasing the local proportion of oxygenated hemoglobin in the blood (11). This reduces local distortions in the magnetic field in and around the blood vessels and relaxation rates ($1/T_2$ and $1/T_2^*$) are reduced (17,18). More specifically, relaxation rates in solutions are proportional to the concentration of magnetically active particles, such as deoxyhemoglobin (19). In T_2- or T_2^*-weighted images, the MR signal is therefore higher at sites of increased neuronal activity because of the reduced relaxation rates.

The magnitude of the BOLD signal change therefore depends on the change in the concentration of deoxyhemoglobin that results from the imbalance between the increase in oxygen consumption ($\Delta CMRO_2$) and the increase in blood flow (ΔCBF). The proportion of oxygen that is taken up by the tissues out of the total amount of oxygen available in the blood is known as the oxygen extraction fraction, E (20):

$$E = \frac{CMRO_2}{4\, S_A[Hb]CBF}$$

In this equation the rate of oxygen consumption, $CMRO_2$, is in units of [μmol/s], S_A is the arterial oxygen saturation (typically 0.98), [Hb] is the concentration of hemoglobin in the blood [μmol/mL], and the cerebral blood flow (CBF) is in units of [mL/s]. The value 4 in the denominator reflects the number of oxygen-binding sites in each hemoglobin molecule. The value of E can range from 0 to 1, with 0 meaning that no oxygen is extracted from the blood and therefore the concentration of deoxyhemoglobin would equal that in the incoming arterial blood. In the non-physiological case of E = 1, all of the oxygen would be extracted from the blood and the concentration of deoxyhemoglobin would match that of the total hemoglobin. The value of E is directly related to the concentration of deoxyhemoglobin in the blood (E = [deoxy-Hb]/[Hb]), and therefore the change in E when a stimulus is applied is also related to the change in relaxation rates ($R_2 = 1/T_2$ and $R_2^* = 1/T_2^*$). In solutions of proteins (19), or with the administration of MR contrast agents such as ultrasmall superparamagnetic iron oxide (USPIO) (21), it has been shown that R_2 is proportional to the concentration of solutes; and similarly for blood vessel networks R_2^* is predicted to be proportional to the magnetic susceptibility of the vessels (22). Therefore R_2 and R_2^* are expected to be proportional to the concentration of deoxyhemoglobin, [deoxy-Hb]. However, models of a distribution of blood vessel sizes, and including the effects of water diffusion around the vessels, predict that $R_2^* \propto [\text{deoxy-Hb}]^\beta$ (23) where β is between 1 and 2, or roughly 1.5 for a mixture of vessel sizes (16). Putting these concepts together we therefore get:

$$R_2 \text{ and } R_2^* \propto (E[Hb])^\beta$$

$$R_2 \text{ and } R_2^* \propto \left(\frac{CMRO_2}{4 \, S_A \, CBF} \right)^\beta$$

The net total relaxation rate observed for MR signal from a voxel also depends on the volume proportion, V, which is occupied by the blood vessels. The complete predicted relationship between physiological parameters and the MR relaxation rates is therefore:

$$R_2 \text{ and } R_2^* \propto V \left(\frac{CMRO_2}{4 \, S_A \, CBF} \right)^\beta$$

Clearly the model becomes complex when we try to account for all of the physiological parameters that can affect the BOLD signal change. Still, the most important physiological features influencing the MR signal when a stimulus is applied are how $CMRO_2$ and CBF change relative to each other in time.

Referring again to Figure 6.1b, we can now explain some of the features of the timing in the response function, at least qualitatively, such as described by the Balloon Model, developed by Buxton et al. (24). The initial lag in the BOLD response after the onset of the stimulus, or even a brief initial decrease in signal (the *initial dip*) (25), reflects a nearly simultaneous increase in $CMRO_2$ and CBF, or possibly a slight lag in the CBF increase. The time taken to reach the peak BOLD response then presumably reflects the time it takes for the total increase in CBF to reach the sites of neural activity that have elevated $CMRO_2$ and also to reach the venules and veins draining the area. This also raises the interesting point that some regions downstream (that is, along the venules and veins) from the area of increased neural activity may not have an elevated $CMRO_2$ but yet have an increased CBF because of the upstream activity. The time taken for the MR signal to decrease to baseline values again after the stimulation is ceased is similarly expected to reflect the time it takes to reestablish the baseline blood flow. The poststimulus undershoot is proposed to occur because of the blood volume (V) settling back to baseline more slowly than the blood flow (CBF) (16). The total hemodynamic response can therefore be explained by the above equations, particularly with focus on the ratio between $CMRO_2$ and CBF. Precise quantitative models have been developed and described in the literature (9,16,23), but the qualitative description used here serves to identify the key physiological changes that are reflected by BOLD signal changes.

The next step is to understand the relationship between changes in neural activity and changes in $CMRO_2$ and CBF (Figure 6.3). A key component of the physiological changes that is not included in the model described above is the role of astrocytes that circumscribe most (if not all) neuronal synapses and more than 99% of the total cerebrovascular surface area (26). Relatively recently, multiple laboratories have independently confirmed that astrocytes control local cerebrovascular microcirculation (27–31), making these cells a key link in the BOLD and other fMRI responses (32). When synaptic input is increased, releasing neurotransmitters into synaptic clefts, the surrounding astrocyte end processes take up the neurotransmitters to serve three important functions. The first is to prevent neurotransmitters (such as glutamate and γ–amino-butyric acid (GABA)) from diffusing to adjacent synapses and producing unwanted cross-excitation (33), the second is to prevent synaptic excitotoxicity (33,34), and the third is to recycle neurotransmitters (glutamate into glutamine) for subsequent reuse by neurons (35). The energy needed by active brain tissue is derived from both oxidative and nonoxidative (i.e., glycolytic) metabolism (11,36). It has been suggested that while neurons primarily use oxidative phosphorylation to meet their increased energy demands, astrocytes may be preferentially

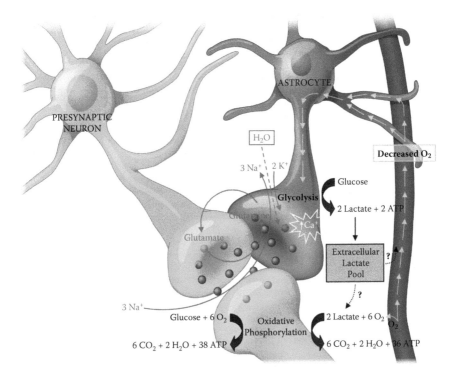

FIGURE 6.3 The interaction between neurons, astrocytes, and blood vessels to produce a neural signaling event. When neurotransmitters are released at the synaptic cleft, they bind to receptors to open ion channels of the postsynaptic neuron before they are taken up by the adjacent astrocyte, where glutamate is converted to glutamine and released for reuse by the neurons. Water cotransports with ions and neurotransmitters into the astrocyte, causing it to swell slightly with water. One of the current competing theories is that the energy needed by the astrocyte can be provided in part by glycolysis, resulting in increased extracellular lactate. This lactate acts as a local vasodilator and, according to some theories, may also serve as an energy substrate for the neurons. (Figure courtesy of C.R. Figley [32].)

glycolytic (37–39). The importance of astrocyte metabolism in hemodynamic changes has been further supported by the observation of activity-related increases in the glycolytic end-product lactate (40), which acting as a vasodilator, has been shown to control the polarity of cerebrovascular coupling mechanisms in an oxygen-dependent manner (30,37). It has also been proposed that the extracellular lactate might subsequently be used by neurons as an energy substrate for oxidative phosphorylation (30,37). However, there are competing theories that glucose is the primary source of energy for neurons during increased activity, and this area is the subject of both active research and debate (41,42). It has also been proposed that the vascular response may be mediated by potassium carried by astrocytes between the active neurons and capillaries (43). Regardless of which of these theories is correct, it appears that astrocytes, as well as neurons, play an important role in the signal changes that are detected with fMRI.

The relationship between presynaptic input and BOLD signal changes has been demonstrated by the observation that hemodynamic changes correlate better with local field potentials (LFPs) than either single-unit or multiunit recordings (44–49). LFPs represent the input to a given region, as well as the local processing. The important consequence of this relationship for fMRI based on BOLD contrast is that the signal changes reflect the net change in presynaptic input, whether excitatory or inhibitory, presumably regardless of the net change in neuronal output. A strong coupling between LFPs and changes in tissue oxygen concentration has even been demonstrated in the absence of spiking output, supporting the conclusion that hemodynamic changes reflect synaptic, more than spiking, activity (47,48).

Key Points

1. When the neuronal firing rate increases, the rate of oxygen consumption is increased because of the added energy demands.
2. The local blood flow is increased at sites of increased neural activity, such that the oxygen supply exceeds the added demand, resulting in more oxygen in the tissues when they are more metabolically active.
3. Iron in hemoglobin in blood is *diamagnetic* (very weak magnetic effect) when bound to oxygen but is *paramagnetic* (stronger magnetic effect) when not bound to oxygen. This property of iron in hemoglobin results in blood oxygenation-level dependent (BOLD) changes in MR signal relaxation times.
4. The relaxation rate ($1/T_2$ or $1/T_2^*$) is proportional to the concentration of deoxy-hemoglobin in the blood.
5. The relaxation rate is therefore proportional to the ratio of $CMRO_2$ to CBF (actually $(CMRO_2/CBF)^\beta$, where β is a constant value between 1 and 2).
6. The signal intensity detected in T_2- or T_2^*-weighted MR images is therefore increased from regions of the CNS that have increased metabolic demand in relation to increased neural activity.
7. The total change in energy required, and therefore oxygen consumption, is most closely linked to presynaptic input, which can be both excitatory and inhibitory, as opposed to the postsynaptic output activity, which results from the net effect of all of the inputs.
8. The predicted BOLD MR signal change response is the convolution of the stimulation pattern and the hemodynamic response function, which is the response measured for a very brief stimulus.

6.2.3 Quantifying the BOLD MR Signal Change

The MRI signal change upon a change in neural activity is determined by the difference in relaxation times between the *baseline* and *stimulation* states. Focusing on the expected change in T_2^* (for gradient-echo images) or T_2 (for spin-echo images), the signal intensity during baseline periods can be expressed as:

$$S = S_0 \exp(-TE/T_2^*) \text{ or } S = S_0 \exp(-TE/T_2)$$

(Recall that $\exp(x)$ is the same as e^x.) In this equation, S is the measured signal, and S_0 is the magnitude of the signal that would be observed in the absence of relaxation (i.e., if TE = 0). The echo time TE and the relaxation times T_2^* and T_2 are as described in Chapter 3. For the rest of this discussion we can just assume we are using gradient-echo images and T_2^* is the relevant relaxation time. The signal intensity during stimulation periods can then be expressed as (3,50):

$$S' = S_0 \exp(-TE/T_2^{*\prime})$$

This equation imposes two assumptions, that (1) the proton density, related to S_0, does not depend on either the baseline or stimulation conditions, and (2) the tissues can be treated as a single relaxation environment. The fractional change in signal intensity between baseline and stimulation periods is $\Delta S/S_0 = (S_{stim} - S_{baseline})/S_{baseline}$, and can be expanded as follows:

$$\frac{\triangle S}{S} = \frac{S_0 \exp\left(-TE/T_2^{*\prime}\right) - S_0 \exp\left(-TE/T_2^{*}\right)}{S_0 \exp\left(-TE/T_2^{*}\right)}$$

$$\frac{\triangle S}{S} = \exp\left(-TE/T_2^{*} + TE/T_2^{*}\right) - 1$$

The difference in relaxation rates $(-1/T_2^{*\prime} + 1/T_2^{*})$ can be replaced with ΔR_2^{*} for clarity:

$$\frac{\triangle S}{S} = \exp\left(-TE \ \triangle R_2^{*}\right) - 1$$

As long as the magnitude of $TE \ \Delta R_2^{*}$ is relatively small (<0.3), then the following approximation can be made with less than 5.5% error:

$$\frac{\triangle S}{S} \cong -TE \ \triangle R_2^{*}$$

This approximation introduces very little error because BOLD signal changes are typically less than 20% (i.e., $\Delta S/S < 0.2$). Measured values of ΔR_2^{*} and ΔR_2 have been reported to be as summarized in Table 6.1.

As a result, the BOLD signal changes that can occur with a change in neural activity are expected to scale linearly with the chosen echo time, TE, up to a reasonable limit. The BOLD signal changes also scale roughly linearly with the magnetic field strength, and are three to four times larger with a gradient echo than with a spin echo at the same echo time (52). However, results of such measurements are also expected to vary, depending on the stimulus used and the brain region(s) that are involved. The other factor that must be considered is the contrast-to-noise ratio. This is the signal change between rest and stimulation compared with the random noise, N, in the image data, and is given by:

$$\frac{\triangle S}{N} = \frac{S_0 \exp\left(-TE/T_2^{*\prime}\right) - S_0 \exp\left(-TE/T_2^{*}\right)}{N}$$

The maximum contrast-to-noise ratio, and therefore the highest sensitivity to BOLD signal changes, is obtained when $TE \approx T_2^{*}$ for gradient-echo images or $TE \approx T_2$ if a spin echo is used.

Table 6.1 Reported Relaxation Rate Changes Arising from the BOLD Effect at 1.5 T and 3 T				
	$\Delta (1/T_2^{*})$ at 1.5 T	$\Delta (1/T_2)$ at 1.5 T	$\Delta (1/T_2^{*})$ at 3 T	$\Delta(1/T_2)$ at 3 T
Stroman et al.	−0.43 s⁻¹	−0.12 s⁻¹	−1.22 s⁻¹	−0.37 s⁻¹
Bandettini et al.	−0.84 ± 0.05 s⁻¹		−1.76 ± 0.05 s⁻¹	
Zhong et al.		−0.11 s⁻¹		

Sources: Bandettini, P.A. et al., *NMR Biomed* 7, 1–2, 12–20, 1994 (52); Stroman, P.W. et al., *Magn Reson Imaging* 19, 6, 827–831, 2001 (51); Zhong, J. et al., *Magn Reson Med* 40, 4, 526–536, 1998 (53).

9. The MR signal change, ΔS, due the BOLD effect between rest and stimulation conditions is given by , $\Delta S/S \cong -\text{TE}\ \Delta(1/T_2^*)$ for gradient-echo images or $\Delta S/S \cong -\text{TE}\ \Delta(1/T_2)$ for spin-echo images, where S is the signal intensity at rest.
10. The changes in relaxation rates increase with higher magnetic fields and are three to four times higher for gradient-echo signals, $\Delta(1/T_2^*)$, than for spin-echo signals, $\Delta(1/T_2)$.
11. The maximum contrast-to-noise ratio is obtained when $\text{TE} \approx T_2^*$ for gradient-echo images or $\text{TE} \approx T_2$ for spin-echo images.

6.3 General BOLD fMRI Methods

The essential properties of the MR imaging method used for BOLD fMRI are as follows: (1) it must be sensitive to the BOLD effect; (2) it must provide a time series of images with adequate spatial and temporal resolution; (3) the images must span the region of the CNS that is being investigated; and (4) the images must have sufficiently high signal-to-noise ratio (SNR) to enable detection of the BOLD signal changes. While these points may be obvious, they constrain the imaging methods that can be used, and impose the need to balance the imaging parameters to meet all of these needs. For example, it is not possible to acquire images that have very high spatial resolution, in a very short amount of time, and with a high SNR. A considerable amount of debate has also occurred over the questions of what are "adequate" and "sufficient" values, as listed above. Typical BOLD fMRI methods provide adequate sensitivity, spatial and temporal resolution, coverage of the anatomy, and SNR, but rarely meet the optimal values that might be desired for all of these parameters, as will be described below.

The most commonly used imaging methods for BOLD fMRI are gradient-echo methods, because with T_2^*-weighting these methods provide higher BOLD sensitivity than the alternative spin-echo methods. The need for very short acquisition times to achieve high temporal resolution makes single-shot echo-planar imaging (EPI) or spiral imaging methods most effective for a wide range of applications.

The theoretical optimal resolution that will provide the highest BOLD contrast is to have 1.5 mm cubic voxels (54). This is the approximate size of the *smallest vascular unit*, which is the extent of the smallest region that is supplied by a single arteriole. It has been proposed, therefore, that using higher resolution (smaller voxels) will not improve the spatial precision of the activity maps obtained with fMRI.

The optimal echo time, TE is equal to T_2^*, in order obtain the highest contrast-to-noise ratio. At 3 tesla (T), for example, this echo time will therefore be approximately 30 msec. Since $\Delta S/S = -\text{TE}\ \Delta(1/T_2^*)$, as described in Section 6.2.1, and $\Delta(1/T_2^*) = -1.2\ \text{s}^{-1}$ to $-1.8\ \text{s}^{-1}$ at 3 T, the BOLD signal changes are expected to be around 4%–5%. The minimum necessary SNR is therefore around 25, although SNR values several times higher are desirable to detect more subtle effects, to quantify differences between regions or responses to different stimuli, and so forth.

The combination of these optimal parameters means that to obtain fMRI data spanning the entire brain, in a 22 cm × 18 cm × 14 cm field of view, would require a matrix size of 147 × 120 for each image and 93 slices, each 1.5 mm thick. Even with a relatively large receiver bandwidth of 250 kHz, the sampling rate is 4 μsec per data point. Sampling 147 × 120 data points in continuous succession for each slice would take 71 msec. Sampling the center of k-space at an echo time of 30 msec is therefore only possible if only a portion of k-space is sampled. Assuming a 3 msec RF pulse is used, and allowing 1.5 msec for gradients to be applied prior to data sampling (refer to Figure 5.22), then 3 msec is needed from the center of the RF pulse to when data sampling can begin. There is a maximum of 27 msec available for sampling data before the echo time is reached. Each line of 147 data points will take 0.588 msec plus 0.300 msec for turning gradients on and off (assuming a typical 150 μsec gradient ramp time), for a total of 0.888 msec. This means that in 27 msec we can sample the center of k-space at the center of the 30th line of data (that is, we can sample 29.5 lines of data by the time we reach 30 msec after the center of the RF pulse). After this, we need to sample the remaining 60.5 lines spanning the other side of k-space, and this will take an additional 53.7 msec. The resulting 90 lines of data will need to be zero-filled to complete the 120 lines of the matrix to be sampled. The total time for collecting the data for each slice is therefore 53.7 msec + TE, or 83.7 msec. Now, to collect data from 93 slices will take 7.8 sec. As one final point, this acquisition would require a frequency-encoding gradient that provides 250 kHz/22 cm and is therefore a gradient of 2.7 gauss/cm. This is a relatively large value but is possible with most modern MRI systems.

The SNR to be expected with this imaging method can be estimated as described in Chapter 5, with SNR $\propto$ *Total imaging volume*$/\sqrt{(N_X N_Y BW)}$. Since this is a proportionality statement, we need a reference for comparison. SNR values have been reported to be 332 at 3 tesla with gradient-echo EPI, TE = 30 msec, 3 mm × 3 mm × 3.5 mm resolution, with a 64 × 64 matrix (55). In this example the bandwidth was not specified but for our purposes is assumed to be 200 kHz based on comparable observations. Using this reference we can estimate the SNR of the acquisition described above, as follows:

$$\frac{SNR_2}{SNR_1} = \frac{Total\ imaging\ volume_2/\sqrt{(N_{X2}\ N_{Y2}\ BW_2)}}{Total\ imaging\ volume_1/\sqrt{(N_{X1}\ N_{Y1}\ BW_1)}}$$

$$\frac{SNR_2}{332} = \frac{(220\ mm \times 180\ mm \times 1.5\ mm)/\sqrt{(147 \times 90 \times 250\ kHz)}}{(192\ mm \times 192\ mm \times 3\ mm)/\sqrt{(64 \times 64 \times 200\ kHz)}}$$

$$SNR_2 = 89$$

Based on this estimate we can expect that the method described above with 1.5 mm cubic voxels will provide an adequate SNR for fMRI at 3 tesla. However, one problem is that the imaging time of 7.8 sec is relatively slow compared with that of commonly used fMRI methods. To achieve high sensitivity for detecting BOLD signal changes, it is desirable to have a large number of data points to describe the signal intensity time series. On the other hand, it is also necessary that the total fMRI acquisition is not exceedingly long because of the need to keep the participant engaged in performing the assigned task(s) while remaining motionless. Using the same process as above to estimate new imaging parameters, we can estimate the SNR and timing that would be needed for a range of selected spatial resolutions, as shown in Table 6.2. These values are

Table 6.2 Summary of Estimated Acquisition Times and Signal-to-Noise Ratios for Various Choices of Imaging Parameters for Gradient-Echo EPI, Based on a 22 cm × 18 cm × 14 cm Field of View to Span the Brain, at 3 Tesla, with TE = 30 msec, and a 250 kHz Bandwidth

Resolution	No. of Slices Needed	Sampling Matrix	Time per Slice	Acquisition Time	SNR (estimate)
1.5 × 1.5 × 1.5	93	147 × 90 (partial k-space)	84 msec	7.8 sec	89
2 × 2 × 2	70	110 × 90	64 msec	4.5 sec	137
3 × 3 × 3	47	73 × 60	48 msec	2.3 sec	308
4 × 4 × 4	35	55 × 45	42 msec	1.5 sec	547

based on a 22 cm × 18 cm × 14 cm field of view to span the brain, at 3 tesla, with TE = 30 msec and a 250 kHz bandwidth.

In practice, lower bandwidth values are often used to further increase the SNR and also to reduce the gradient strengths that are needed. Using lower gradients will increase the imaging time but can improve image quality because electrical currents may be induced in components of the MRI system by rapidly changing magnetic fields. These electrical currents are termed *eddy currents* and can cause image distortions, particularly in EPI and spiral imaging methods with long acquisition times. A common site of eddy currents is the wall of the helium-filled dewar, although most modern MRI systems are designed to reduce and/or compensate for eddy currents. A related problem is small electrical currents that can be induced in peripheral nerves of the person being imaged, if their arms or legs form a closed loop, such as if their hands are clasped together or if their ankles are crossed and bare skin is touching. Peripheral nerve stimulation can cause brief sensations of tingling or slight pressure.

Another factor that may cause actual imaging parameters to be different from the estimates above is limitations in the options available on the MRI system. For example, some systems may limit the choice of acquisition matrix sizes to be multiples of powers of 2, such as 256, 192, 128, 64, and so on. The imaging sequence may also require small additional amounts of time for each slice if additional gradients are used, or different durations of RF pulses if gradient ramp times are used, and so forth. As a result, the above examples cannot be taken as exact but are good estimates of the effects of using different parameters for fMRI acquisitions.

There is no single "optimal" fMRI method because the ideal values of speed, resolution, SNR, and coverage of the anatomy cannot be delivered simultaneously. The best balance of what can be achieved depends on the particular application. For example, for an investigation of function in the visual cortex, high spatial and temporal resolution may be achieved by imaging only the particular region of the brain that is of interest as opposed to whole-brain coverage. Alternatively, detection of all of the brain regions involved with a distributed system, such as with the perception of pain, may benefit much more from whole-brain coverage and high temporal resolution than from high spatial resolution. There are also options such as whether or not to leave a gap between adjacent imaging slices to span more of the anatomy. However, leaving a gap can cause greater sensitivity to small body movements, unless very long repetition times (TR) are used, because the magnetization in the imaging slices will not have the same T_1-weighting as the magnetization in the gaps. Even slight movement could cause tissues that were previously in the gap between slices to move into an imaging slice, resulting in higher signal intensity in the subsequent acquisitions of that image location, until a new steady-state T_1-weighting is reached. As a result, the signal intensity time course could include signal changes in time that are not related to neural activity, potentially causing errors in the fMRI results. Another imaging option is whether or not to have cubic voxels to facilitate spatial normalization or reslicing the imaging data to other orientations, or, more importantly, to have consistent partial-volume effects in all regions of the anatomy. *Partial-volume effects* refer to the signal intensity observed in every voxel being the sum of the signals from all of the different tissues within the voxel. When voxels are not cubic, partial volume effects may vary systematically, depending on the position in the brain, because at different locations the largest dimension of the voxels may align parallel to the cortical gray-matter layer, whereas at other locations it may be transverse to the gray-matter layer (Figure 6.4).

Finally, considering all of these factors, a common range of imaging parameters emerges. A survey of recently published fMRI studies done at 3 tesla (to represent an intermediate range) indicates that the most commonly used imaging method and parameters for human brain fMRI are listed in Table 6.3. Among the examples surveyed, the bandwidth (BW) was not specified. However, the BW can be adjusted to be the lowest possible while still achieving the desired imaging speed.

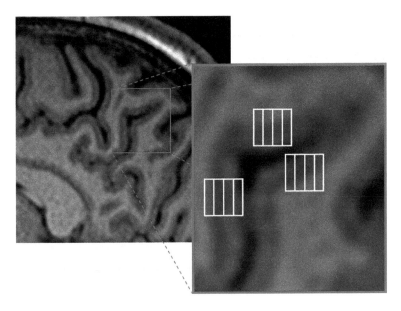

FIGURE 6.4 Illustration of the systematic variation of partial volume effects that may occur depending on position in the brain with noncubic voxels. The magnified view depicts three simulated examples of sets of four voxels with larger vertical than horizontal dimensions, at various locations relative to the gray matter (which appears darker than white matter in this T_1-weighted image). The average signal that would be contained within each voxel can be seen to vary with the orientation of the boundary between gray matter and white matter.

Table 6.3 Commonly Used Imaging Parameters for Human Brain fMRI at 3 Tesla	
Imaging Method	**Gradient-Echo EPI**
Echo time (TE)	30 to 40 msec
Repetition time (TR)	2 to 4 sec
Slice thickness	3 to 5 mm
Field of view (FOV)	192 mm × 192 mm to 240 mm × 240 mm
Sampling matrix	64 × 64 is most often used, but 128 × 128 is also common
In-plane image resolution	1.8 mm × 1.8 mm to 3.8 mm × 3.8 mm

Key Points

12. Imaging methods for BOLD fMRI must balance the requirements of BOLD sensitivity, rapid and repeated imaging of the region of interest with spatial precision, and a high SNR.

13. The highest BOLD sensitivity is achieved with gradient-echo imaging, with the echo time, TE, set equal to the T_2^* of the tissues in the region of interest.

14. The optimal balance of spatial resolution, imaging speed, BOLD sensitivity, and SNR, appears to be obtained with single-shot fast imaging methods, such as EPI or spiral sampling of k-space, to image multiple contiguous 2D image slices, with approximately 3 mm × 3 mm × 3 mm cubic voxels.

15. Typical repetition times that are needed to acquire a complete set of images spanning the region of interest are 2 to 4 sec.
16. The optimal imaging parameters depend strongly on the specific needs of the fMRI application and on the MRI system capabilities and magnetic field strength. There is no single standard method that is suitable for every application.

6.4 Special Regions

The challenges presented by magnetic susceptibility differences between air, tissues, and bone and by physiological motion can significantly affect the sensitivity and reliability of fMRI results and have a strong influence on the optimal methods for any given application. There are particular areas of the CNS that are much more affected by these challenges than other areas. For example, the frontal lobe and temporal lobes of the brain are near the air spaces in the sinuses and ear canals, and are prone to signal loss and image distortion. The brainstem and inferior portions of the basal ganglia are similarly near air spaces; and the spinal cord is surrounded by the bone in the spine, and its entire length is near the lungs. As a result, these areas of the CNS are much less frequently discussed in the scientific literature than are other areas of the brain. Nonetheless, with appropriate adaptations of fMRI methods, it is possible to map activity in all of these challenging areas.

6.4.1 Causes of Image Distortion and Signal Loss

The volume magnetic susceptibilities, χ, of air, bone, and tissue (gray matter and white matter) are respectively 0.4, –8.4, and –8.8 to –9.0 ppm (repeated from Table 3.2). The volume susceptibility of intervertebral discs in the spine have been measured to be similar to water at around –9.8 ppm, whereas the lipid component of marrow in the vertebrae increases the susceptibility value by about 0.15 ppm (56). Recall that when the body is in a magnetic field, B_0, inside the various materials in the body the magnetic field is altered and is given by $B = (1 + \chi)B_0$. The magnetic field is continuous across boundaries between tissues and so at air/tissue and bone/tissue interfaces there are spatial field variations that are, in effect, small magnetic field gradients. When echo-planar imaging is used with a field of view of 22 cm and a receiver bandwidth of 220 kHz, then a gradient of 10 kHz/cm, or 2.35 gauss/cm, is needed. The magnetic field difference of 0.25 gauss between air and tissues (in a field of 3 tesla) might seem insignificant in comparison. However, the field difference at such interfaces produces two important effects. The first is its contribution to T_2^*, since:

$$\frac{1}{T_2^*} = \frac{1}{T_2} + \gamma \Delta B$$

where ΔB is the total field variation across a voxel. Typical values of T_2 and T_2^* in the brain at 3 tesla are 90 msec and 30 msec, respectively, indicating that the value of $\gamma \Delta B$ is 22.2 sec^{-1}, or $\Delta B = 0.83 \times 10^{-3}$ gauss. If the effect of an air/tissue interface spans across 10 voxels, then the variation in each voxel is 0.025 gauss, which results in an additional $\gamma \Delta B$ of 669 sec^{-1}. The resulting value of T_2^* in the voxels within the effect of the interface is therefore reduced to 1.4 msec. The field variation effect diminishes with distance from the interface, and this number is just an estimate, but it serves to demonstrate the cause of the reduced signal intensity observed in T_2^*-weighted images in the vicinity of such interfaces (Figure 6.5).

The second effect of the field variations at interfaces is to cause image distortions by adding to the magnetic field gradients that are applied for spatially encoding the MR signal. An important difference between the gradients and the field variations at interfaces is that the gradients can be applied for brief periods and turned off and can be reversed if needed. The field variations at interfaces are constant. The echo-planar imaging method (as described in Chapter 5.5 and shown in Figure 5.22) samples across k-space in the $+k_X$ direction, the phase-encoding gradient

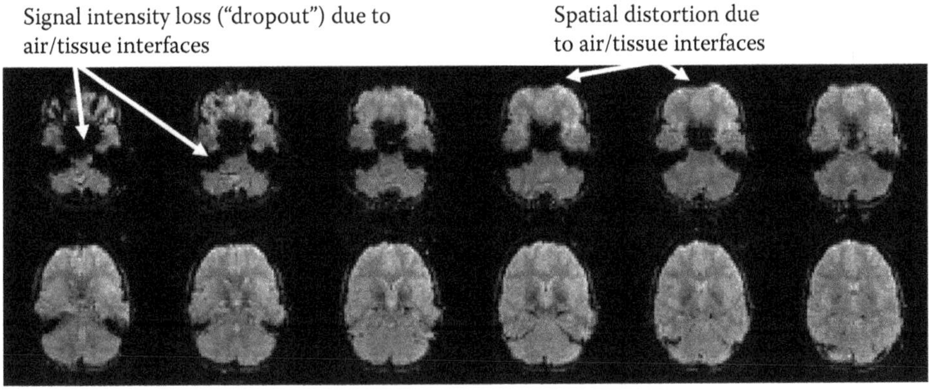

Signal intensity loss ("dropout") due to air/tissue interfaces

Spatial distortion due to air/tissue interfaces

FIGURE 6.5 Axial images acquired with single-shot EPI showing the loss of signal and spatial distortion caused by magnetic susceptibility differences at air/tissue interfaces.

is applied briefly to step the value of k_Y by a small amount, and then the frequency-encoding gradient is reversed to sweep back across k-space in the $-k_X$ direction, and so on. The field variations at interfaces will therefore add to the gradient when sampling across k-space in one direction and will subtract from the gradient when sampling in the opposite direction. The values of k_X and k_Y are expected to be $k_X = \gamma G_X t$ and $k_Y = \gamma G_Y T$ and the phase of the signal in a voxel at a location (x,y) at a fixed point in time is expected to be $\phi = k_X x + k_Y y$. This two-dimensional encoding is, of course, applied after a slice-selective RF pulse has tipped the magnetic moments within a slice away from alignment with B_0. The effect of the interface will depend on the position of the slice that is selected as well. We can use the term $\Delta B_{interface}$ to refer to the field variation caused by the interfaces at any three-dimensional (x,y,z) location, and for this discussion the x, y, and z coordinates correspond with the G_X, G_Y, and slice directions. The actual phase of the signal at a specific (x,y,z) location is therefore given by:

$$\phi = k_X x + k_Y y + \gamma \ B(x,y,z)_{interface} t$$

An important point here is that t' is the time elapsed since the center of the RF excitation pulse (the time when the magnetic moments were rotated away from alignment with B_0). For the signal from some localized regions (i.e., where $\Delta B_{interface} \neq 0$), the actual position in k-space that is being measured does not match the expected position. Keep in mind that during this time the values of k_X and k_Y are being varied systematically to sweep across all of the value of k-space that we want to measure, and that every point in k-space represents data from the entire imaging slice. It is not possible to simply correct the values of k-space for the spatially localized magnetic field distortions, because this would require prior knowledge of what the ideal image would look like. However, it is possible to correct for larger scale or global magnetic field distortions that alter the effective values of G_X or G_Y by measuring the field distortions, and then accounting for the systematic shifts in the actual values of k_X and k_Y these would cause. The effect of $\Delta B_{interface}$ on the resulting images is simulated and shown in Figure 6.6.

The simulation shown in Figure 6.6 is approximate and takes into consideration only the tissues that are in the plane of the imaging slice. Actual distortions are equally caused by interfaces outside of the imaging slice as well, the important factor being the distance from the interface. Actual distortions and signal losses due to reductions in T_2^*-values are shown in Figure 6.7 for three different imaging methods applied at the exact same location and slice orientation in one person. The HASTE method (half-Fourier single-shot fast-spin echo), shown in the leftmost frame of Figure 6.7, is a fast spin echo method, with slightly more than half of k-space sampled, whereas the middle and rightmost frames show spin-echo and gradient-echo acquisitions, respectively, and both of these were acquired with EPI spatial encoding. Significant

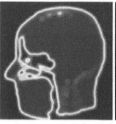
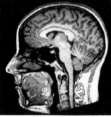
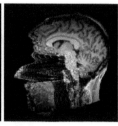
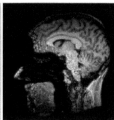

| Field distortion map | Original image | Simulated distortion with EPI | Simulated distortion and signal loss |

FIGURE 6.6 Simulation of the effects of magnetic susceptibility differences, $\Delta B_{interface}$, as shown in the left-most frame, if the image data shown in the original image were to be acquired with an EPI method. The resulting distortion is shown (simulated) in the third frame, and the additional effect of signal loss due to changes in T_2^*, is shown in the rightmost frame. Significant distortion and signal loss around the frontal lobe, brainstem, and spinal cord are clearly visible.

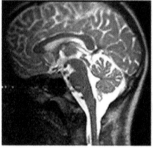
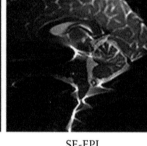
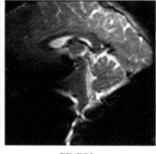

| HASTE | SE-EPI | GE-EPI |

FIGURE 6.7 Actual examples of the image distortion and signal loss caused by magnetic susceptibility differences at air/tissue and bone/tissue interfaces. HASTE is a half-Fourier fast spin-echo method, whereas the middle and rightmost frames show results obtained with EPI spatial-encoding schemes, combined with spin-echo (SE) and gradient-echo (GE) methods, respectively.

image distortion can be seen in regions of the brain, brainstem, and spinal cord, and can be seen to depend primarily on the spatial-encoding method (i.e., HASTE versus EPI) and very little on the differences between spin-echo and gradient-echo methods. Spatial distortion is expected to occur in any images with long acquisition times, such that values of $\gamma\Delta B_{interface}t'$ are significant relative to $k_X x + k_Y y$, including spiral methods, EPI, or any encoding scheme.

To apply fMRI in regions near interfaces with magnetic susceptibility differences, it is therefore necessary to avoid long acquisitions. The question then is, What is a sufficiently short acquisition time in order to avoid distortions? A distortion will be apparent whenever one part of the image appears to be shifted in position with respect to its surroundings. We can consider different regions of an object separately and treat the resulting image as the sum of the images of the different regions. If the images to be constructed are $N_X \times N_Y$ points, then the image of one region will be shifted by the width of 1 pixel, if there is a phase increment of $2\pi/N_Y$ between successive points across k-space in the k_Y direction. The phase difference, φ, between two regions with a magnetic field difference of ΔB, after a time interval t, is given by $\varphi = \gamma \Delta B\, t$. With an EPI spatial-encoding scheme, successive data points in the k_X direction are sampled at the dwell time, DW, apart (recall that DW = 1/BW). Successive lines in the k_Y direction, however, are sampled at intervals (N_X DW + 2 T_{ramp}) apart, where T_{ramp} is the time taken to turn the gradients on or off (i.e., to *ramp* the gradients). This means that to get the phase increment of $2\pi/N_Y$ in the time elapsed between successive points in the k_Y direction, we need a magnetic field difference of:

$$B = \frac{2\pi/N_Y}{\gamma\left(N_X\ \mathrm{DW} + 2\mathrm{T_{ramp}}\right)}$$

Having this field difference between two regions would cause the image of one region to be shifted by 1 pixel in the y direction with respect to the image of the other region. When the images are summed (to produce the actual image), then there may be an area where the shifted and nonshifted regions overlap, resulting in much higher signal, and another area where the shifted region leaves nothing but a signal void in the image, as can be seen in Figures 6.6 and 6.7.

Actual values for an EPI acquisition could be an acquisition matrix of 64 × 64, with a sampling rate of 4 μsec per point and a ramp time of 150 μsec. Using these values, the field difference needed to produce a shift of 1 pixel width is:

$$B = \frac{2\pi/64}{\gamma\left(64\times4\quad\mathrm{sec} + 2\times150\quad\mathrm{sec}\right)}$$

$$B = 0.0066\ \mathrm{gauss}$$

A field shift of 0.025 gauss near an air/tissue interface (i.e., 10% of the magnetic field difference between air and tissues) could therefore cause the signal from that region to be shifted by 3.8 pixels in the resulting image. Any increase in N_X, N_Y, or DW would reduce the magnetic field difference needed to cause a shift of 1 pixel width. However, increasing N_X or N_Y also reduces the size of each pixel.

Another effect of long acquisition times is that the transverse relaxation that occurs while the data are being acquired becomes significant and reduces the spatial resolution of the resulting images, as discussed in Chapter 5. In particular, between sequential values of k_y, the signal decays by the amount e^{-t_s/T_2^*}, where t_s is equal to $(N_X\ \mathrm{DW} + 2\ \mathrm{T_{ramp}})$. Using the values for N_X, DW, and $\mathrm{T_{ramp}}$ from the previous example, t_s is therefore equal to 0.556 msec. Since T_2^* is approximately 30 msec in the brain at 3 tesla, then the signal decays by a factor of 0.982 between each successive line. While this may not seem very large, the cumulative effect across 64 lines is a reduction to 0.305. The progressive exponential decay of the signal across k_Y lines results in the same effect as the resulting image being convolved with the Fourier transform (FT) of the exponential decay, that is, a Lorentzian function. Recall from Chapter 2 the Fourier convolution theorem that $\mathrm{FT}(A \times B) = \mathrm{FT}(A) \otimes \mathrm{FT}(B)$. The decay of the signal depends on the time interval between successive k_Y points, and the full width at half of the peak height of the corresponding Lorentzian function is equal to $1/(2T_2^*)$. Using the same frequency units, each pixel in the resulting image spans a width equal to $2\pi/(N_Y\ t_s)$. The full width at half height can therefore be expressed as $((N_Y\ t_s)/4\pi\ T_2^*)$ pixel widths. Depending on how the signal decay is modeled, whether uniformly across the data sampling or symmetrically from the center of k-space as with some imaging methods, there are slightly different forms of this expression in the literature. For example, Jesmanowicz et al. (57) expressed the width as $((N_Y\ t_s)/2\pi\ T_2^*)$ and abbreviated the duration $(N_Y\ t_s)$ as $\mathrm{T_{RO}}$ for the *read-out time*. Nonetheless the key point is the same, being that longer sampling times with larger values of either N_Y or t_s will result in lower resolution. If we continue the example above and use values of 64 for N_Y, 0.556 msec for t_s, and 30 msec for T_2^*, then the effect of the transverse decay across k-space is to convolve the image with a Lorentzian function with a width of 0.0944 pixels. This would cause only a very slight reduction of spatial resolution, such that if the acquired pixel size was 3 mm, then the actual resolution would be approximately 3.14 mm (i.e., multiplied by 1 + 0.0944/2). However, if instead we acquired data in a 128 × 128 matrix, the width of this Lorentzian function would be 0.378 pixels, and the 3 mm pixel size would be spread out to an image resolution of approximately 3.57 mm.

It is possible to avoid the distortions described above by using conventional encoding schemes in which the phase encoding is applied individually for each line of k-space, as opposed to being accumulated across successive lines, as with EPI or spiral imaging. Another alternative is to use *segmented* EPI acquisitions, in which, for example, every second line of k-space is acquired after one RF excitation pulse, and then the remaining lines are acquired after a second RF excitation pulse. Similarly, every fourth line could be acquired in four separate acquisitions to sample all of the lines of k-space. In this case, the phase difference between successive lines is spread across four lines, and the resulting shift caused by ΔB is decreased by a factor of 4. That is, instead of a shift of 3.8 pixel widths in the above example, we would see a shift of 0.95 pixel widths, and therefore much less distortion.

Signal loss due to reduction of T_2^*-values at interfaces between materials with different magnetic susceptibilities can only be avoided by significantly reducing the echo time, TE, or by using spin-echo imaging methods. As described earlier, spin-echo methods can be used to detect the T_2 changes produced by the BOLD effect, although with less sensitivity than would be obtained with detecting T_2^* changes with gradient-echo methods (52). In some regions, the improved image quality may be a necessary trade-off for the reduced BOLD sensitivity.

Specific methods for fMRI of the frontal lobe or temporal lobes, in order to obtain high sensitivity to BOLD signal changes and also spatial accuracy, can be theorized (actual examples are shown in Section 6.5) based on this knowledge of how the distortions and signal loss are caused. Limiting the choices to BOLD-sensitive T_2^*-weighted acquisitions, then a gradient-echo method must be used. Conventional spatial-encoding methods (i.e., one line or k-space per RF excitation pulse) can be used with a small flip angle and short repetition time in an effort to avoid distortions while keeping the imaging time short. Continuing the example of working with a 3 tesla MRI system, then we want the echo time to be equal to 30 msec. This will affect the lower limit of how quickly the data for one slice can be acquired. If we use similar parameters as those for EPI, as in the previous examples, then for now we can assume we have a 64×64 matrix, 250 kHz receiver bandwidth (i.e., 4 μsec sampling rate or DW), and 22 cm × 22 cm field of view to provide 3.4 mm × 3.4 mm in-plane resolution. The time to sample one line of 64 points with different k_X values is 0.256 msec (4 μsec/point × 64 points). If the echo time is set at the center of each line of data, then the minimum time to acquire one line of data from one slice is 31.9 msec. This time allows an additional 1.5 msec for half of the RF pulse not included in the echo time, half of the data sampling time after the echo time, and 0.15 msec for ramping down the gradients. As a result, the minimum time needed to sample 64 lines of data is 2.04 sec, and this does not allow time to acquire data from multiple slices, and very little time for longitudinal relaxation between RF excitation pulses. Even a relatively small number of slices, such as 10, would take a minimum TR of 319 msec and a total acquisition time of 20.4 sec, while still only covering a 34 mm span if we used cubic voxels. As a result, this approach is not practical for most applications of fMRI unless a very small number of slices is sufficient.

An alternative approach is therefore to use a segmented EPI acquisition (58,59). As indicated above, to acquire one line of 64 points at a sampling rate of 4 μsec/point, plus 150 μsec each to turn the gradients on and off, adds up to 0.556 msec. Acquiring 16 of the 64 lines with different phase-encoding (k_Y) values would therefore take 8.9 msec. The point in time when the center of k-space is sampled is the echo time. If the center of k-space is sampled at the center of the ninth line, then another 8.5 lines or 4.7 msec is needed to complete the sampling. The minimum time needed to acquire one segment of k-space data for one slice is 36.2 msec (1.5 msec for half of the RF pulse, 30 msec echo time, plus 4.7 msec). The time needed to acquire one segment of data for 40 slices is therefore 1.45 sec, and to acquire all of the data needed to construct the images for 40 slices is 5.8 sec. Because the repetition time will be 1.45 sec, the flip angle for excitation needs to be reduced from 90° to reduce the T_1-weighting of the resulting images. The optimal angle is the Ernst angle (described in Chapter 3) at 74° for a TR of 1.45 sec, assuming a T_1-value of 1.1 sec. At this flip angle the signal intensity will still be 96% of that obtained with a 90° flip angle (i.e., sin(74°) = 0.96). This segmented-EPI method will therefore produce the same image resolution as a typical single-shot EPI acquisition for fMRI, with approximately 96% of the SNR, with image distortions reduced by a factor of 4, at the expense of a longer, but still reasonable, total acquisition time of 5.8 sec (although with more sensitivity to movement of the person being imaged).

A degree of reduction of spatial distortions and signal loss due to magnetic susceptibility differences can also be achieved by other means or a combination. For example, the TE could be reduced below the typical optimal value, causing a slight reduction in BOLD sensitivity. While this does not decrease the total time span over which the data are acquired, it can reduce the signal loss from regions where T_2^* is reduced. The slice thickness can also be reduced, thereby decreasing the amount of signal variation across the voxel, or similarly the in-place resolution could be increased. Both of these changes would result in a smaller voxel volume and therefore lower SNR. The time taken to acquire the data, and over which the errors accumulate, can be reduced by only sampling part of k-space and using zero filling, or parallel imaging methods can be applied (both of the methods are described in Chapter 5). These methods also result in a reduction of SNR. As always, the improvements that are gained with these approaches must be balanced with the costs.

Additional challenges that may be encountered with fMRI of regions such as basal ganglia or regions of the brainstem are the small physical dimensions of some regions (such as specific regions of the thalamus) (60), and the motion of the brainstem that occurs with each heartbeat (61). The image distortion due to air/tissue interfaces, as discussed above, is also a problem in these regions, and so must still be considered as well. The movement of the brainstem in the head-foot direction has been measured to be about 0.6 mm, and there may be an anterior–posterior component to the motion as well, as observed in the cervical spinal cord (61). The impact of this movement on fMRI data is expected to depend on the image resolution. For example, if image data are obtained with 1 mm transverse slices, so that the movement is through the plane of the slice, then 0.6 mm movement may cause significant signal variations. One component of the effect is the variation due to the tissues within the slice changing, as the brainstem moves, and the second component is that the amount of T_1-weighting may vary due to changing times between repeated excitations of the same slice. To clarify, recall that imaging slices are typically acquired in an interleaved order, such as 1, 3, 5 …, then 2, 4, 6 …, in order to avoid the effects of slight cross-excitation between adjacent slices. While each slice is excited once every TR period, its immediate neighbor is not excited immediately after, but rather approximately ½ TR after (and ½ TR before as well, since the images are repeated). If the brainstem moves 0.6 mm, compared with a 1 mm slice thickness, then some tissue will be excited twice in the time 1/2 TR, and other tissue will be excited twice in 3/2 TR, resulting in variable T_1-weighting of these tissues. The resulting signal variation could obscure the neuronal-activity–related signal changes that we are trying to detect. Allowing gaps between the imaging slices would not improve the situation but may actually compound it by adding an additional possibility for how often the tissues are excited. Possible solutions, or at least ways of reducing the signal variation, include (1) using thicker slices, so that the amount of movement is proportionately smaller; (2) using a longer TR and/or smaller flip angle, so that the amount of T_1-weighting even in ½ TR is relatively small; or (3) using a cardiac-gated acquisition. This last option will typically require more time to acquire the image data but can provide the additional benefit of reducing the artifacts that are produced in the phase-encoding direction when image data are acquired from moving tissues (as described in Chapter 5).

The challenge presented by small physical dimensions of some regions, such as the brainstem, requires adaptations to acquire image data with higher resolution and careful consideration of the slice orientation that is best suited to the particular application. While one option to achieve higher spatial resolution is to reduce the FOV, this will substantially decrease the SNR of the resulting images. Recall that SNR $\propto$ *Total imaging volume*/$\sqrt{(N_X N_Y \text{ BW})}$. If in the previous example the FOV is reduced from 220 mm × 220 mm to 110 mm × 110 mm to decrease the resolution to 1.7 mm × 1.7 mm, then the SNR will be reduced by a factor of 4. However, if instead the FOV is not changed but the matrix is increased from 64 × 64 to 128 × 128, then the SNR will be reduced by a factor of 2. However, the trade-off is that the acquisition time will be longer, potentially increasing the total imaging time and being more prone to image distortions. Continuing the example above, if we used a single-shot EPI method, the magnetic field shift needed to produce a distortion of 1 pixel width would be:

$$B = \frac{2\pi/128}{\gamma\left(128 \times 4 \quad \text{sec} + 2 \times 150 \quad \text{sec}\right)} = 0.0023 \text{ gauss}$$

If there is a field shift of 0.025 gauss near an air/tissue interface, the pixels in the resulting image would be shifted from this location by 10.9 pixel widths. The time needed to acquire a 128 × 128 matrix of data with single-shot EPI would be approximately 104 msec (again assuming a 4 μsec sampling rate and 150 μsec ramp time). As described above (Table 6.1) we would have to resort to a partial-Fourier acquisition to sample the center point of k-space at an echo time of 30 msec. Considering the long acquisition and high expected distortion, the single-shot method is not well-suited to this application.

Once again, a segmented EPI method may provide an appropriate solution to this imaging problem. By acquiring the data in eight segments, for example, the image distortion would be reduced from a displacement of 10.9 pixel widths to 1.4 pixel widths. The time needed to acquire the data for one segment from one slice would be 13.0 msec, and the total time needed for each segment for each slice would be 37.6 msec with an echo time of 30 msec. We could therefore acquire the data from 1 segment for each of 20 slices in 0.75 sec, or the entire set of image data in 6 sec. The optimal flip angle, the Ernst angle, in this case would be 60°, which would reduce the SNR further because the signal intensity will be multiplied by sin(60°) or 0.87. The end result is a trade-off of higher spatial resolution and less image distortion for slower imaging time, although the time of 6 sec for each complete set of images is still reasonable for fMRI.

For comparison, we can also consider acquiring the data in four segments. This would require an acquisition time for each segment of 26.0 msec, and the total time needed to acquire one segment for each slice would be 44.1 msec, for an echo time of 30 msec. One segment from each of 20 slices could be acquired within a TR of 882 msec, and a complete set of image data would therefore take 3.5 sec. In this case, with the longer TR compared to the eight-segment example above, the Ernst angle would be 63° and only a very slight increase in SNR would be expected. The drawback of the four-segment approach would be greater image distortion compared with the eight-segment method, but the advantage would be a shorter imaging time.

Specific adaptations of BOLD fMRI for the spinal cord can also be made as described above, using segmented EPI to reduce distortion (62). However, imaging the spinal cord presents the greatest challenges, with both very large magnetic field distortions and very small physical dimensions of the spinal cord cross-sectional anatomy. Efforts to reduce the image distortions and signal losses that are caused by nonuniform magnetic fields also tend to reduce the sensitivity to the BOLD effect, which itself depends on local distortions of the magnetic field around deoxyhemoglobin (63). The requirements of sensitivity to the BOLD effect and also high image quality are therefore at odds.

In the relatively extreme case of the challenges presented by fMRI of the spinal cord, the alternative contrast mechanism signal enhancement by extravascular water protons, or SEEP, is needed (described in detail in Section 6.6). Because SEEP contrast can be detected with proton density–weighted images, this method provides the option of using spin-echo imaging methods as opposed to gradient-echo imaging, which is preferred for BOLD contrast. In addition, spin-echo methods have options for fast imaging methods that do not require the use of long acquisition periods as with EPI or spiral methods. It is important to keep in mind that EPI and spiral are methods for sampling k-space and can be used with either gradient-echo or spin-echo imaging methods. In spite of this fact, it has become relatively common in scientific papers on fMRI to simply use the term EPI as though it also implies a gradient-echo imaging method unless otherwise specified. As described in Chapter 5 (Figure 5.21), with a fast spin-echo method, each line of data at a single value of k_Y is sampled after a 180° refocusing pulse, with the phase encoding applied specifically for each line of data. That is, the phase encoding does not rely on the cumulative effect of repeated gradients as with EPI, and therefore the resulting image does not have the same sensitivity to image distortion. Producing each successive line of data requires the time to apply the 180 pulse°, then the phase-encoding (k_Y) gradient, then turn on the frequency-encoding (k_X) gradient, sample the data over a range of k_X values, and then apply the reverse phase-encoding gradient to return the k_Y value to zero. If we sample data for a 128×128 matrix with a receiver bandwidth of 100 kHz, for example, then the sampling rate is 10 μsec/point and requires 1.28 msec to sample each line of k_X values. Other timing parameters may depend on the MRI system capabilities, but reasonable values are 3 msec to apply each RF pulse, 1 msec for each phase-encoding gradient, including 0.15 msec to ramp the gradients on or off. Using these timing values, the total time required for each line of data is therefore 6.58 msec. As a result, sampling the 128×128 matrix for a single-shot fast spin echo will require a 90° pulse followed by a train of 128 separate 180° pulses and will take approximately 846 msec for each complete image. The fast spin-echo method therefore cannot approach the speed provided by EPI methods and deposits considerably more energy into the tissues due to the RF pulses, as described in Chapter 5. However, this method can provide high resolution (1 mm × 1 mm (64)), high SNR, and high image quality with little to no distortion, as shown with the HASTE method in Figure 6.7.

As demonstrated above, although certain areas of the CNS present additional challenges for fMRI, function can nonetheless still be mapped in these areas. Typical fMRI acquisition methods must be adapted for each specific application, and trade-offs of slower imaging speed must occasionally be tolerated.

Key Points

17. Magnetic susceptibility differences between air, bone, and tissues create magnetic field distortions that can cause image distortion and signal loss, particularly with fast imaging methods.
18. Imaging methods with longer data acquisition periods (i.e., more time spent actually sampling the data), such as with EPI and spiral k-space sampling, are more sensitive to the effects of magnetic field distortions.
19. Progressive signal attenuation across k-space as a result of transverse relaxation while data are being sampled results in a reduction of spatial resolution in the resulting image.
20. Image distortions in areas near air/tissue or bone/tissue interfaces can be reduced by using imaging methods with conventional (line-by-line) k-space sampling or by using *segmented* or *multishot* EPI or spiral k-space sampling.

6.5 Specific Examples of fMRI Applications— Setting the Acquisition Parameters

As demonstrated in the previous section, there is no single optimal method that suits all applications of fMRI, but rather the method must be adapted for each specific case. However, the range of choices is limited by the capabilities of the MRI system and the imaging method and by the needs for high sensitivity to neuronal-activity–related signal changes. Many published fMRI studies therefore use very similar imaging parameters, but it is very useful to understand when and how these parameters can be adjusted to improve the sensitivity and reliability of the resulting maps of neural activity. Likely, the best guides available for demonstrating optimal imaging parameters are the published fMRI studies that have achieved the desired sensitivity and spatial precision for mapping neural activity changes in response to a stimulus or task. While understanding the underlying theory is extremely useful, the differences between theory and practice are revealed by the experience that has been gained by practitioners of fMRI over the two decades that have passed since it was first introduced.

One such example is by Greene et al. (65), who aimed to study the neural correlates of moral judgment. In such a study, it is expected that a number of brain areas will be involved, and the design must account for the possibility that some areas may be unexpected or may have relatively brief responses. The fMRI method in this example was applied at both 1.5 T and at 3 T and employed BOLD contrast. Although it was not specified in the paper, we can presume that gradient-echo imaging methods were used. For the experiments carried out at 1.5 T, single-shot spiral k-space encoding was used with a TE of 45 msec to sample a 240 mm × 240 mm FOV, to produce image data with 3.75 mm isotropic voxels. The repetition time was set at 2 sec and the flip angle was 80°. A total of 20 slices were imaged in each TR period, and these were oriented parallel to the AC-PC line (the line between the anterior commissure and the posterior commissure). The experiments done at 3 T employed single-shot EPI k-space encoding with a TE of 25 msec, a 192 mm × 192 mm FOV, and 3 mm isotropic voxels. In this case the authors left a 1 mm gap between slices, and image data were acquired from 22 slices parallel to the AC-PC line. Again, a TR of 2 sec was used but the flip angle was set at 90°.

The imaging parameters used in this study at both field strengths correspond very closely with those predicted by the theory in the previous sections as being optimal. An effective balance between the primary needs of high temporal resolution, relatively high spatial resolution, and whole-brain coverage is demonstrated by the choice of imaging parameters.

Another example is a study by Barry et al. aimed at investigating the effectiveness of data processing methods for reducing distortions in fMRI data acquired with gradient-echo EPI. The data in this example were acquired at 4 T and used a two-shot EPI sequence (i.e., data were acquired in two segments) "to improve image quality by decreasing T_2^* blurring" (66, p. 237). Images of 17 slices were oriented parallel to the calcarine sulcus and were acquired with a 192 mm × 192 mm FOV, in a 64 × 64 matrix, to produce image data with 3 mm isotropic voxels. The TE was set at 15 msec, and the TR was 1 sec, with a flip angle of 40°, and the resulting time for each complete set of images was 2 sec.

A more application-based example of fMRI is the study of emotional and cognitive aspects of pain by Wiech et al. (67). The goal of this study was to investigate the perception and neural processing of pain and therefore, as in the first example above, required whole-brain imaging to demonstrate the distributed networks involved with pain perception, as well as relatively high temporal resolution. In this example, fMRI data were acquired at 3 T using gradient-echo EPI with an echo time of 30 msec. Images were acquired in 33 axial slices with a thickness of 3 mm and a 1-mm gap between slices, with a 192 mm × 192 mm FOV, and a 64 × 64 matrix. Again, the resulting data represented 3 mm isotropic voxels. The repetition time was set at 2.38 sec for each complete set of images, and the flip angle was 90°. In this case, whole-brain imaging with cubic voxels was balanced with the need for high speed.

A related study of pain processing was carried out in the brainstem by Fairhurst et al., also employing BOLD contrast and a gradient-echo EPI method. In this case, however, the imaging method was "designed to optimise functional sensitivity in the brainstem" (60, p. 103). Images were acquired of 24 coronal slices, each 2 mm thick in order to span the entire brainstem, with a TR of 3 sec. The FOV was set at 192 mm × 224 mm, with a 64 × 64 matrix, to produce images with a resolution of 3 mm × 3.5 mm. This study was carried out at 3 T and the TE was again set at 30 msec, as in previous examples. Here, relatively thin slices were imaged in coronal planes in order to match the anatomy of interest.

Finally, a recent example of fMRI applied in the spinal cord (68) employs SEEP contrast (see Section 6.6). In this example a number of adaptations are made to accommodate the imaging challenges presented by the spinal cord. Image data were acquired at 3 T in nine sagittal slices, each 2 mm thick, using a HASTE method. The FOV was set at 200 mm × 100 mm, and a 192 × 96 matrix was sampled to produce approximately 1 mm × 1 mm in-plane spatial resolution. The TE was set at 38 msec to produce primarily proton density–weighted images, and the TR was limited to 1 sec per slice for a total of 9 sec to acquire each complete set of images. In comparison with the previous examples, this method employs the less common SEEP contrast, at the cost of much slower imaging speed, to achieve much higher than usual spatial resolution and little to no spatial distortion to accommodate the anatomy of the spinal cord.

There are many more such examples in the published literature, in which deviations from what might be considered the "standard" fMRI imaging method have been adopted to suit the needs of the function of interest and/or the anatomy of interest. The examples above serve to illustrate the point that there is no real standard method for fMRI, although even the variations are within reasonable boundaries. More examples of fMRI applications, specifically clinically related studies, are provided in Chapter 9.

Key Points

21. There is no single optimal method that suits all applications of fMRI, but rather the method must be adapted for each specific case.
22. Many published fMRI studies use very similar imaging parameters nonetheless.
23. The most commonly varied parameters are the spatial resolution, slice thickness, slice orientation, numbers of slices, and the repetition time and corresponding flip angle.

6.6 Alternative Contrast Mechanisms

The same physiological changes that give rise to BOLD signal changes, as described above, can also produce MR signal changes by other mechanisms and can be used for fMRI. However, the contrast mechanisms, and sensitivity to neural activity, can be quite distinct from the BOLD effect. These alternative contrast mechanisms are summarized in Table 6.4 and are discussed in detail in the following sections.

Table 6.4 Summary of Alternative Contrast Mechanisms That Can Be Used for fMRI, and the Underlying Physiological Mechanism That Is Used to Detect Changes in Neural Activity	
Contrast Mechanism	Physiological Property That Is Detected at Sites of Neural Activity
Signal enhancement by extravascular water protons (SEEP)	Activity-dependent changes in tissue water content due to swelling of neurons and glial cells
Vascular space occupancy (VASO)	Blood volume changes
Perfusion-weighted imaging (PWI)	Blood flow changes
Diffusion-weighted imaging (DWI)	Activity-dependent changes in water self-diffusion due to swelling of neurons and glial cells

6.6.1 Signal Enhancement by Extravascular Water Protons (SEEP)

The SEEP contrast mechanism was discovered during an attempt to verify that the BOLD signal change occurs in the human spinal cord as it does in the brain (51,69–72). These early studies demonstrated that the BOLD effect does indeed occur in the spinal cord, but there is also another process contributing to MR signal changes as well that cannot be attributed to a change in relaxation times.

As described above for the BOLD effect, astrocytes make contact with both blood vessels and neurons and play a role in providing metabolites to neurons, maintaining the extracellular concentration of glutamate (73–75), and have been shown to play an important role in effective neuronal signaling (76,77). When glutamate is released from vesicles in the axon terminal and travels across the synaptic cleft to trigger the depolarization of the adjoining neuron, the remaining glutamate is rapidly absorbed by astrocytes by means of high-affinity sodium-dependent transporters (76). This process helps to maintain the extracellular glutamate concentration at the essentially low level required for effective neuronal function. Astrocytes also convert the glutamate they absorb into glutamine, which is then rereleased into the extracellular space and absorbed by neurons. Neurons then convert the glutamine back into glutamate and package it into vesicles ready for release at the synapse. When glutamate is actively taken up by astrocytes, each molecule is accompanied by three Na^+ and one H^+ transported into the cell, and one K^+ transported out, and so is also accompanied by water entering the cells (76). In addition, astrocytes have been shown to be depolarized during neuronal activity as a result of uptake of potassium in proportion to the number of active neurons in the vicinity and the frequency at which they fire (74). As a result of these effects, the extracellular/intracellular volume ratio has been shown to change significantly in isolated tissue samples of rat spinal cord tissue during and following neuronal stimulation (78,79). During electrical stimulation of a dorsal nerve root at 10 Hz for 1 minute, the extracellular volume was observed to decrease by roughly 5%. However, these measurements were done in isolated tissue in the absence of blood flow where water could only move between intracellular and extracellular spaces. *In vivo* the situation is more complex because of the added contribution from water moving out of the blood vessels into the extracellular space. The outcomes of the cellular volume changes are that (1) the astrocyte processes wrap around the synapse more closely, making it more effective at taking up glutamate; and (2) the extracellular space becomes more tortuous, restricting the diffusion of glutamate to neighboring synapses (77). The function of the astrocyte therefore plays a role in the effective functioning of the neuron as well by inhibiting propagation of neuronal signaling.

Increases in blood flow to sites of neural activity are driven by increased perfusion pressure due to upstream relaxation of arterial smooth muscle. Water continually crosses the blood vessel walls in the CNS, primarily in capillaries, to produce extracellular fluid (ECF) (80,81), and some may be reabsorbed near the venous end of the capillaries. The production of extracellular fluid from capillaries arises from the well-known Starling Law of Capillaries. Some of the ECF moves through the interstitial space until it is absorbed across the glial-pial membrane and enters the cerebrospinal fluid (CSF). Approximately 30% of CSF obtained from a spinal tap originates from the capillary bed (82). This is a normal but important physiological function because it compensates for the absence of lymphatics in the CNS. At sites of neuronal activity, the increased perfusion pressure is expected to alter the fluid balance and increase the rate of production of ECF. Functional imaging studies carried out with positron emission tomography (PET) using radio-labeled water as a tracer support this idea as they have demonstrated increased unidirectional clearance of water from the blood into the brain parenchyma (83,84).

The net result of the water movements out of the blood vessels and into the astrocytes is a local increase in tissue water content, or spin density, and can be detected with MRI to demonstrate sites of neural activity (51,69,85,86). Hence, the term *signal enhancement by extravascular water protons*.

Evidence of activity-dependent changes in water content, and therefore proton density, has been provided in a series of studies. Initial evidence was obtained in the spinal cord, which demonstrated roughly equal signal changes with spin-echo and gradient-echo EPI at the same echo time, in contrast to the BOLD model (51,52,69,70). Studies of signal changes as a function of echo time with gradient-echo and spin-echo images in the spinal cord demonstrate that the signal change extrapolated to TE = 0 for spin-echo data is consistently around 2.5%, again in contrast with BOLD theory. The intercept value is not significantly different from zero with the gradient echo, showing that the water compartment contributing to the signal change is more easily detected with a spin echo than with a gradient echo. Therefore, this compartment is expected to have a short T_2^* (< 10 msec) and cannot be attributed to blood. A detailed study in the spinal cord has demonstrated that there are two components contributing to the signal changes, one that demonstrates the BOLD effect and another than demonstrates a proton-density change, and signal changes of 3.3% were observed in spin-echo data with an echo time of 11 msec (69). In the same study, the proton-density change in one component was determined to be approximately 5.6% upon a change in neuronal activity. Based on this work, an effective method for fMRI of the spinal cord (discussed further in Sections 6.4.2 and 9.2.7) has been developed based on single-shot fast spin-echo imaging with the echo time set as short as possible (TE = 35–40 msec).

A comparison of spin-echo (T_2-weighted) and gradient-echo (T_2^*-weighted) data in the brain over a range of echo times shows that the intercept value extrapolated to TE = 0 is approximately 1% with spin echo and is significant, but is lower in the brain than in the spinal cord (51). A similar comparison of SEEP and BOLD fMRI in the brain at 3 T demonstrates the relative locations of activity with each, as well as the fact that the SEEP signal changes are around 2% with spin-echo EPI at an echo time of 22 msec (85). Areas of activity with SEEP are in close proximity to those observed with BOLD, but the two are adjacent and overlap only slightly. This evidence suggests that SEEP signal changes may occur near capillaries and near the site of neural activity, whereas the BOLD changes are known to occur primarily in venules and veins. More evidence for the existence of a proton-density change contribution has also been shown by fMRI studies done with fast spin echo with an echo time of 24 msec at a low field of only 0.2 T (86,87). Even in a field of 0.2 T the signal changes were detected at sites of neuronal activity with a two-hand motor task, with a magnitude of 2%. The SEEP signal changes detected in the brain have therefore been seen to be around 2% at 0.2 T, 1.5 T, and 3 T and so do not demonstrate the same field dependence as the BOLD effect.

The best evidence to date for the SEEP effect has come from recent studies combining light-transmittance microscopy and functional MRI of superfused cortical tissue slices from rats. Tissue slices were stimulated with 26-mM potassium (K+) at room temperature to elicit neuronal activity, followed by perfusion with artificial CSF to allow recovery (88,89). Light-transmittance microscopy confirmed the occurrence of cell swelling. In separate experiments, the same preparation was used in combination with fMRI at 3 T, using imaging methods established for SEEP fMRI of the spinal cord (90,91). The results of fMRI also revealed signal changes in the cortical gray matter that occurred upon stimulation with K+ (Figure 6.8). Experiments with hyper- and hypoosmotic perfusion solutions also confirmed that tissue swelling or shrinking due to osmotically driven water movements result in decreases, and increases, respectively, in both light transmittance and MRI signal intensities. These findings confirm a mechanism of MR signal change arising from changes in tissue water content that is independent of blood or blood oxygenation and that arise concomitant to neuronal activity.

Unlike the BOLD contrast mechanism, SEEP does not rely on a change in relaxation time and can therefore be detected without any relaxation-time weighting of the data. That is, it can be detected with proton density–weighted imaging. This weighting provides the distinct advantage of producing images with the highest possible SNR. In addition, SEEP fMRI data can be acquired with spin-echo methods, again to provide higher SNR and better image quality,

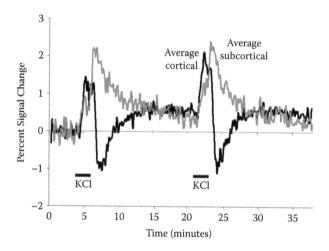

FIGURE 6.8 Signal intensity time course data observed in fMRI experiments with live cerebral slices from rats, superfused with oxygenated artificial cerebrospinal fluid (CSFa) at room temperature during baseline periods. Slices were briefly depolarized twice by changing the control superperfusate to iso-osmotic aCSF containing 26 mM potassium chloride (KCl) for 2 minutes. Periods of superfusion with high KCl are indicated with black bars at the bottom of each plot. Signal intensity is plotted for regions containing predominantly cortical gray matter (black line) or subcortical gray and white matter (gray line). Each trace represents the averaged signal from six cerebral slices.

in terms of lower spatial distortion and less sensitivity to image artifacts. SEEP has also been shown to provide a greater contrast-to-noise ratio than BOLD (92). The ability to acquire data with spin-echo methods makes it possible to acquire fMRI data in areas of poor magnetic field homogeneity near air/tissue and bone/tissue interfaces, as discussed in Section 6.4.

The timing of the SEEP response has been determined by characterizing the response function, similar to the hemodynamic response function for BOLD (Figure 6.9) (90). The SEEP response function demonstrates a slightly slower response than BOLD, taking approximately 7 sec to reach the peak response, and a slow return to baseline after the stimulus, taking approximately 14 sec with no poststimulus undershoot.

The drawbacks of the SEEP contrast mechanism are that (1) it is therefore slower than BOLD, and (2) the areas of activity are more highly localized. While this latter point may reflect greater precision, it presents a challenge because the localized areas of activity are more difficult to detect than the relatively broad areas of activity demonstrated by BOLD contrast. The smaller areas of activity with SEEP therefore require higher spatial resolution, are more affected by motion, and require greater spatial precision when aligning and combining or comparing results from different people.

6.6.2 Perfusion-Weighted Imaging (PWI)

An important feature that is common to BOLD contrast and SEEP contrast, discussed above, is the change in blood flow at sites of increased neural activity, and this blood flow change can be measured directly in order to reveal a change in activity. The delivery of arterial blood to the capillaries where gas exchange occurs with the tissues is termed *perfusion* and is typically quantified in terms of milliliters of blood, per gram of tissue, per unit time. Normal values in the gray matter regions of the brain are around 1.2 to 1.4 mL/g/min, as summarized in Table 6.5. Over the entire brain with an average mass of 1.4 kg, the average rate of perfusion is 750 mL blood/min, resulting in 0.54 mL/g/min. The overall rate of O_2 consumption is 46 mL/min or 0.033 mL/g/min.

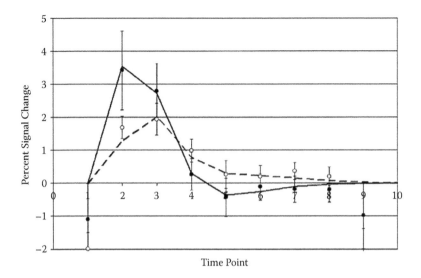

FIGURE 6.9 Measured response functions for brain fMRI data obtained with BOLD contrast (solid line) and with SEEP contrast (dashed line). Data were measured at 3 sec intervals, and therefore each time point represents a 3 sec span.

Table 6.5 Rates of Blood Flow to Various Regions of the Brain	
CNS Region	**Mean Perfusion (mL/g/min)**
Whole brain (average)	0.54
Sensorimotor cortex	1.38
Auditory cortex	1.30
Visual cortex	1.25
Caudate nucleus	1.10
Thalamus	1.03
Association cortex	0.88
Cerebellar nuclei	0.87
Cerebellar white matter	0.24
Cerebral white matter	0.23

Sources: Data from Kety, S.S., *Brain Res Bull* 50, 5–6, 415–416, 1999 (93); Ganong W.F., *Review of Medical Physiology,* 13th ed, Appleton & Lange, Norwalk, CT, 1987, Table 32 (94).

Functional imaging based on detecting changes in perfusion has been developed to supplement BOLD fMRI, or in some cases has been argued to be a more sensitive and more direct measure of brain function (95). By quantifying changes in CBF, and also knowing the BOLD signal changes, it is possible to estimate changes in $CMRO_2$ at sites of neural activity (96). The perfusion information itself can be used for as well as for characterizing blood flow in tumors or other tissues.

Several different methods can be used for perfusion imaging, but all are based on the general idea that the blood flowing into a region can be made to be differently T_1-weighted than the stationary water in the tissues and so forth in the region being imaged. Variations of the method are called *arterial spin labeling* (ASL), for reasons described below, or *pulsed arterial spin labeling* (PASL), *continuous arterial spin labeling* (CASL), EPISTAR, PULSAR, QUIPSS, and many others. Perfusion MRI is an in-depth topic on its own, and for this brief introduction it is best to focus on the common features of this family of methods (97–99). The three different basic approaches that are used in essentially all perfusion imaging methods include the following:

1. Inject a bolus of paramagnetic contrast agent, so that the transit time of the bolus through the region of interest can be characterized with time-series images (100).

2. Apply a 180° inversion pulse over a region containing the major arteries feeding the region of interest, and then ~1–2 sec later image the region of interest. This method is referred to as *tagging* or *labeling* the blood and is used in variations of the ASL method (101). The results are calibrated by comparison with images obtained in the absence of an inversion pulse (Figure 6.10).

3. Apply a 180° inversion pulse over the region of interest, and image the region after allowing time for blood to flow into the region that has not been affected by the inversion pulse. A variation of this method is to acquire images of the region of interest preceded by a slice-selective 180° inversion pulse, and then acquire a second set of images preceded by a nonspatially selective 180° inversion pulse. The time allowed between the inversion pulses and the imaging is set to be equal to the T_1-value of blood, to allow time for the blood to flow into the imaging region. This method is called *flow-sensitive alternating inversion-recovery*, or FAIR (102,103).

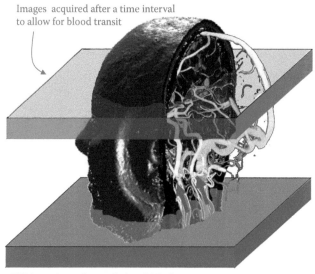

Images acquired after a time interval
to allow for blood transit

180° inversion pulse applied to thick slice
for "labeling" the arterial blood

FIGURE 6.10 Arterial spin-labeling method of perfusion imaging. An inversion pulse is applied over the neck, causing the blood flowing into the brain to be T_1-weighted. Images of the brain therefore have reduced signal intensity depending on the rate of perfusion, the time elapsed since the inversion pulse, and the T_1-value of the blood.

The last method list above, FAIR, provides examples of the common features of all of the perfusion imaging methods, and so we can focus on it. The contrast between the stationary tissue and the amount of blood that moves either into or out of the region is provided by the time allowed for perfusion, given that the T_1 of blood is relatively long at 1200–1350 msec at 1.5 T, and around 1600 msec at 3 T (104). The blood that has been affected by the inversion pulse is replaced by blood that was not affected, causing the net observed MR signal to be altered, in proportion to the perfusion. That is, if fully relaxed blood flows into a slice that has experienced an inversion pulse, then the MR signal is greater in the subsequent image of that slice. The effective longitudinal relaxation time appears to be reduced, because the signal appears to have recovered more quickly because of the perfusion. On the other hand, if blood that has been affected by an inversion pulse flows into an unaffected region, then the total signal is reduced when that region is imaged, again in proportion to the rate of perfusion.

The effective change in the observed relaxation rate (in the case of fully relaxed blood flowing into a region that has experienced an inversion pulse), is given by:

$$\frac{1}{T_1^*} = \frac{1}{T_1} + \frac{f}{\lambda}$$

where T_1^* is the observed longitudinal relaxation time, T_1 is the longitudinal relaxation time of the tissues, f is the rate of perfusion (typically around 0.02 mL/g tissue/sec in gray matter in the brain), and λ is the tissue partition fraction (0.9 mL water/g tissue). The signal obtained with nonselective inversion at a time interval TI before a gradient-echo EPI acquisition, for example, would be:

$$S_{NS} = S_0 \left(1 - 2e^{-TI/T_1}\right) e^{-TE/T_2^*}$$

With an inversion pulse that is not spatially selective, the in-flowing blood receives the same excitation as the blood in the slice and there is no apparent reduction in T_1. On the other hand, when the slice-selective inversion is used, then the apparent T_1-value, T_1^*, more accurately reflects the properties of the signal that is observed:

$$S_{SS} = S_0 \left(1 - 2e^{-TI/T_1^*}\right) e^{-TE/T_2^*}$$

With these two equations the assumption is made that the proton density, S_0, is constant between the two scans. T_2^* effects are also included in these equations because they may not be constant if there is a change in blood oxygenation (i.e., the BOLD effect). The difference between the two signals is:

$$S_{SS-NS} = S_0 \left(1 - 2e^{-TI/T_1^*}\right) e^{-TE/T_2^*} - S_0 \left(1 - 2e^{-TI/T_1}\right) e^{-TE/T_2^*}$$

$$S_{SS-NS} = 2S_0 e^{-TE/T_2^*} \left(e^{-TI/T_1} - e^{-TI/T_1^*}\right)$$

Now we can put in the full expression for T_1^*, and rename the signal S_{FAIR}:

$$S_{FAIR} = 2S_0 e^{-TE/T_2^*} \left(e^{-TI/T_1} - e^{-TI/T_1} e^{-TI f/\lambda}\right)$$

$$S_{FAIR} = 2S_0 e^{-TE/T_2^*} e^{-TI/T_1} \left(1 - e^{-TI f/\lambda}\right)$$

Because the value of TI f/λ is always much less than 1, the last term can be expanded with a first-order approximation:

$$S_{FAIR} = 2S_0 e^{-TE/T_2^*} e^{-TI/T_1} \, TI \, f/\lambda$$

Based on this signal difference, S_{FAIR}, the rates of perfusion can be mapped and quantified, and differences in perfusion resulting from neuronal activity, exercise, respiratory challenge, and so forth can be measured. The ratio of the FAIR signals measured during rest and stimulation conditions can be expressed as (103):

$$S_{FAIR} = \frac{S_{FAIR}^{(stim)}}{S_{FAIR}^{(rest)}} - 1$$

$$S_{FAIR} = \frac{2S_0 e^{-TE/T_{2,stim}^*} e^{-TI/T_1} \; TI \, f_{stim}/\lambda}{2S_0 e^{-TE/T_{2,rest}^*} e^{-TI/T_1} \; TI \, f_{rest}/\lambda} - 1$$

$$S_{FAIR} = \frac{f_{stim}}{f_{rest}} \; e^{-TE\left(1/T_{2,stim}^* - 1/T_{2,rest}^*\right)} - 1$$

$$S_{FAIR} = \frac{f_{stim}}{f_{rest}} \; \beta_{BOLD} - 1$$

In this last equation, β_{BOLD} accounts for possible variations in the BOLD effect, given by the $\exp(-TE/T_2^*)$ term, and recall that f is the perfusion, or local CBF, for a given image voxel. As a result, it is possible to quantify the relative changes in perfusion between stimulation and baseline conditions. If the tissue T_1-value is accurately determined, then absolute perfusion rates can also be estimated.

The challenges of perfusion imaging methods are primarily the poor SNR that results from applying the inversion recovery, and then subtracting two images with selective and nonselective inversion. The ratio of signals between rest and stimulation conditions is therefore quite sensitive to noise. Another challenge arises with trying to obtain multi-slice data with the same inversion times for each slice when nonselective inversion is used, while still maintaining a reasonable repetition time for functional imaging. Particularly with multi-slice acquisitions, RF pulses used to invert the magnetization can cause magnetization transfer effects (see Section 4.7) in neighboring regions, and some variations of the method are specifically to compensate for this effect. Nonetheless, a number of different approaches have been proposed to overcome these challenges, and sensitive perfusion-weighted images of the human brain have been demonstrated and can provide important information to support fMRI data.

6.6.3 Vascular Space Occupancy (VASO)

A variation of the perfusion imaging methods described above can demonstrate changes in blood volume that may occur in relation to neuronal activity. Using a nonspatially selective inversion pulse with the inversion time, TI, set to null the signal from blood, the image signal is reduced according to how much blood is in each voxel. An increase in blood volume therefore results in a decrease in MR signal. Changes in blood volume between rest and stimulation conditions can therefore be detected and quantified using similar procedures as with other functional imaging methods. For example, at 3 tesla the T_1-values of gray matter, white matter, blood, and CSF are, respectively, 1209 msec, 758 msec, 1627 msec, and 4300 msec (values used here are from Donahue et al. 2006 (105), although there is some variation in the values reported in the literature (105–107)). This means that the MR signal that would be measured at 1128 msec after a 180° inversion pulse would be multiplied by a factor of 0 for blood, 0.21 for gray matter, 0.55 for white matter, and −0.54 for CSF. The signal attenuation is due to the longitudinal recovery of the magnetization after the inversion pulse, according to the expression:

$$S = S_0\left(1 - 2e^{-TI/T_1}\right)$$

If the fraction of volume that is occupied by blood increases in a given voxel, then the signal intensity that is measured will decrease accordingly.

There are a number of challenges with this method, including obtaining an accurate estimate of the T_1 of blood and acquiring images of multiple slices with the same inversion time, with a repetition time that is suitable for functional imaging. There are also confounding effects from blood flow through the imaging region between repeated scans, the attenuation of the signal from gray matter and white matter, and the potential for contribution from CSF (105,108). Nonetheless, it has been reported that VASO-based fMRI reveals areas of activity that are more closely localized to sites of neuronal activity than are observed with BOLD fMRI, because VASO is less sensitive to the veins draining active regions (109).

6.6.4 Diffusion-Weighted Imaging (DWI)

Although diffusion-weighted imaging is more commonly used to detect changes in tissue structure at the cellular level, it has also been shown to reveal changes in tissue properties in relation to neuronal activity (110,111). Diffusion arises from random thermal motion of molecules and is characterized as a *random walk*. The average net displacement of a molecule in time is described by the diffusion coefficient, D, by the expression derived by Einstein: $<d> = \sqrt{2D\ t}$, where $<d>$ is the expectation value (i.e., average) of the displacement over the time interval Δt. Structural components of tissues can have a significant effect on the observed diffusion, however, by restricting the distance that can be traveled, or simply by presenting more barriers to diffusion (Figure 6.11).

The method of diffusion weighting the MRI signal relies on magnetic field gradients to highlight the displacement of hydrogen nuclei in time. By applying a strong magnetic field gradient, followed by an equal but opposite gradient, the effects of the gradients are canceled out only if the nuclei are stationary. Even slight amounts of movement result in a residual change in the phase of the MR signal from each nucleus. For a large number of nuclei (such as in even a one microliter (μL) of water), the phases are distributed across a range of values depending on how much, and in which direction, the nuclei have moved. Because the movement is generally random, the observed effect is a reduction in signal intensity proportional to the diffusion coefficient, D:

$$S = S_0\ e^{-bD}$$

In this expression the value of b is the diffusion-weighting coefficient that is determined by the amplitude, duration, and spacing of the gradients, and will be described below.

(a)   (b)

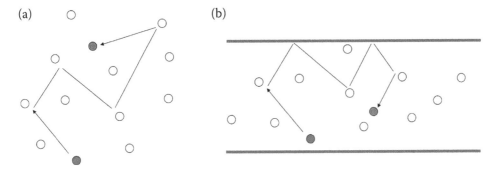

FIGURE 6.11 Representations of water self-diffusion by means of *random walk* movement driven by thermal motion, in both the unrestricted (left) and restricted (right) cases. In each case, one particle (such as a water molecule) is shown as starting at the blue dot, and traveling through several interactions to arrive later at the red dot.

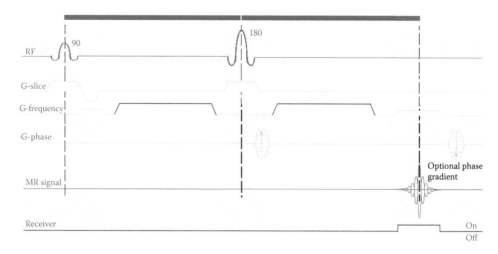

FIGURE 6.12 Schematic of a spin-echo sequence with a diffusion-weighting gradient added in only the frequency-encoding direction (shown in black).

With a gradient-echo sequence, the diffusion-weighting gradients must be applied between the RF excitation pulse and the MR signal acquisition and must employ two equal and opposite gradients. A more common method is to use a spin-echo method with one gradient applied between the 90° excitation pulse and the 180° pulse, and a second gradient of equal magnitude and duration between the 180° pulse and the acquisition, as indicated in Figure 6.12. This allows more time for applying the diffusion-weighting gradients and therefore stronger diffusion weighting. Similarly, stimulated echo sequences (see Chapter 4) are commonly used with the equal diffusion-weighting gradients applied before the second 90° RF pulse and after the third 90° RF pulse. The stimulated echo allows even more time to impose the diffusion weighting while maintaining a relatively short effective echo time.

Using this scheme of two balanced gradients, the diffusion-weighting factor is given by the expression (112,113):

$$b = \gamma^2 G^2 \delta^2 \left(\quad - \delta/3 \right)$$

The value of δ is the duration of each gradient pulse, and the value of Δ is the time between the onsets of the two pulses. The gradient amplitude is indicated by G. The typical water self-diffusion coefficient, D, in the brain has a value of 0.95 to 1.76×10^{-3} mm²/s. However, the diffusion can depend significantly on the orientation of structural elements and has been observed to be two to four times higher along the direction of white matter tracts than perpendicular to the tracts and is even higher in some regions of the spinal cord. The directional dependence of the diffusion coefficient is known as *diffusion anisotropy* and has been proven to be very sensitive for detecting disruption of normally ordered tissues. (See the review of DWI by Beaulieu [114].) The typical magnitude of the diffusion coefficient demonstrates the need for very strong diffusion weighting to achieve sensitivity to changes in tissues. For example, with diffusion weighting of 1000 s/mm², the attenuation from diffusion in the brain is roughly $\exp^{(-1000 \text{ s/mm}^2 \times 10^{-3} \text{ mm}^2/\text{s})} = 0.37$.

Diffusion-weighted images therefore reflect the movement of water, due to diffusion, in a particular direction, as determined by the direction of the gradient that is applied. Bulk movement of the body, in which all of the hydrogen nuclei within a voxel move together, will result in a phase change that is the same for all nuclei and will therefore not contribute to the distribution of phases that causes the reduction in the total signal. Blood flow tends to have a distribution of flow speeds across the vessel cross-section, and so generally tends to cause a very large

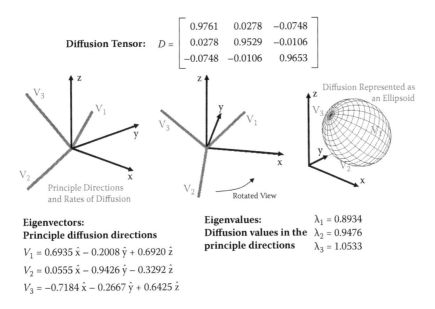

Diffusion Tensor: $D = \begin{bmatrix} 0.9761 & 0.0278 & -0.0748 \\ 0.0278 & 0.9529 & -0.0106 \\ -0.0748 & -0.0106 & 0.9653 \end{bmatrix}$

Eigenvectors:
Principle diffusion directions

$V_1 = 0.6935\,\hat{x} - 0.2008\,\hat{y} + 0.6920\,\hat{z}$

$V_2 = 0.0555\,\hat{x} - 0.9426\,\hat{y} - 0.3292\,\hat{z}$

$V_3 = -0.7184\,\hat{x} - 0.2667\,\hat{y} + 0.6425\,\hat{z}$

Eigenvalues:
Diffusion values in the
principle directions

$\lambda_1 = 0.8934$
$\lambda_2 = 0.9476$
$\lambda_3 = 1.0533$

FIGURE 6.13 Example of a 3 × 3 diffusion tensor (from Basser et al. 1994 (96)) showing the form of the tensor and how it can be expressed as the eigenvectors or characteristic vectors that are along the principal directions of diffusion. That is, the eigenvectors show the directions of maximum and minimum diffusion. The eigenvalues indicate the magnitude of the diffusion coefficient along each of the principal directions. The units of the numbers are 10^{-3} mm²/s.

reduction in signal, similar to that expected for extremely rapid diffusion. The signal from large blood vessels can be almost entirely eliminated by applying diffusion weighting with a very small b-value of only 2 s/mm² (96). Larger (although still relatively small) b-values of around 250 s/mm² (111,115) are needed to eliminate the signal from smaller blood vessels, due to the slower flow and presumably smaller distribution of flow speeds. To achieve good sensitivity to water self-diffusion in the brain, b-values of around 1000 s/mm² are commonly used. As a result, in spite of the constant presence of blood and CSF flow, and efforts to avoid bulk movement of the body, the MR signal weighting can be dominated by the effects of diffusion, due to its random nature and the relatively small movement it produces.

In most cases, when characterizing the tissues in the body, there is not a single known direction of water diffusion that is of interest, and we want to observe diffusion in many directions and make comparisons. It is possible to simply acquire several images, each with diffusion weighting in a different direction, to obtain this information. However, it can be more descriptive to construct the diffusion tensor (Figure 6.13), which is a 3 × 3 matrix that is somewhat like a 3D vector describing the diffusion. However, the tensor describes the diffusion in each direction in 3D, as well as the relationships between the different diffusion directions. The tensor can be rotated to describe the diffusion relative to any set of 3D coordinate system, or similarly the tissues being measured can be oriented in any direction and the tensor will be rotated accordingly, showing the diffusion in different directions in the tissues.

By acquiring diffusion-weighted images with diffusion gradients in a sufficient number of different directions, a diffusion tensor can be constructed for each voxel. This is termed *diffusion-tensor imaging*, or DTI, and is therefore an extension of DWI. The number of diffusion directions needed is at least three, as long as they are all 90° apart, but the accuracy of the diffusion tensor increases with more directions. In many current applications of DTI, diffusion weighting is applied in six different directions, but measurements in 20 or 30 different directions are also common. A limitation of this method is that the tensor approach describes the water movement due to diffusion as an ellipsoid (as indicated in Figure 6.13). To achieve greater accuracy the method has been extended with *high angular resolution diffusion imaging* (HARDI), or with *Q-ball imaging* with as many as 252 directions. These are being developed with the understanding that tissues may be complex, and may not be fully described by a tensor with three principal diffusion directions (116).

The value of constructing the tensor (or more complex forms) is to reveal detailed structural information about tissues at the cellular level. Tissues such as white matter in the CNS have a structure that is highly oriented, and water can diffuse relatively freely parallel to the direction of long axons, but not across the axons because the cellular walls, myelin layers, and so forth, restrict the movement of the water. The diffusion tensor therefore reveals a high degree of diffusion anisotropy, which refers to the ratio of diffusion coefficients in the direction of most rapid diffusion, compared with the direction of slowest diffusion. The magnitudes of the diffusion coefficients in the each of the principle directions are typically indicated with values λ_1, λ_2, and λ_3 (as in Figure 6.13), and in this example are ordered from smallest to largest. Several different methods are used to quantify the anisotropy, depending on the situation. One method is to quantify the deviation from the *spherical* case in which diffusion is equal in all three directions, with

$$C_a = \frac{\lambda_3 + \lambda_2 - 2\lambda_1}{\lambda_1 + \lambda_2 + \lambda_3}$$

Using the diffusion coefficients from the example in Figure 6.13 would give a value of 0.07, indicating a relatively low degree of diffusion anisotropy. For comparison, in normal appearing white matter in the brain, anisotropy values have been measured at 0.30 in healthy control subjects and 0.25 in people with multiple sclerosis (117).

Gray matter, on the other hand, has no preferred direction of structural orientation and has a very low degree of (or zero) diffusion anisotropy. Differences in diffusion tensors between voxels in the brain or spinal cord can therefore provide detailed information about the tissues in the voxels and reveal important changes as a result of tissue damage and loss of white matter structure after trauma, or demyelination as a result of multiple sclerosis (for example).

Another application of DTI is to use the information it provides about the orientation of structures at the cellular level to estimate the paths of long white matter tracts. This process is called *fiber tracking* and is demonstrated in Figure 6.14. Limits of fiber tracking are imposed by

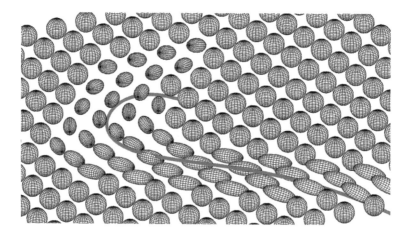

FIGURE 6.14 Simulated example of diffusion tensor information in a plane. A long white-matter tract passes through the plane, with the center of the tract indicated by the red line. The diffusion anisotropy reveals the fiber orientation and can be used to trace the white-matter path. The surrounding region is simulated to contain tissues or fluids with isotropic diffusion, as indicated by the spherical diffusion distribution.

the precision of the diffusion tensors that can be constructed for each voxel and partial-volume effects (i.e., a voxel may contain more than one type of tissue). Another significant challenge is created by regions of converging or diverging white-matter tracts and crossing fibers. To overcome these challenges, methods are being developed using HARDI or Q-ball imaging to detect multiple fiber directions within a single voxel (116,118).

Diffusion-weighted imaging methods are used in relation to fMRI in a number of different ways. Some investigations of the physiological changes underlying the BOLD effect, as well as attempts to improve the spatial precision of fMRI, have used diffusion weighting to eliminate the MR signal from large blood vessels (96,111,115). Another application is to combine fMRI with fiber tracking to attempt to trace the white matter pathways that connect physically separated areas of neural activity (119). Finally, diffusion changes have been described at sites of neural activity (55,120) and are detected with time-series DWI while neuronal activity is systematically varied. The diffusion changes occur outside of blood vessels and are believed to arise from cellular swelling resulting in a change in the relative proportion of water that is in contact with cellular membrane surfaces.

Key Points

24. Astrocytes and neurons swell slightly with water at sites of increased neural activity, changing the tortuosity of the extracellular space and causing astrocyte end-feet to wrap around synapses.

25. *Signal enhancement by extravascular water protons*, or SEEP, arises from water being exuded, or seeping, out of capillaries into the extracellular space and being taken up by metabolically active neurons and glial cells at sites of increased neural activity.

26. MR signal changes due to SEEP arise from a change in proton density (i.e., water content) and are not due to changes in relaxation times.

27. *Perfusion-weighted imaging*, PWI, demonstrates changes in blood flow, or perfusion, through tissues and can therefore reveal neuronal-activity-dependent changes in blood flow.

28. Changes in perfusion can be detected by means of *arterial spin-labeling* (ASL) in which an RF pulse is first applied to a region to label or tag the blood. Because blood has a long T_1-value, the blood that has experienced the labeling pulse can affect the MR signal if it moves into or out of the region of interest, which is then imaged ~1–2 sec later.

29. *Vascular space occupancy*, or VASO, demonstrates changes in blood volume in relation to neuronal activity. The method is based on an inversion-recovery method to null the signal from blood.

30. *Diffusion-weighted imaging*, or DWI, can be used to quantify rates of water self-diffusion in tissues. Apparent water diffusion coefficients are strongly influenced by cellular structures that restrict water movements and can therefore reveal details about tissue structure at the cellular level, including the orientation of axons in white matter.

31. Diffusion changes have also been demonstrated in relation to neuronal activity and are believed to arise from changes in the cellular surface area interacting with water, as a result of neuronal and/or glial swelling.

32. *Diffusion-tensor imaging*, or DTI, is an extension of DWI by making use of the fact that the MRI sensitivity to diffusion is direction dependent, and so the diffusion characteristics in each voxel can be described as a 3D tensor, and preferred directions of diffusion can be revealed.

33. *Fiber tracking* is an application of DTI in which the preferred diffusion parallel to long axons in white matter is used to attempt to trace the paths of long sections of white-matter tracts.

34 The underlying physiological changes that are used to detect neuronal activity with the methods of BOLD, SEEP, PWI, VASO, and DWI are all related.

References

1. Ogawa S, Lee TM. Magnetic resonance imaging of blood vessels at high fields: *In vivo* and *in vitro* measurements and image simulation. *Magn Reson Med* 1990;16(1):9–18.
2. Pauling L, Coryell CD. The magnetic properties and structure of hemoglobin, oxyhemoglobin and carbonmonoxyhemoglobin. *Proc Natl Acad Sci USA* 1936;22:210–216.
3. Menon RS, Ogawa S, Kim SG, Ellermann JM, Merkle H, Tank DW, Ugurbil K. Functional brain mapping using magnetic resonance imaging. Signal changes accompanying visual stimulation. *Invest Radiol* 1992;27 Suppl 2:S47–S53.
4. Ogawa S, Tank DW, Menon R, Ellermann JM, Kim SG, Merkle H, Ugurbil K. Intrinsic signal changes accompanying sensory stimulation: Functional brain mapping with magnetic resonance imaging. *Proc Natl Acad Sci USA* 1992;89(13):5951–5955.
5. Kwong KK, Belliveau JW, Chesler DA, Goldberg IE, Weisskoff RM, Poncelet BP, Kennedy DN, Hoppel BE, Cohen MS, Turner R. Dynamic magnetic resonance imaging of human brain activity during primary sensory stimulation. *Proc Natl Acad Sci USA* 1992;89(12):5675–5679.
6. Bandettini PA, Wong EC, Hinks RS, Tikofsky RS, Hyde JS. Time course EPI of human brain function during task activation. *Magn Reson Med* 1992;25(2):390–397.
7. Collins CM, Yang B, Yang QX, Smith MB. Numerical calculations of the static magnetic field in three-dimensional multi-tissue models of the human head. *Magn Reson Imaging* 2002;20(5):413–424.
8. Spees WM, Yablonskiy DA, Oswood MC, Ackerman JJ. Water proton MR properties of human blood at 1.5 Tesla: Magnetic susceptibility, $T_{(1)}$, $T_{(2)}$, $T_{(2)}^{*}$, and non-Lorentzian signal behavior. *Magn Reson Med* 2001;45(4):533–542.
9. Ogawa S, Menon RS, Tank DW, Kim SG, Merkle H, Ellermann JM, Ugurbil K. Functional brain mapping by blood oxygenation level-dependent contrast magnetic resonance imaging. A comparison of signal characteristics with a biophysical model. *Biophys J* 1993;64(3):803–812.
10. Gjedde A, Marrett S, Vafaee M. Oxidative and nonoxidative metabolism of excited neurons and astrocytes. *J Cereb Blood Flow Metab* 2002;22(1):1–14.
11. Fox PT, Raichle ME. Focal physiological uncoupling of cerebral blood flow and oxidative metabolism during somatosensory stimulation in human subjects. *Proc Natl Acad Sci USA* 1986;83(4):1140–1144.
12. Friston KJ, Jezzard P, Turner R. Analysis of functional MRI time series. *Hum Brain Mapp* 1994;1:153–171.
13. Frahm J, Kruger G, Merboldt KD, Kleinschmidt A. Dynamic uncoupling and recoupling of perfusion and oxidative metabolism during focal brain activation in man. *Magn Reson Med* 1996;35(2):143–148.
14. Kruger G, Kleinschmidt A, Frahm J. Dynamic MRI sensitized to cerebral blood oxygenation and flow during sustained activation of human visual cortex. *Magn Reson Med* 1996;35(6):797–800.
15. Worsley KJ, Friston KJ. Analysis of fMRI time series revisited—Again. *NeuroImage* 1995;2(3):173–181.
16. Buxton RB, Uludag K, Dubowitz DJ, Liu TT. Modeling the hemodynamic response to brain activation. *NeuroImage* 2004;23 Suppl 1:S220–S233.
17. Ogawa S, Lee TM, Kay AR, Tank DW. Brain magnetic resonance imaging with contrast dependent on blood oxygenation. *Proc Natl Acad Sci USA* 1990;87(24):9868–9872.
18. Ogawa S, Lee TM, Nayak AS, Glynn P. Oxygenation-sensitive contrast in magnetic resonance image of rodent brain at high magnetic fields. *Magn Reson Med* 1990;14(1):68–78.
19. Menon RS, Allen PS. Solvent proton relaxation of aqueous solutions of the serum proteins alpha 2-macroglobulin, fibrinogen, and albumin. *Biophys J* 1990;57(3):389–396.
20. Kocsis L, Herman P, Eke A. Mathematical model for the estimation of hemodynamic and oxygenation variables by tissue spectroscopy. *J Theor Biol* 2006;241(2):262–275.

21. Simon GH, Bauer J, Saborovski O, Fu Y, Corot C, Wendland MF, drup-Link HE. T1 and T2 relaxivity of intracellular and extracellular USPIO at 1.5T and 3T clinical MR scanning. *Eur Radiol* 2006;16(3):738–745.

22. Yablonskiy DA, Haacke EM. Theory of NMR signal behavior in magnetically inhomogeneous tissues: The static dephasing regime. *Magn Reson Med* 1994;32(6):749–763.

23. Davis TL, Kwong KK, Weisskoff RM, Rosen BR. Calibrated functional MRI: Mapping the dynamics of oxidative metabolism. *Proc Natl Acad Sci USA* 1998;95(4):1834–1839.

24. Buxton RB, Wong EC, Frank LR. Dynamics of blood flow and oxygenation changes during brain activation: The balloon model. *Magn Reson Med* 1998;39(6):855–864.

25. Buxton RB. The elusive initial dip. *NeuroImage* 2001;13(6 Pt 1):953–958.

26. Agulhon C, Petravicz J, McMullen AB, Sweger EJ, Minton SK, Taves SR, Casper KB, Fiacco TA, McCarthy KD. What is the role of astrocyte calcium in neurophysiology? *Neuron* 2008;59(6):932–946.

27. Zonta M, Angulo MC, Gobbo S, Rosengarten B, Hossmann KA, Pozzan T, Carmignoto G. Neuron-to-astrocyte signaling is central to the dynamic control of brain microcirculation. *Nat Neurosci* 2003;6(1):43–50.

28. Mulligan SJ, MacVicar BA. Calcium transients in astrocyte endfeet cause cerebrovascular constrictions. *Nature* 2004;431(7005):195–199.

29. Schummers J, Yu H, Sur M. Tuned responses of astrocytes and their influence on hemodynamic signals in the visual cortex. *Science* 2008;320(5883):1638–1643.

30. Gordon GR, Choi HB, Rungta RL, Ellis-Davies GC, MacVicar BA. Brain metabolism dictates the polarity of astrocyte control over arterioles. *Nature* 2008;456(7223):745–749.

31. Petzold GC, Albeanu DF, Sato TF, Murthy VN. Coupling of neural activity to blood flow in olfactory glomeruli is mediated by astrocytic pathways. *Neuron* 2008;58(6):897–910.

32. Figley CR, Stroman PW. The role(s) of astrocytes and astrocyte activity in neurometabolism, neurovascular coupling, and the production of functional neuroimaging signals. *Euro J Neurosci* 2011; 33(4):577–588.

33. Kullmann DM, Asztely F. Extrasynaptic glutamate spillover in the hippocampus: evidence and implications. *Trends Neurosci* 1998;21(1):8–14.

34. Rothstein JD, Dykes-Hoberg M, Pardo CA, Bristol LA, Jin L, Kuncl RW, Kanai Y, Hediger MA, Wang Y, Schielke JP, Welty DF. Knockout of glutamate transporters reveals a major role for astroglial transport in excitotoxicity and clearance of glutamate. *Neuron* 1996;16(3):675–686.

35. Sibson NR, Dhankhar A, Mason GF, Behar KL, Rothman DL, Shulman RG. *In vivo* 13C NMR measurements of cerebral glutamine synthesis as evidence for glutamate-glutamine cycling. *Proc Natl Acad Sci USA* 1997;94(6):2699–2704.

36. Fox PT, Raichle ME, Mintun MA, Dence C. Nonoxidative glucose consumption during focal physiologic neural activity. *Science* 1988;241(4864):462–464.

37. Pellerin L, Magistretti PJ. Glutamate uptake into astrocytes stimulates aerobic glycolysis: A mechanism coupling neuronal activity to glucose utilization. *Proc Natl Acad Sci USA* 1994;91(22):10625–10629.

38. Kasischke KA, Vishwasrao HD, Fisher PJ, Zipfel WR, Webb WW. Neural activity triggers neuronal oxidative metabolism followed by astrocytic glycolysis. *Science* 2004;305(5680):99–103.

39. Pellerin L, Magistretti PJ. Neuroscience. Let there be (NADH) light. *Science* 2004;305(5680):50–52.

40. Prichard J, Rothman D, Novotny E, Petroff O, Kuwabara T, Avison M, Howseman A, Hanstock C, Shulman R. Lactate rise detected by 1H NMR in human visual cortex during physiologic stimulation. *Proc Natl Acad Sci USA* 1991;88(13):5829–5831.

41. Mangia S, Giove F, Tkac I, Logothetis NK, Henry PG, Olman CA, Maraviglia B, Di SF, Ugurbil K. Metabolic and hemodynamic events after changes in neuronal activity: Current hypotheses, theoretical predictions and *in vivo* NMR experimental findings. *J Cereb Blood Flow Metab* 2009;29(3):441–463.

42. Chih CP, Lipton P, Roberts EL, Jr. Do active cerebral neurons really use lactate rather than glucose? *Trends Neurosci* 2001;24(10):573–578.

43. Paulson OB, Newman EA. Does the release of potassium from astrocyte endfeet regulate cerebral blood flow? *Science* 1987;237(4817):896–898.

44. Logothetis NK, Pauls J, Augath M, Trinath T, Oeltermann A. Neurophysiological investigation of the basis of the fMRI signal. *Nature* 2001;412(6843):150–157.

45. Logothetis NK. The underpinnings of the BOLD functional magnetic resonance imaging signal. *J Neurosci* 2003;23(10):3963–3971.

46. Logothetis NK, Pfeuffer J. On the nature of the BOLD fMRI contrast mechanism. *Magn Reson Imaging* 2004;22(10):1517–1531.

47. Viswanathan A, Freeman RD. Neurometabolic coupling in cerebral cortex reflects synaptic more than spiking activity. *Nat Neurosci* 2007;10(10):1308–1312.

48. Logothetis NK. The ins and outs of fMRI signals. *Nat Neurosci* 2007;10(10):1230–1232.

49. Logothetis NK. What we can do and what we cannot do with fMRI. *Nature* 2008;453(7197):869–878.

50. Gati JS, Menon RS, Ugurbil K, Rutt BK. Experimental determination of the BOLD field strength dependence in vessels and tissue. *Magn Reson Med* 1997;38(2):296–302.

51. Stroman PW, Krause V, Frankenstein UN, Malisza KL, Tomanek B. Spin echo versus gradient echo fMRI with short echo times. *Magn Reson Imaging* 2001;19(6):827–831.

52. Bandettini PA, Wong EC, Jesmanowicz A, Hinks RS, Hyde JS. Spin-echo and gradient-echo EPI of human brain activation using BOLD contrast: A comparative study at 1.5 T. *NMR Biomed* 1994;7(1–2):12–20.

53. Zhong J, Kennan RP, Fulbright RK, Gore JC. Quantification of intravascular and extravascular contributions to BOLD effects induced by alteration in oxygenation or intravascular contrast agents. *Magn Reson Med* 1998;40(4):526–536.

54. Hyde JS, Biswal BB, Jesmanowicz A. High-resolution fMRI using multislice partial k-space GR-EPI with cubic voxels. *Magn Reson Med* 2001;46(1):114–125.

55. Aso T, Urayama S, Poupon C, Sawamoto N, Fukuyama H, Le BD. An intrinsic diffusion response function for analyzing diffusion functional MRI time series. *NeuroImage* 2009;47(4):1487–1495.

56. Schick F, Nagele T, Lutz O, Pfeffer K, Giehl J. Magnetic susceptibility in the vertebral column. *J Magn Reson B* 1994;103(1):39–52.

57. Jesmanowicz A, Bandettini PA, Hyde JS. Single-shot half k-space high-resolution gradient-recalled EPI for fMRI at 3 tesla. *Magn Reson Med* 1998;40(5):754–762.

58. Hoogenraad FG, Pouwels PJ, Hofman MB, Rombouts SA, Lavini C, Leach MO, Haacke EM. High-resolution segmented EPI in a motor task fMRI study. *Magn Reson Imaging* 2000;18(4):405–409.

59. Menon RS, Thomas CG, Gati JS. Investigation of BOLD contrast in fMRI using multi-shot EPI. *NMR Biomed* 1997;10(4–5):179–182.

60. Fairhurst M, Wiech K, Dunckley P, Tracey I. Anticipatory brainstem activity predicts neural processing of pain in humans. *Pain* 2007;128(1–2):101–110.

61. Figley CR, Stroman PW. Investigation of human cervical and upper thoracic spinal cord motion: Implications for imaging spinal cord structure and function. *Magn Reson Med* 2007;58(1):185–189.

62. Bouwman CJ, Wilmink JT, Mess WH, Backes WH. Spinal cord functional MRI at 3 T: gradient echo echo-planar imaging versus turbo spin echo. *NeuroImage* 2008;43(2):288–296.

63. Gauthier C, Cohen-Adad J, Brooks J, Rossignol S, Hoge RD. Investigation of venous effects in spinal cord fMRI using hypercapnia. International Society for Magnetic Resonance in Medicine 2009; 17th Annual Meeting, Honolulu, HI, May 18–25:3185.

64. Stroman PW, Figley CR, Cahill CM. Spatial normalization, bulk motion correction and coregistration for functional magnetic resonance imaging of the human cervical spinal cord and brainstem. *Magn Reson Imaging* 2008;26(6):809–814.

65. Greene JD, Sommerville RB, Nystrom LE, Darley JM, Cohen JD. An fMRI investigation of emotional engagement in moral judgment. *Science* 2001;293(5537):2105–2108.

66. Barry RL, Williams JM, Klassen LM, Gallivan JP, Culham JC, Menon RS. Evaluation of preprocessing steps to compensate for magnetic field distortions due to body movements in BOLD fMRI. *Magn Reson Imaging* 2010;28(2):235–244.

67. Wiech K, Farias M, Kahane G, Shackel N, Tiede W, Tracey I. An fMRI study measuring analgesia enhanced by religion as a belief system. *Pain* 2008;139(2):467–476.

68. Stroman PW. Spinal fMRI investigation of human spinal cord function over a range of innocuous thermal sensory stimuli and study-related emotional influences. *Magn Reson Imaging* 2009;27:1333–1346.

69. Stroman PW, Krause V, Malisza KL, Frankenstein UN, Tomanek B. Extravascular proton-density changes as a non-BOLD component of contrast in fMRI of the human spinal cord. *Magn Reson Med* 2002;48(1):122–127.

70. Stroman PW, Krause V, Malisza KL, Frankenstein UN, Tomanek B. Characterization of contrast changes in functional MRI of the human spinal cord at 1.5 T. *Magn Reson Imaging* 2001;19(6):833–838.

71. Stroman PW, Ryner LN. Functional MRI of motor and sensory activation in the human spinal cord. *Magn Reson Imaging* 2001;19(1):27–32.

72. Stroman PW, Nance PW, Ryner LN. BOLD MRI of the human cervical spinal cord at 3 tesla. *Magn Reson Med* 1999;42(3):571–576.

73. Bouzier-Sore AK, Merle M, Magistretti PJ, Pellerin L. Feeding active neurons: (Re)emergence of a nursing role for astrocytes. *J Physiol Paris* 2002;96(3–4):273–282.

74. Nicholls JG, Martin AR, Wallace BG. *Properties and Functions of Neuroglial Cells. From Neuron to Brain.* 3rd ed. Sunderland, MA: Sinauer Associates, Inc.; 1992, 146–183.

75. Pellerin L, Magistretti PJ. Food for thought: Challenging the dogmas. *J Cereb Blood Flow Metab* 2003;23(11):1282–1286.

76. Nedergaard M, Takano T, Hansen AJ. Beyond the role of glutamate as a neurotransmitter. *Nat Rev Neurosci* 2002;3(9):748–755.

77. Piet R, Vargova L, Sykova E, Poulain DA, Oliet SH. Physiological contribution of the astrocytic environment of neurons to intersynaptic crosstalk. *Proc Natl Acad Sci USA* 2004;101(7):2151–2155.

78. Sykova E, Vargova L, Kubinova S, Jendelova P, Chvatal A. The relationship between changes in intrinsic optical signals and cell swelling in rat spinal cord slices. *NeuroImage* 2003;18(2):214–230.

79. Sykova E. Diffusion properties of the brain in health and disease. *Neurochem Int* 2004;45(4):453–466.

80. Fitzgerald MJT, Folan-Curran J. *Clinical Neuroanatomy and Related Neuroscience.* 4th ed. Toronto: W.B. Saunders; 2002.

81. Abbott NJ. Evidence for bulk flow of brain interstitial fluid: Significance for physiology and pathology. *Neurochem Int* 2004;45(4):545–552.

82. Fitzgerald MJT. *Neuroanatomy, Basic and Clinical.* 2nd ed. London: Bailliere Tindall; 1992.

83. Fujita H, Meyer E, Reutens DC, Kuwabara H, Evans AC, Gjedde A. Cerebral [15O] water clearance in humans determined by positron emission tomography: II. Vascular responses to vibrotactile stimulation. *J Cereb Blood Flow Metab* 1997;17(1):73–79.

84. Ohta S, Meyer E, Fujita H, Reutens DC, Evans A, Gjedde A. Cerebral [15O] water clearance in humans determined by PET: I. Theory and normal values. *J Cereb Blood Flow Metab* 1996;16(5):765–780.

85. Stroman PW, Tomanek B, Krause V, Frankenstein UN, Malisza KL. Functional magnetic resonance imaging of the brain based on signal enhancement by extravascular protons (SEEP fMRI). *Magn Reson Med* 2003;49(3):433–439.

86. Stroman PW, Malisza KL, Onu M. Functional magnetic resonance imaging at 0.2 tesla. *NeuroImage* 2003;20:1210–1214.

87. Wong KK, Ng MC, Hu Y, Luk DK, Ma QY, Yang ES. Functional MRI of the spinal cord at low field. Proceedings of the International Society for Magnetic Resonance in Medicine 2004; 12th Annual Meeting, Kyoto, Japan, May 15–21, 2004:1534.

88. Anderson TR, Andrew RD. Spreading depression: Imaging and blockade in the rat neocortical brain slice. *J Neurophysiol* 2002;88(5):2713–2725.

89. Andrew RD, MacVicar BA. Imaging cell volume changes and neuronal excitation in the hippocampal slice. *Neuroscience* 1994;62(2):371–383.

90. Stroman PW, Kornelsen J, Lawrence J, Malisza KL. Functional magnetic resonance imaging based on SEEP contrast: Response function and anatomical specificity. *Magn Reson Imaging* 2005;23(8):843–850.

91. Stroman PW. Discrimination of errors from neuronal activity in functional magnetic resonance imaging in the human spinal cord by means of general linear model analysis. *Magn Reson Med* 2006;56:452–456.

92. Stroman PW, Kornelsen J, Lawrence J, Malisza KL. Functional magnetic resonance imaging based on SEEP contrast: Reproducibility, hemodynamic response function, and anatomical specificity. Proceedings of the International Society for Magnetic Resonance in Medicine 2004; 12th Annual Meeting, Kyoto, Japan, May 15–21, 2004:2543.

93. Kety SS. Circulation and metabolism of the human brain. *Brain Res Bull* 1999;50(5–6):415–416.

94. Ganong WF. *Review of Medical Physiology.* 13th ed. Norwalk, CT: Appleton & Lange; 1987. Table 32.

95. Silva AC. Perfusion-based fMRI: Insights from animal models. *J Magn Reson Imaging* 2005;22(6):745–750.

96. Leontiev O, Buxton RB. Reproducibility of BOLD, perfusion, and CMRO2 measurements with calibrated-BOLD fMRI. *NeuroImage* 2007;35(1):175–184.

97. Golay X, Petersen ET, Hui F. Pulsed star labeling of arterial regions (PULSAR): A robust regional perfusion technique for high field imaging. *Magn Reson Med* 2005;53(1):15–21.

98. Edelman RR, Chen Q. EPISTAR MRI: Multislice mapping of cerebral blood flow. *Magn Reson Med* 1998;40(6):800–805.

99. Petersen ET, Zimine I, Ho YC, Golay X. Noninvasive measurement of perfusion: A critical review of arterial spin labelling techniques. *Br J Radiol* 2006;79(944):688–701.

100. Ostergaard L. Principles of cerebral perfusion imaging by bolus tracking. *J Magn Reson Imaging* 2005;22(6):710–717.

101. Wong EC, Buxton RB, Frank LR. Quantitative imaging of perfusion using a single subtraction (QUIPSS and QUIPSS II). *Magn Reson Med* 1998;39(5):702–708.

102. Detre JA, Leigh JS, Williams DS, Koretsky AP. Perfusion imaging. *Magn Reson Med* 1992;23(1):37–45.

103. Kim SG. Quantification of relative cerebral blood flow change by flow-sensitive alternating inversion recovery (FAIR) technique: Application to functional mapping. *Magn Reson Med* 1995;34(3):293–301.

104. Lu H, Clingman C, Golay X, van Zijl PC. Determining the longitudinal relaxation time (T1) of blood at 3.0 tesla. *Magn Reson Med* 2004;52(3):679–682.

105. Donahue MJ, Lu H, Jones CK, Edden RA, Pekar JJ, van Zijl PC. Theoretical and experimental investigation of the VASO contrast mechanism. *Magn Reson Med* 2006;56(6):1261–1273.

106. Bottomley PA, Foster TH, Argersinger RE, Pfeifer LM. A review of normal tissue hydrogen NMR relaxation times and relaxation mechanisms from 1–100 MHz: Dependence on tissue type, NMR frequency, temperature, species, excision, and age. *Med Phys* 1984;11(4):425–448.

107. Hyman TJ, Kurland RJ, Levy GC, Shoop JD. Characterization of normal brain tissue using seven calculated MRI parameters and a statistical analysis system. *Magn Reson Med* 1989;11(1):22–34.

108. Scouten A, Constable RT. Applications and limitations of whole-brain MAGIC VASO functional imaging. *Magn Reson Med* 2007;58(2):306–315.

109. Lu H, Golay X, Pekar JJ, van Zijl PC. Functional magnetic resonance imaging based on changes in vascular space occupancy. *Magn Reson Med* 2003;50(2):263–274.

110. Darquie A, Poline JB, Poupon C, Saint-Jalmes H, Le Bihan D. Transient decrease in water diffusion observed in human occipital cortex during visual stimulation. *Proc Natl Acad Sci USA* 2001;98(16):9391–9395.

111. Le Bihan D, Urayama S, Aso T, Hanakawa T, Fukuyama H. Direct and fast detection of neuronal activation in the human brain with diffusion MRI. *Proc Natl Acad Sci USA* 2006;103(21):8263–8268.

112. Stejskal EO, Tanner JE. Spin diffusion measurements—Spin echoes in presence of a time-dependent field gradient. *J Chem Phys* 1965;42(1):288–292.

113. Le Bihan D, Breton E, Lallemand D, Grenier P, Cabanis E, Laval-Jeantet M. MR imaging of intravoxel incoherent motions: Application to diffusion and perfusion in neurologic disorders. *Radiology* 1986;161(2):401–407.

114. Beaulieu C. The basis of anisotropic water diffusion in the nervous system—A technical review. *NMR Biomed* 2002;15(7–8):435–455.

115. Duong TQ, Yacoub E, Adriany G, Hu X, Ugurbil K, Kim SG. Microvascular BOLD contribution at 4 and 7 T in the human brain: Gradient-echo and spin-echo fMRI with suppression of blood effects. *Magn Reson Med* 2003;49(6):1019–1027.

116. Tuch DS. Q-ball imaging. *Magn Reson Med* 2004;52(6):1358–1372.

117. Agosta F, Valsasina P, Caputo D, Stroman PW, Filippi M. Tactile-associated recruitment of the cervical cord is altered in patients with multiple sclerosis. *NeuroImage* 2008;39(4):1542–1548.

118. Tuch DS, Reese TG, Wiegell MR, Wedeen VJ. Diffusion MRI of complex neural architecture. *Neuron* 2003;40(5):885–895.

119. Hattingen E, Rathert J, Jurcoane A, Weidauer S, Szelenyi A, Ogrezeanu G, Seifert V, Zanella FE, Gasser T. A standardised evaluation of pre-surgical imaging of the corticospinal tract: Where to place the seed ROI. *Neurosurg Rev* 2009;32(4):445–456.

120. Kohno S, Sawamoto N, Urayama S, Aso T, Aso K, Seiyama A, Fukuyama H, Le BD. Water-diffusion slowdown in the human visual cortex on visual stimulation precedes vascular responses. *J Cereb Blood Flow Metab* 2009;29(6):1197–1207.

7

Functional MRI Study Design

The theory underlying how functional magnetic resonance imaging (fMRI) data are obtained has been described in the preceding chapters, and the next essential element to consider is the design of the fMRI study such that the neural function(s) of interest can be detected with the highest degree of sensitivity and specificity. The design strategy consists of deciding which tasks or stimuli can be compared or contrasted to reveal the function(s) of interest, whether these are transient functions or can be sustained for a period of time, and whether or not interactions between functions are of interest. It is also necessary to decide how long the fMRI time series will be, and how many time points or volumes will be imaged to describe it, and how the tasks or stimuli will be applied in time. The wide range of flexibility in fMRI study designs therefore makes this method a powerful tool but can make the design task seem daunting. However, as will be described in this chapter, a considerable number of studies have been carried out to help identify the features of optimal design strategies.

7.1 Basic Principles of fMRI Study Design

The previous chapters showed how a magnetic resonance (MR) image is created (Chapter 5), how the signal intensity in each voxel can be made to depend on relaxation times and the proton density (Chapter 4), and how the relaxation times and proton density can be made to depend on the level of neuronal activity (Chapter 6) (more specifically, metabolism, blood flow, etc.). The net effect is that we can acquire MR images with signal intensities that depend, to some degree, on the local level of neuronal activity, or at least on physiological changes that are related to neuronal activity. Now we can bring all of these ideas together to design an fMRI study. The following sections will discuss only the blood oxygenation–level dependent (BOLD) fMRI method because this is by far the most widely used, and the general principles can be extended to the use of other contrast mechanisms as described in Chapter 6.

The key concept that is common to every fMRI study is that the neuronal activity of interest must be varied systematically so that any voxels that change signal intensity in relation to this activity can be detected. The second key concept is therefore implied, that we must acquire a time series of images in order to record the changes in signal intensity. It is important to keep in mind that we can only detect a change in neuronal activity. The focus of fMRI study design is therefore to determine which tasks or stimuli can be compared to reveal the neuronal activity of interest, and the most effective timing for applying these tasks or stimuli. The choices of tasks or stimuli and how they will be compared must be guided by a clear hypothesis, which is a statement of the proposed result, formed in a way that it can be tested, and shown to be either true or false. By having a clearly defined question to answer, the expected result can often be stated as the hypothesis.

For example, an fMRI study to detect the regions of the brain involved with fine motor skills of the fingers could compare a predetermined movement pattern of the fingers as the *task* condition, and no finger movement as the *baseline* condition. This comparison would reveal any areas of the brain that are involved in any way with the motor task. Alternatively, the baseline condition could be another motor task of the hand, but in which all of the fingers are moved together. Now the comparison would show the areas of the brain involved specifically with fine motor movements of the fingers. The hypothesis in this case could be that fine motor movement, in contrast with bulk movement, involves significantly more activity in the supplementary motor area. Another option is to alternate all three conditions, and make comparisons to reveal both the total motor response and specifically the areas involved with individual finger control. Such tasks could be used in a study of concert pianists or violinists, compared with a similar age and gender group of nonmusicians, and the study could be repeated after each group has an opportunity to practice the tasks. Alternatively, the same tasks could be used to compare a group of healthy control subjects with a group of patients who have suffered strokes affecting their hand motor skills. Each of the three conditions described in this example can be sustained for a period of time to allow the BOLD signal changes to reach a peak and maintain a certain level, or to return to baseline. The activity shown would therefore be the sustained activity over the whole period of each condition, which could be different from transient activity involved with performing each movement. Another alternative would therefore be to design the motor tasks to involve a movement of one finger at a time, or all fingers, every several seconds, and then to compare the responses to each type of movement (i.e., which finger or fingers moved).

Clearly, with even an apparently simple finger movement task, the design of an fMRI experiment can become complex, because of the almost infinite combination of comparisons that can be made. Regardless of which comparisons are made, it is necessary to design the study to detect specifically the function(s) of interest, unambiguously from other neuronal functions, or other sources of signal change such as random noise and physiological motion. At the same time, it is clear that some form of communication with the person being studied is necessary, because they need to be told when to perform the task and when to rest or change to some other task. It is also desirable to record some form of response or feedback from the person being studied so that we have a measure of how they performed the task. The behavioral data are often essential to be able to analyze the fMRI data with the optimum sensitivity and to appropriately interpret the results.

Key Points

1. The neuronal activity of interest must be varied systematically so that any voxels that change signal intensity in relation to this activity can be detected.
2. A time series of images must be acquired in order to record the changes in signal intensity.
3. Only differences or changes in neuronal activity can be detected.

7.2 Choice of Stimulation Method or Task

While clearly the method of stimulation or the cognitive task used for an fMRI study must relate to the neuronal function of interest, there are often options for the choice of stimulus or task and how it is applied, and important considerations and limitations imposed by the environment of the MRI system. The choice of stimulation method, or the task to have the person perform, is often the first decision made when designing an fMRI study. It is also common to have to modify the stimulus or task to adapt to the limitations of the MRI system environment and the imaging parameters, once they are decided.

The person being studied with fMRI must typically lie on their back, within the cylindrical space inside the MRI system, which has a diameter around 65 cm and a typical length of 1.5 to 2 meters. While it is necessary for the person to remain as motionless as possible during the fMRI study, studies of motor functions and proprioception require that the person perform some kind of motor task, and in many other types of studies some form of response or feedback is required from the person being studied, which may involve some movement. A common response method is to have the person press buttons on a keypad. The enclosed environment of the MRI system presents a constraint on how much the person can move or how they can position their hands or arms to respond with a keypad or other device. Any movement can also produce motion artifacts, and changes in position can alter the local magnetic field homogeneity, because of different materials (bone, muscle, air spaces, etc.) having different magnetic susceptibilities. However, some studies in the literature involve fairly complex reaching tasks (1) or even verbal responses (2) with highly reliable fMRI results.

Visual displays are commonly used for fMRI studies because they can effectively elicit emotional responses, evoke memories, or present cognitive tasks such as language or math and so forth, or can be used to present written instructions. The shape of the MRI system and the magnetic field again imposes limits on the presentation of visual stimuli. One immediate factor to consider is that people cannot wear glasses to correct their eyesight while in the MRI system, unless the glasses are specifically made to contain no magnetic or electrically conductive materials. Two common methods for visual presentation are (1) to use MRI-compatible goggles to present images to each eye via fiber-optic cables, and (2) to have a projector positioned some distance from the magnet and project images onto the back of a translucent screen either just outside of or within the MRI system (Figure 7.1) while the person views the front of the screen via a mirror positioned in front of their eyes. The MRI-compatible goggles typically include adjustments to correct the display for people needing glasses, and the visual angle spanned by the display can be very large. The use of a projector and rear-projection screen requires less specialized equipment but may require the use of MRI-compatible glasses and may be limited by the ~65 cm diameter bore of the magnet. However, some projection systems use a screen that is

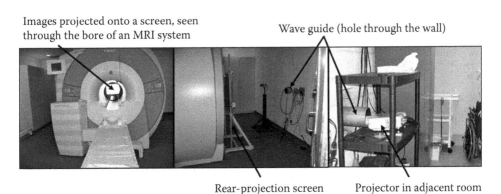

Images projected onto a screen, seen through the bore of an MRI system

Wave guide (hole through the wall)

Rear-projection screen   Projector in adjacent room

FIGURE 7.1 An example of a video projection system used for fMRI, using a projector and rear-projection screen. In this case, the projector is in an adjacent room, to avoid the risk of placing it within the magnet room, and the projected light passes through a waveguide that provides a hole through the wall, while maintaining the shielding of the MRI system from radio-frequency noise. When a person is lying supine inside the MRI system, they can view the screen via a mirror positioned in front of their eyes.

placed very near the person's head while lying in the magnet and may be near enough to view without corrective lenses. Very detailed images and videos can typically be presented with both types of systems.

The enclosed space of the MRI system can be a distraction or even produce anxiety for some people and can therefore affect the performance of the tasks for the fMRI study. At the same time, the rapid gradient switching during imaging can produce sound levels as high as 120 dB, or about the same as a very loud car stereo, so the person being studied needs to wear earplugs and/or headphones for hearing protection and to help with two-way communication. The loud sounds produced by the MRI system therefore present yet another effect that can interfere with the detection of some neural functions and can be particularly problematic for studies of audio processing. However, it is possible to carry out tests of task performance, measure reaction times, and so forth, and compare values measured inside the MRI system with those measured in a more hospitable environment (3,4), to determine whether or not the MRI environment affects the performance.

In spite of these potentially confounding effects, most participants in fMRI studies adapt quite quickly, within only a few minutes, to the environment and sounds of the MRI system. The first images acquired after the person has been positioned inside the MRI system are generally fast images that are used for subsequent selection of slice positions. After these initial images, verbal reassurance over the two-way communication system of the MRI system that everything is going well and that the images look good, can help an anxious study participant to relax. Contrary to what might be expected from the previous paragraph, a fairly common problem is the participant becoming bored or even falling asleep during an fMRI study. Efforts to make the participant comfortable, dim lighting within the MRI system, and rhythmic sounds of the MRI system during repeated imaging can lull the participant to sleep. The solution is often to try to make the tasks or conditions imposed for the fMRI study as engaging and interesting as possible, and limiting the duration of each experiment.

The choice of stimulation method or task to use for an fMRI study is therefore determined by the neural function to be studied, and the ideal choice would likely be the same method that would be used for behavioral testing or clinical tests of function. The environment imposed by the MRI system, however, can impose severe limitations on the choice of task and the like, so the choice must be adapted and a compromise reached for use with fMRI.

Key Points

4. The choice of stimulation method has to evoke the function(s) of interest and must be possible within the environment of the MRI system.
5. Communication is needed with person being studied, such as visual and/or audio presentations, to provide instructions and information to the person, and some form of feedback from the person to determine responses and/or recording of physiological parameters.

7.3 Choice of the fMRI Study Design

The options for the fMRI study design range from *block* designs (to detect sustained activity), in which conditions are applied for periods of time, to *event-related* designs (to detect transient activity) in which the responses to individual events are detected. Any combination between these two limits is possible as well, in the form of *fast event-related* designs, where individual events are carried out in succession so that the BOLD responses to individual events overlap in time; *mixed* designs, where blocks and events are both used allowing both sustained and transient responses to be detected; and *behaviorally driven* designs, which require detailed monitoring of the performance of a task or recording of physical parameters, or can be used with resting-state studies (5). The best choice of design depends on the neural function of interest.

7.3.1 Block Designs

Block designs in which stimuli or tasks are applied for sustained periods of time of ~10 to ~30 sec are probably the most commonly used designs for fMRI (see examples in Chapter 9). The sustained activity produces a relatively strong BOLD response that reaches a plateau after about 6 sec (see Chapter 6, Section 6.2.2), and images can be obtained at multiple time points during the response so that it can be detected with high sensitivity. The drawbacks of the block design are that not all neural functions can be sustained; some are transient by nature. Blocks often consist of multiple repeated trials, such as the finger-movement task described above. Errors in the task performance during the block may make the fMRI data invalid for comparison with other blocks. Over the course of a block the neural function of interest can also change due to practice, habituation, or expectation, or simply because of fatigue or shifting attention. If any of these effects during the block result in changes in the BOLD response, then the analysis to detect the expected sustained response may have reduced sensitivity. Shorter blocks can therefore produce more consistent BOLD responses, resulting in greater sensitivity. However, there are other factors that will be discussed below that support the use of longer blocks for greater statistical power in the analysis, and optimal designs will balance these effects.

7.3.2 Event-Related Designs

Event-related designs are an important alternative to block designs because they enable transient effects to be studied, as opposed to neural functions that can be sustained for several seconds or more. For example, individual finger movements could be studied instead of many repeated movements, or events consisting of reading math problems might be distinguished from thought processes to solve the math problem after it is read and understood. A very useful application is to compare and contrast activity detected between correct and incorrect subject responses, or to sort the trials in other ways depending on the subject responses. Yet another benefit is that stimuli or tasks can be randomized in time so that the person being studied cannot plan responses, anticipate sensations, and so on. However, the primary disadvantages of event-related designs are that (1) they produce much smaller, and more brief, BOLD responses; and (2) a large number of event trials are typically needed, resulting in long fMRI acquisitions. The first point is illustrated in Figure 7.2, with two theoretical BOLD responses calculated for stimuli applied for 1 sec or for 20 sec. The predicted responses are based on the convolution of the BOLD hemodynamic response function (HRF) and the pattern of stimulation in time (6). As a consequence, event-related designs are much more sensitive to accurate models of the HRF for analysis but can also provide more accurate measures of the HRF in the results that are obtained. In practice, the event-related responses are observed to be lower in magnitude than those in the block design, but published examples indicate that the difference may not be as large as suggested by Figure 7.2, and the event-related response may be only about 35% lower than the block design (7–9). However, the observed signal changes depend on a number of factors such as the fMRI study design and the area of the brain that is involved.

The second disadvantage listed above arises due to the fact that the BOLD response must be allowed to return to the baseline level between each event. Here, the *baseline* level refers to the constant image intensity during the condition that is used as a reference for comparison to the stimulation or task conditions. The canonical BOLD HRF discussed in Chapter 6, and as shown in Figure 7.2, takes about 25 sec to return to baseline after a brief event. This means that if the fMRI acquisition takes 10 minutes, there is only time to detect the responses to 24 brief events. As discussed below in Section 7.5.4, this number of events might be sufficient, but a number of factors can affect the sensitivity of an fMRI study.

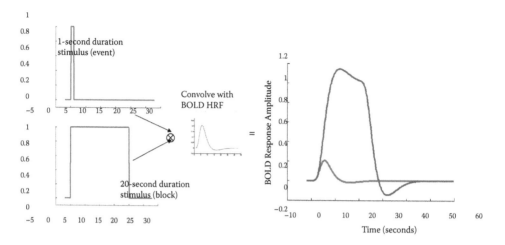

FIGURE 7.2 Theoretical BOLD responses calculated for a 1 sec duration (red line) stimulus or task event, and for a 20 sec duration (blue line) block. The stimulus or task pattern in time is convolved with the BOLD hemodynamic response function (HRF) to predict the total BOLD response patterns in time. The relative magnitudes of the two responses are shown in arbitrary units. This theoretical prediction likely overestimates the relative difference between the magnitudes of the block and event-related responses, and they are actually closer in magnitude with the event-related response being only about 35% lower than the block response. (Bandettini, P.A., Cox, R.W., *Magn Reson Med* 43, 4, 540–548, 2000 [8].)

7.3.3 Fast Event-Related Designs

Fast event-related fMRI study designs (and mixed designs discussed in the next section) provide a combination of the features of block and event-related designs. Fast event-related designs are useful for situations when a fully event-related design is not necessary or is not practical and responses to individual events are required results. The key difference between event-related and fast event-related designs is that there is not enough time allowed for the BOLD response to return to baseline between successive events with fast designs (Figure 7.3). The BOLD responses to successive events therefore overlap. It is assumed that the combined response at any point in time can be modeled as a linear sum of the two overlapping responses (10). There is evidence that BOLD responses to two events, one occurring immediately after the other, do not always sum linearly (the combined response may depend on several factors such as the tasks or stimuli, timing, etc.) (9,10). However, it has been shown that as long as the two events are separated in time by 4 sec or more, the BOLD response to the first event begins to subside by the time the second response is growing, and the linear model is adequate (9).

Even though the BOLD responses to individual events overlap in time with fast designs, differences in activity based on subject responses (such as correct or incorrect answers) can still be detected. Differences in activity between different tasks or stimuli can also be determined. For example, if two different tasks are performed (A and B) and the person being studied responds either correctly or incorrectly, then there are four possible responses to be detected (A_{inc}, A_{corr}, B_{inc}, B_{corr}). Each individual BOLD response may overlap with the responses that precede and follow it, resulting in 64 possible combinations (combinations such as A_{corr} A_{corr} A_{corr}, A_{corr} A_{corr} A_{inc}, A_{corr} A_{corr} B_{corr}, and so on). It may be possible to measure the responses to all 64 combinations several times, if enough events are observed. This would enable average responses to each of the four events, and interactions between them, to be determined. However, in practice, correct and incorrect responses may not occur at equal rates and will depend on the person being studied. Some of the 64 possible permutations may not be observed except infrequently, preventing reliable determination of specific average responses or interactions, and may need to be ignored. As

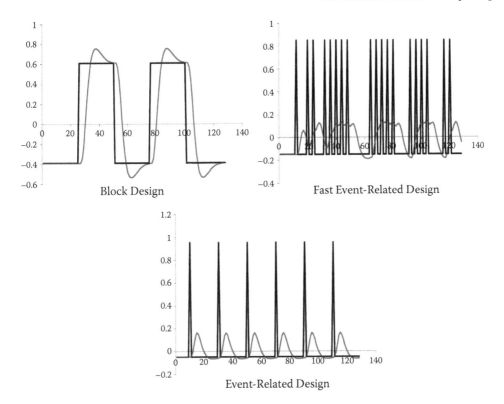

FIGURE 7.3 Examples of fMRI design approaches, ranging from block designs (left), through fast event-related designs (center), to purely event-related designs (right). The black lines indicate the timing for each block or event, and the red lines indicate the expected BOLD response timing determined by convolving the timing pattern with the hemodynamic response function.

long as the important responses of interest, and/or the interactions of interest, are adequately determined, then the study design can be considered to be effective. It may not be necessary to observe multiple combinations of every possible interaction.

The details of data analysis methods are discussed in Chapter 8, but it is worthwhile to understand here, in the context of fMRI study design choices, how the responses to different conditions, and their interactions, would be determined. In the example used above, the analysis of the responses would begin with identifying when each event occurred in the fMRI time series and distinguishing the event based on (1) correct or incorrect response and (2) which events preceded and followed it (Figure 7.4). The fMRI time-series data would be analyzed to detect when BOLD signal changes consistently occurred during each response type. In total, it would be possible to identify up to 64 different types of events (as described above for this particular example) and to determine the BOLD response to each type. To clarify, the results could show the time course of the BOLD response to each event type, as well as where in the image data the BOLD responses occurred. Interactions between events could be determined by contrasting one type of event (such as type A_{corr}, for example) with different types of events preceding them (e.g., such as types A_{corr} and A_{inc}).

7.3.4 Mixed Designs

Mixed designs consist of blocks of stimuli or tasks, but within each block are multiple types of events (11,12). The purpose of this design is to enable both transient and sustained components of neural responses to be detected at the same time. An important feature of the design is that

167

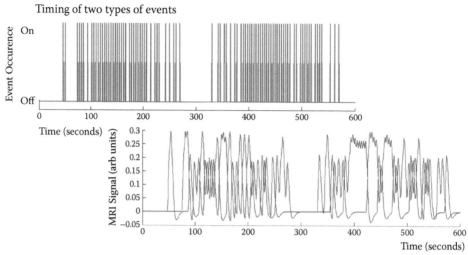

Predicted BOLD responses to each event type

FIGURE 7.4 Example of a rapid event-related design with two types of trials (such as different tasks or stimuli, or correct or incorrect responses to one task type, etc.), with the timing of events of one type indicated in red and the other type in blue. The events occur only at time points spaced 4 sec apart, but with a probability of occurring that varies slowly during the time series. The BOLD response that is predicted by convolving the event timing with the canonical hemodynamic response function is shown in the lower half of the figure.

the events within the block of stimuli must be applied at variable intervals (i.e., the timing is *jittered*), so that the events do not produce a steady-state response that might be indistinguishable from the sustained components of the neural activity (11).

As a recent example of the use of a mixed design, Madden et al. (13) investigated age-related differences in executive control during task switching. During task blocks, the participant was given a cue word for how to categorize the succeeding target word. The participant pressed one of two buttons to indicate to which category the target word belonged. During the task block, the categorization was repeated or was switched unpredictably. In addition, for 25% of trials there was no target word presented. During baseline periods the participant viewed a row of three fixation crosses. The sustained activity during the task blocks was therefore demonstrated by consistent signal changes between task and baseline periods, whereas transient responses were detected in response to a switch of the categorization instruction. Events in which the cue word was presented without a target word also permitted detection of the response to the cue word alone.

The analysis of mixed designs requires the assumption that BOLD responses to simultaneous or successive events are linear (i.e., they can be added or subtracted to produce the total response). Nonetheless, the published examples of the use of this design option demonstrate that it is highly effective and is essential for detecting both transient and sustained components of neural processes (12).

7.3.5 Behaviorally Driven Designs

A special case of fMRI analysis arises in *resting-state* studies in which the participant is often asked to do nothing or may be asked to simply fixate on a visual display. In this situation, there is no predictable BOLD response to model for the analysis. The analysis instead consists of detecting spatially separated areas of the brain, or a network of specific regions of interest that have synchronous signal intensity changes over the fMRI time series. Alternatively, even in more common fMRI studies in which a task is performed, participants may be allowed to function at their own pace. The timing of responses or actions can be recorded precisely and used as a model for subsequent fMRI analysis. In this latter case, the fMRI study design could be very much like the event-related, fast event-related, or mixed designs described above, depending on the timing chosen by the person being studied and on the neural functions that are of interest.

Key Points

6. The best choice of design depends on the neural function of interest.

7. *Block* designs consist of applying stimuli or tasks for sustained periods of time of ~10 to ~30 sec and enable study of sustained functions.

8. *Event-related* designs consist of brief events and the BOLD response to each event is detected. These designs enable transient effects or functions that cannot be sustained to be studied.

9. *Fast event-related* fMRI study designs consist of events applied in succession so that BOLD responses to each event overlap in time. These designs provide a combination of the features of block and event-related designs and allow study of interactions between trials.

10. *Mixed* designs consist of blocks of stimuli or tasks, but within each block are multiple types of events. The purpose of this design is to enable both transient and sustained components of neural responses to be detected at the same time.

11. *Behaviorally driven* designs are used for resting-state studies or self-paced tasks in which there is no predetermined pattern of task or stimulation.

7.4 Order and Timing of Presentation of Tasks or Stimuli

The duration of each stimulus or event, the timing between them, and the order in which different tasks or stimuli are presented are key elements of the fMRI study design. These factors have a strong influence on the BOLD signal changes, and consequently on how well different neural functions and interactions between them can be detected. As mentioned at the beginning of this chapter, the key concept that is common to every fMRI study is that the neuronal activity of interest must be varied systematically so that any voxels that change signal intensity in relation to this activity can be detected. If two different conditions are applied, such as a finger-movement task and a rest condition, as in the example used earlier, then only the differences in function between these two conditions can be detected. If the finger-movement task involves a pattern of individual movements of each finger to press a sequence of buttons on a keypad, then the results may show areas of the brain involved with understanding the cue to move; planning the movement; the motor control of each finger, the hand, the wrist; and so on. If it is of interest to determine the function that is specifically involved with coordinating movements of individual fingers, then a second task could be used consisting of moving all of the fingers together to push several buttons at once. Now, the differences between the two tasks and the rest condition would be expected to have some common features such as understanding the cue to move, planning the movement, and so on. However, there are also differences expected because of the different movements that were performed. As a result, the contrast between the two movement tasks would show the signal intensity changes, and areas of the brain, that differ between moving each finger individually and moving all fingers together.

In this example, since the finger or hand movements can be performed repeatedly, a block design will provide the greatest sensitivity for detecting function (as described below). Now it is important to consider whether there should be a rest condition between each movement task, or if the task should be changed without a rest. Will the change of task alter the function that is detected, even if only transiently? Is the activity involved with the change of task important to detect? Will the order of the two tasks matter in this situation? With the different fMRI study designs described in the previous section, we can see that it could be possible to detect transient effects or to look for interactions between successive stimuli, and these factors must be considered in the study design. For this example, we could decide that we do not want to detect transient switching-related activity or look for interactions between tasks, so we can put a rest period between each task. If the neural functions of interest were different, then a different combination of tasks and rest periods might be better. The example of the finger-movement tasks used here might seem simplistic, but in many regards it is no different to detect the activity involved with simple finger movements than it is to detect activity involved with complex cognitive processes. In both of these examples, neuronal activity changes, altering the local rate of metabolism, thereby producing the net change in MR signal that we detect. The main difference is that it is easy to verify that a simple finger movement has been performed, and can be difficult to confirm that the desired cognitive process has been evoked. In either situation, monitoring of the task performance and feedback from the person are necessary for the data to be analyzed with the greatest sensitivity.

The example used above illustrates the importance of the different tasks or stimuli that are employed and how they are compared or contrasted to reveal the neural function of interest. Even with the apparently simple finger motor task, there are a number of options for how the task blocks are compared, and each can produce different results because they answer different questions. The comparison strategies that are most often used are summarized by Amaro and Barker (5) as *subtraction, factorial, parametric,* and *conjunction* (Figure 7.5).

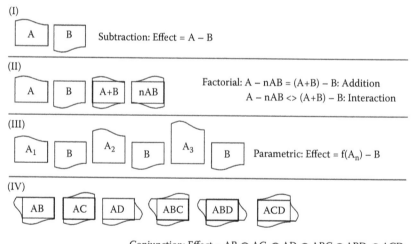

FIGURE 7.5 The original caption reads: "Cognitive comparison strategies: (I) subtraction, based on 'pure insertion' principle; (II) Factorial, which provides a framework for testing 'pure insertion' theory; (III) Parametric, in which the 'nature' of the cognitive process is maintained, but its intensity is modulated; (IV) Conjunction, in which the conditions sharing the same cognitive component can the further analysed using an 'intersection' approach. Symbols: A, B, C, and D represent cognitive components in a given experimental condition in the experiment; nAB represents a condition where the cognitive components 'A' and 'B' are absent; A_1, A_2, A_3 represent the 'A' cognitive component of a condition with three different cognitive 'loads.'" (Reproduced from Amaro Jr., E., and Barker, G.J., Study Design in fMRI: Basic Principles, in *Brain and Cognition*, Volume 60, 220–232, 2006, with permission from Elsevier.)

7.4.1 Subtraction Method

Comparisons based on subtraction demonstrate the BOLD signal intensity differences between two tasks, indicated by A and B in Figure 7.5. This was one of the earliest methods used in fMRI (14–16) and is probably still the most commonly used. The underlying assumption is that neural processes and the BOLD signal changes that are produced are additive, and so one can be subtracted from the other. Even though this assumption is not expected to be valid in most cases, this method nonetheless serves to demonstrate where BOLD signal differences consistently occur between two conditions and has been shown to reliably identify brain areas involved with a task or stimulus. In spite of the name given to this approach, the method does not involve subtraction of the image data obtained during one condition from the other condition, but rather the name refers to the contrast of the BOLD responses between the two conditions, as will be discussed in Chapter 8.

7.4.2 Factorial Method

The factorial approach employs individual tasks or conditions, A and B, a reference condition not involving A or B (indicated by "nAB" in Figure 7.5), and also the combination of conditions (A + B). This method enables interactions between the conditions A and B to be studied. Again, it is necessary to assume that BOLD responses from the two conditions A and B sum linearly to produce the response to A + B.

For example, Michielsen et al. (17) used a factorial design to investigate the neural basis for the effects of mirror therapy in patients with stroke. Mirror therapy consists of showing a patient, after stroke or amputation affecting one arm, a reflection of their unimpaired arm in a mirror, so that they receive visual feedback of having both arms intact and functioning. It is used to treat chronic pain and for rehabilitation of motor deficits. In order to investigate how this therapy works, Michielsen et al. carried out an fMRI study with two separate paradigms. In one study, participants performed a unimanual movement task (opening and closing the hand) with their unaffected hand, and task blocks consisted of either viewing their hand directly (no mirror condition) or viewing their hand in a mirror (mirror condition) while performing the task for 30 sec. In the second experiment, patients moved both hands while looking directly at the affected hand (no mirror condition) or while observing their unaffected hand in the mirror (mirror condition), also for 30 sec. Patients could see two hands in all four conditions. In each experiment 5 task blocks were interleaved with 5 rest blocks of 30 sec duration. The task blocks were then contrasted in the analysis to determine the main effects of movement (task condition versus rest), mirror (mirror condition versus no mirror condition), and the interaction effect between movement and mirror for both the unimanual and bimanual experiments. In this example, the factorial approach provided the necessary means to investigate how observing a reflection in a mirror can interact with motor function.

7.4.3 Parametric Method

Parametric comparisons involve modulating the intensity of one of the tasks or stimuli and applying at least one other condition for comparison. Some areas of the brain might have BOLD signal changes that increase with the intensity of a stimulus or complexity of a task, whereas other areas may show more constant BOLD responses, such as those involved with maintaining attention to the task or processing instructions and so forth. With this approach, neural activity involved with different aspects of a cognitive task, for example, can be investigated. However, not all tasks or stimuli can be applied with consistently varying intensity or complexity, and increasing the task may not result in an increased BOLD response but may instead involve recruitment of other areas of the brain.

As an example of the parametric approach for fMRI study design Westerhausen et al. (18) investigated the attentional and cognitive control required when hearing two competing syllables of varying sound intensities, and listening to either the louder or the quieter of the two, as instructed. The fMRI study design consisted of presenting different sounds simultaneously to the right and left ears (dichotic), with five different levels of sound intensity difference between the two ears. The participants were instructed to focus their attention to and to report either the right-ear stimulus (forced-right condition, FR) or the left-ear stimulus (forced-left condition, FL). Which ear to attend to was varied and indicated to the participant before each trial by an arrow pointing in the respective direction. The participants responded orally after each trial by naming the syllable they heard, and their responses were recorded. A control condition was also applied in which no attention instruction was given (nonforced attention, NF). In total there were therefore 15 experimental conditions composed of 5 interaural intensity differences with 3 attention manipulations (NF, FR, and FL). Each condition consisted of 18 stimulus presentations, resulting in 270 stimulus presentations that were intermixed with 135 *null events* with no acoustic stimulation, to obtain a stochastic event-related fMRI design. The comparisons of responses, however, was a parametric approach, and the magnitude of BOLD responses detected for each of the 15 experimental conditions was analyzed using an analysis of variance (ANOVA) with the 5 levels of interaural differences and 3 attention manipulations. The results showed that the BOLD responses varied in magnitude according to the interaural intensity differences in a number of relevant brain regions (including the inferior parietal lobe, supramarginal gyrus, precentral gyrus, presupplementary motor area, inferior frontal gyrus, insula, and anterior cingulate cortex).

7.4.4 Conjunction Method

The conjunction approach to comparing responses is described by Amaro and Barker (5) as being a subtle deviation from the factorial design. Stimulus blocks consist of combinations of task or stimulus conditions, and the study design and comparisons focus on determining common features of the responses to the conditions.

In a recent example of the use of the conjunction approach, Mier et al. (19) employed both a facial emotion recognition and an emotional intention recognition task, in an event-related fMRI study. Participants were shown pictures of faces with varying emotional expressions (joy, anger, fear, disgust, or neutral). Each trial started with the presentation of a statement, followed by a facial picture, with the choice yes or no below, and participants indicated their choice with a button press. The statements indicated (1) the person's emotion ("This person is angry"), (2) their intention ("This person is going to run away"), or (3) a control condition ("This person is female"). A conjunction analysis was carried out to reveal common activation between the recognition of emotion and intent.

Key Points

12. The duration of each stimulus or event, the timing between them, and the order in which different tasks or stimuli are presented are key elements of the fMRI study design.

13. Comparisons based on *subtraction* demonstrate the BOLD signal intensity differences (contrasts) between two tasks (A and B).

14. The *factorial* approach enables interactions between the conditions A and B to be studied (conditions A, B, A + B, and not A or B).

15. *Parametric* comparisons differentiate cognitive components of responses (B, and varying intensities of A).

16. The *conjunction* approach determines common features of responses to the conditions with different trials (AB, AC, AD, ABC, ABD, ACD).

7.5 Timing of Tasks or Stimuli, Duration, Sampling Rate

The previous discussions about the choice of stimulus conditions and how they can be compared or contrasted have hinted at yet another critical issue in fMRI design, which is the duration, timing, and numbers of tasks or events to apply. The effectiveness of the fMRI study design can depend critically on the answers to questions regarding how many time points should be measured, how many blocks or events should be applied, and what should be the duration and interval between successive blocks or events.

The two key factors that influence the timing of an fMRI experiment are the rate at which images can be acquired, in order to describe the time series of changes, and the rate of change of the BOLD effect itself. Each image, or set of images, is acquired over the period of a few seconds and represents one time point in the time series. Because images are typically acquired from multiple slices in order to span a three-dimensional (3D) volume of the brain, spinal cord, and so forth, each time point is often called a *volume* in the time series as well. As described in Chapter 6, the rate at which the images can be acquired will depend on the imaging parameters, the spatial encoding method used, the number of slices acquired, and so on. In general, functional MRI data spanning the entire brain can be acquired in 2 to 4 sec, with T_2^*-weighting to be sensitive to the BOLD effect.

7.5.1 The Sampling Rate

Most fMRI acquisitions employ single-shot imaging, typically with EPI or spiral k-space encoding, and so the time between repeated images of the same region (the sampling rate) is equal to the repetition time, TR. As discussed in Chapter 5 (Section 5.4), there is an important balance between the TR and the amount of image data (sampling matrix, number of slices) that can be acquired. To reduce the TR to attain more speed, it may be necessary to reduce the matrix size or the number of slices. Reducing the matrix will increase the voxel size (reduced spatial resolution) unless it is balanced with a corresponding reduction in the field of view (FOV), which may reduce the coverage of the anatomy. Reducing the number of slices will similarly reduce the coverage of the anatomy, unless balanced by an increase in the slice thickness or the introduction of gaps between the slices. Regardless of the solution that is chosen, increasing the sampling rate (reducing the TR) will have some cost of reduced spatial information.

It has been shown that lower TR values produce fMRI results with higher statistical power, and that optimally, the TR should be relatively short at less than 1.5 sec (20). The authors of this study further recommended that the number of slices be kept to the minimum required, in order to allow for the shortest TR possible. However, they point out the limitation that some 3D motion correction methods require whole-brain imaging. Also, with short TR values some components of noise in the time-series image data may be temporally correlated, as opposed to being random with a Gaussian distribution, as is assumed in many statistical analyses that are applied to fMRI data. Nonetheless, the authors show fMRI results with a selection of different tasks all of the same length, but sampled at different TR values, that support the conclusion that TR values are optimal at less than 1.5 sec. In many cases this may not be achievable, and so the next best option is to use the shortest TR possible, given the other requirements for the study design.

7.5.2 The Order and Timing of Blocks or Events—Design Efficiency

The primary factor limiting the timing for fMRI studies is therefore not the image acquisition but the speed of the BOLD response itself. The BOLD response takes approximately 5 to 6 sec to reach its peak at the onset of a stimulus and takes approximately 12 sec to return to baseline values after the stimulus is removed. The response to a very brief stimulus, as in event-related design (discussed in detail below), has a total duration from beginning to end of roughly 25 sec. With time points (i.e., volumes) sampled 2 to 4 sec apart, the transitions of the BOLD signal changes will be detected, with at least one sampled point on the upward transitions at the onset

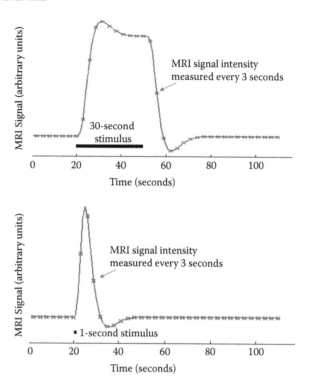

FIGURE 7.6 Examples of sampling the expected BOLD response to a stimulus that is applied for 30 sec, and for 1 sec, with the MR signal intensity measured every 3 sec (in the absence of noise). The blue line represents the signal change time course whereas the red X symbols indicate the discrete measurement time points, when each MR image in the time series is acquired.

of the stimulus, and likely three or more sampled points on the downward transitions after the stimulus is removed (Figure 7.6).

One common analysis approach to detect the BOLD response to a task or stimulus is to use the general linear model (GLM) (21) (as described in Chapter 2):

$$y = X\beta + e$$

Here, β is the magnitude of the response that we are trying to determine, y is the time series of measured data points for a single voxel, X is a time series describing the pattern of signal changes that we are looking for (we choose this prior to doing the analysis), and e is the residual error, or the difference between the fit values and the measured values at each time point. In many papers describing analysis methods X is called a matrix of response variables, or the *design matrix*. For example, X could be defined to be −1−1−1 1 1 1..., meaning that the signal is expected to be at some lower value (i.e., off) for the first three time points, and then at some higher value (on) for the next three time points, and so forth, or could contain more than one pattern in time, as shown in Figure 7.7. This on/off pattern would typically be convolved with the shape of the hemodynamic response function to account for the time it takes for the BOLD response to occur and to return to baseline. The solution to the equation above gives the value of β, which is the magnitude of X as a fit to the measured data, y. (β will contain one value for each pattern in X, but for this discussion we are considering X with only one pattern of signal change, for simplicity.)

For this discussion the important value is the variance of β, which is equal to $\sigma^2(X^TX)^{-1}$, where σ^2 is the fit variance given by the sum of the squares of the values in e, divided by the number of degrees of freedom (i.e., approximately the number of values in y minus the number of β values

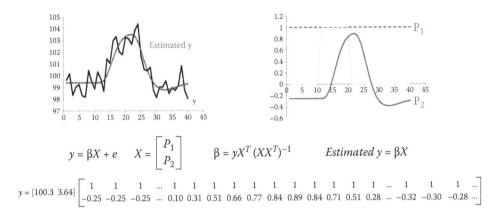

$$y = \beta X + e \qquad X = \begin{bmatrix} P_1 \\ P_2 \end{bmatrix} \qquad \beta = yX^T (XX^T)^{-1} \qquad \text{Estimated } y = \beta X$$

$$y = [100.3 \ 3.64] \begin{bmatrix} 1 & 1 & 1 & \cdots & 1 & 1 & 1 & 1 & 1 & 1 & 1 & 1 & 1 & 1 & \cdots & 1 & 1 & 1 & \cdots \\ -0.25 & -0.25 & -0.25 & \cdots & 0.10 & 0.31 & 0.51 & 0.66 & 0.77 & 0.84 & 0.89 & 0.84 & 0.71 & 0.51 & 0.28 & \cdots & -0.32 & -0.30 & -0.28 & \cdots \end{bmatrix}$$

FIGURE 7.7 Example of a general linear model (GLM) used to fit a sum of predetermined patterns, P_1 and P_2, to a measured time series, y. Note that the pattern P_2 is the convolution of the BOLD hemodynamic response function (HRF), with a block pattern (shown in green in the right-hand plot) that models the signal increasing to a high value for 10 sec. The mean value of P_2 has also been set to zero, and the constant function, P_1, accounts for the nonzero average value of the measured data, y. In this example, the GLM equation is written as $y = \beta X + e$, instead of $y = X\beta + e$, as in the text, only so that X can be shown as a long horizontal matrix with two rows instead of a tall vertical matrix with two columns. (Remember that matrices are multiplied by multiplying row elements with column elements and summing the results.) Note that this arrangement also changes the order of the terms in the equation to determine the values in β.

being determined). The notation used here, X^T, indicates the transpose of X, which means that X is flipped so that the rows and columns are swapped, and $(X^TX)^{-1}$ means the inverse of (X^TX). The fit variance depends on the noise, or other spurious fluctuations in the measured signal, and on how well the chosen design matrix, X, matches the measured pattern of values. The other factor in the estimate of the variance of β, however, is $(X^TX)^{-1}$, which depends entirely on our choice of design matrix. The *efficiency* of the design is described as being inversely related to the variance of the β value(s), (which is also equal to the standard error squared) (22,23):

$$\left(standard\ error\right)^2 = \frac{1}{efficiency} = \left(\sigma^2 \left(X^T X\right)^{-1}\right)$$

The value of (X^TX) is a square matrix with diagonal elements that are equal to the sum of the squares of the values in the patterns in X. A number of papers have been published on this topic, and several provide a detailed analysis of the design efficiency, showing that we can characterize the efficiency regardless of what type of study design is used (i.e., block design, event-related, or a combination) (22–24).

As a side note, the reader might notice that in comparison with the equation used by Mechelli et al. (22), the equation above omits the *contrast* term, c, which makes the right-hand side of the equation $\sigma^2\ c^T(X^TX)^{-1}\ c$. The contrast term indicates the weighting to give each pattern in the design matrix, so that different responses can be compared, summed, contrasted, or just ignored. With more than one pattern in the design matrix, $X = [P_1\ P_2\ P_3\ ...]$ where each pattern is a different column in X. For example, if the fMRI time-series data contain 100 time points, then each pattern to model the response contains 100 time points; and if there are three patterns, then X will be a 100×3 matrix, and c will be a column of three numbers (i.e., a 3×1 matrix). So, the contrast, c, makes the equation more general, as follows (24):

$$\frac{1}{efficiency} = trace\left(\sigma^2 c^T \left(X^T X\right)^{-1} c\right)$$

Here *trace* indicates the sum of the diagonal elements of the resulting matrix. However, for now we will consider only one pattern, P (i.e., a single trial type), in the design matrix X.

The design efficiency for a particular timing pattern, P, can therefore be expressed as ΣP^2, meaning the sum of the square of each time point in P. Keep in mind that this does not describe the total variance of the fit parameters, β, that will be determined with a GLM but accounts for the component of the variance that we can control with our choice of design. To observe the effects of choosing different durations of stimulus blocks and baseline periods, and different numbers of stimulus blocks, and so forth, a wide range of block designs were simulated and the design efficiencies are shown in Figure 7.8. The duration of the time courses was kept constant at 600 sec, and the repetition time, TR, was set at 3 sec, resulting in a total of 200 time points. The variations in efficiency in this figure therefore do not include the effects of different sampling rates or durations, as discussed below. The effects of modeling the signal intensity patterns with and without convolution with the BOLD HRF are also shown. These comparisons consistently show that regardless of the number and duration of stimulus blocks, the highest efficiency is achieved when the stimulation periods account for approximately 50% of the total duration. Again, it is worth noting that this is for the case of one trial type (compared with some baseline or reference condition).

A detailed discussion of the design efficiency with event-related designs was presented by Dale (23) and by Friston et al. (24) and demonstrated the importance of varying the interstimulus interval (ISI) as opposed to setting it at one fixed value. The same dependence is shown in Figure 7.9, although the exact conditions used by Dale were not replicated. For comparison with Figure 7.8, the estimated efficiencies are also plotted against the proportion of the time series that is during stimulation conditions, for a range of event-related designs. However, this comparison is only shown for the case when the timing pattern was not convolved with the HRF. Again it can be seen that the highest efficiency is achieved when the stimulation periods account for approximately 50% of the total duration of the time series. Friston et al. point out the considerable advantages of using shorter ISIs for event-related designs, such as allowing the person being studied to maintain a particular cognitive or attention state and reducing the time available to develop alternative strategies for carrying out the tasks. By varying the stimulus presentation rate, the person's ability to anticipate the stimulus or performance of a response is also reduced.

The stimulus onset asynchrony (SOA) is the amount the ISI is varied between stimuli, and there is a wide range of options for how the SOA can be manipulated during an fMRI acquisition. Even with the SOA varying throughout the time series, the events may occur at predetermined times, and these fMRI designs are called *deterministic*; whereas the SOA can be varied by setting the probability that an event will occur at a series of time points, and these designs are termed *stochastic*. To allow even more variability in the fMRI design, the probability that

---

FIGURE 7.8   *(See facing page.)* The *design efficiency* is estimated by the value of (P^T P) or ΣP^2, where P is the pattern in time (i.e., the *paradigm*) used to model the expected fMRI signal changes. Here P is set to a constant duration of 200 time points and is modeled with a repetition time of 3 sec. Each paradigm is modeled initially as a block paradigm, consisting of alternated baseline and stimulation periods, with values of 0 for lower signal periods (baseline) and 1 for higher signal periods (stimulation). The block paradigm is convolved with the hemodynamic response function (HRF) for the plot on the left, but is left as a block pattern for the plot on the right. The average value of the time series is subtracted from each point in order to set the average of P to zero. The value n in the legend indicates the number of stimulation periods that are modeled in the paradigm, and the estimated design efficiency is plotted as a function of the proportion of the time points that are modeled as being during stimulation conditions. The top frame of the figure shows four selected examples of patterns in time, P, which have the same number of time points modeled during stimulation conditions, but with different numbers of stimulation periods. This estimate of design efficiency indicates that the most efficient design has approximately 50% of the time spent in stimulation conditions and 50% in baseline periods.

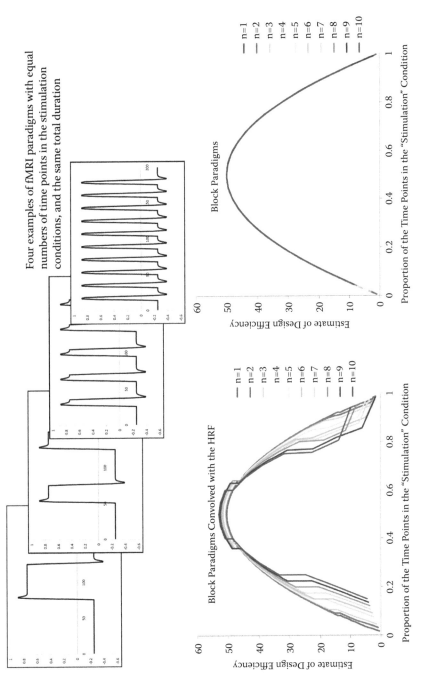

Four examples of fMRI paradigms with equal numbers of time points in the stimulation conditions, and the same total duration

The relative efficiency of the time pattern used for fMRI study design is estimated for a range of combinations of numbers of points in each of the two conditions, and the number of times the stimulation condition is applied, for a fixed total duration of the stimulation paradigm.

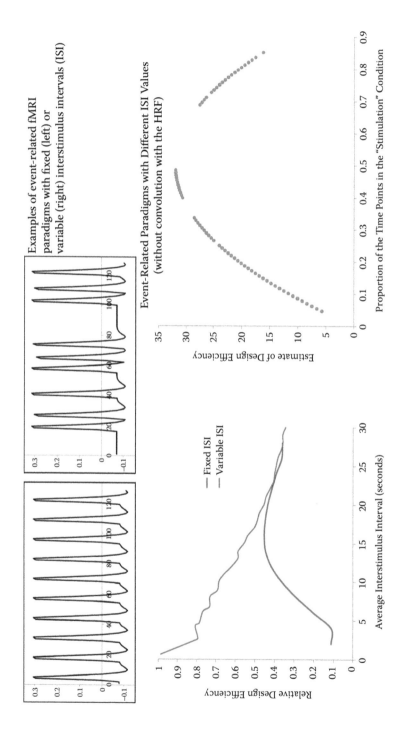

FIGURE 7.9 The design efficiency estimated for a range of event-related designs with fixed and variable interstimulus intervals (ISIs). The plot at bottom left shows the dependence of the design efficiency on the ISI, and on whether it is fixed or variable, as shown by Dale 1999 (23). The plot at the lower right shows that, as with the block designs, the efficiency depends on the proportion of time that is during stimulation conditions, with the optimal design having about equal time spent during stimulation and baseline periods. The upper frames show two examples of event-related designs, with fixed and variable ISIs.

an event will occur can be fixed (stationary) or can be varied in time (nonstationary). In their analysis of design efficiency, Friston et al. (24) show that nonstationary stochastic designs are more efficient than stationary designs for event-related fMRI, and they show that block designs (the same as events presented in very rapid succession) are the most efficient. The results of their analysis show that long SOA designs (requiring longer times between events or trials) are less efficient than designs in which the events or trials are presented rapidly. Also, slowly modulating the probability of an event or trial occurring at any point in time during the fMRI acquisition provides high design efficiency. Varying the ISI also provides *jittered* sampling of the BOLD response time course, meaning that the shape of the response is sampled at different time points across repeated events, and a detailed response curve can be constructed. The authors conclude that the slowly modulated SOA design "may represent a useful compromise between the high efficiency of block designs and the psychological benefits and latitude of stochastic designs" (p. 615). For stationary stochastic designs they show that the high design efficiency is achieved with 50% occurrence probability at each time point, when one trial type is used. This is similar to the result above, showing that with block designs, equal timing of stimulus/task conditions and rest conditions provide the highest efficiency. For stationary stochastic designs with two trial types, the optimum efficiency is with a roughly 30% probability of an event occurring. However, the authors of this study point out that the efficiency of an fMRI study design depends on the trial types, as well as on whether the effect of interest is the observed response to each trial, or the differences between responses, and so there is no single optimum design for all studies.

7.5.3 The Number of Time Points (Volumes)

The number of time points to acquire for an fMRI study (or in other words, the number of volumes to image) is another key choice that goes along with the choice of sampling rate discussed in the previous section. The total duration of the fMRI study is determined by the combination of the number of time points and the sampling rate. When considering the comfort of the person being studied, and the expectation that they can maintain their performance of a task, remain motionless, and so on, it would seem appropriate the keep the duration of fMRI studies as short as possible. However, it is also clear that more time points to describe the fMRI time series will provide higher sensitivity for detecting of BOLD signal changes. An analysis of this question by Murphy et al. (25) showed that there is a minimum number of required time points that depends on the effect size (the percent signal change to be detected), the signal-to-noise ratio of the fMRI data, and on the desired significance level. This analysis was based on the correlation between a model time series and the measured data, to test for a significant matching pattern in the data. A theoretical analysis showed that

$$N = \frac{2}{R(1-R)} \left(\frac{erfc^{-1}(P)}{(TSNR)(eff)}\right)^2$$

where N is the number of time points required, R is the ratio of time points in the on period (task/stimulation period), $erfc^{-1}$ indicates the inverse complementary error function, P is the statistical significance p-value, $TSNR$ is the temporal signal-to-noise ratio, and eff is the effect size. This equation shows, for example, that if a significance level of $p \le 10^{-6}$ is desired and the $TSNR$ of the fMRI data is 100 (or a 1% fluctuation due to random noise across time points), there are equal amounts of time spent during stimulation and rest conditions ($R = 0.5$), and the smallest BOLD response that is to be detected is 1%, then:

$$N = \frac{2}{0.5(1-0.5)} \left(\frac{erfc^{-1}(10^{-6})}{(100)(0.01)}\right)^2 = \frac{2}{0.25}\left(\frac{3.4589}{(100)(0.01)}\right)^2 = 96$$

In this situation then it is expected that at least 96 time points should be measured to achieve the desired sensitivity of the fMRI results. Although this is an estimate, Murphy et al. (25) showed by means of simulated data with various forms of noise, and data from six healthy volunteers, that this estimate is useful for guiding an appropriate choice of the minimum number of time points to acquire.

There are other ways of estimating the effect of varying the number of time points in fMRI data, such as by looking at the design efficiency discussed above. Regardless of the method used, it can be seen that higher numbers of time points produce higher sensitivity for fMRI. However, increasing the number of time points has diminishing benefits as the time series gets longer. This is illustrated by the example above, since theoretically we could detect a 3% signal change response with 11 time points, or a 2% response with 24 time points, and 1% with 96 time points. One goal of fMRI study design is therefore to maximize the number of volumes that are acquired to describe the time series in order to achieve the highest sensitivity, balanced by practical factors such as image acquisition parameters and the comfort of the person being studied. The number of volumes can be increased by sampling as quickly as possible by reducing the acquisition time, and by measuring a longer time series. As discussed above, the minimum acquisition time is determined by the image resolution, field of view, and so forth, and therefore has a practical lower limit. The total duration of the fMRI time series also has practical limits, given that the person being studied must remain motionless and be able to carry out the required tasks for a sustained period of time. It is also necessary that the conditions that are applied (tasks, sensations, etc.) do not change over the duration of the time series because of adaptation or learning of the task, onset of fatigue or boredom, wandering attention, and so on. Such changes over time in the person's cognitive state could alter the function being studied or evoke a time series of changes that interferes with the detection of the functions of interest. Variations in attention, fatigue, and the like may be avoided by designing the stimulus or tasks to be interesting and engaging for the person being studied. Still, practical upper limits on the length of each fMRI time series are around 10 minutes, although the limit depends considerably on the nature of the tasks.

Applying the estimation of the design efficiency as in the previous section, the dependence on the duration of stimulus/task blocks can be seen (Figure 7.10). Assuming the canonical BOLD hemodynamic response function as described above (and as shown in Figure 7.2), the efficiency of block designs increases with increasing block lengths, up to a length of around 15 sec. With blocks longer than 15 sec the efficiency is roughly constant or drops off slowly with increasing long blocks. With a fixed duration of the fMRI time series it can be seen that shorter TR values provide higher efficiency, because more time points are sampled. If instead the number of time points is held constant, so that the duration of the time series is longer with higher TR values, then the design efficiency changes only marginally with different TR values.

7.5.4 The Number of Events or Blocks

The number of stimulus/task blocks to use for block designs, or the number of events to use for event-related designs, is yet another important factor to consider for fMRI study design. Many of the factors discussed already in this chapter impact the choices of numbers of blocks, events, trials, and so forth (where a *trial* can be one event or can be applied repeatedly to create a block). For example, the numbers of blocks or events will certainly impact the duration of the fMRI time series, and this has already been discussed. For this discussion, we could consider that the number of time points to be acquired, the sampling rate, and the total duration of the fMRI acquisition have already been decided so that we do not need to rediscuss these factors.

For block designs, the solution becomes fairly simple given that the optimum design efficiency has been shown (Figures 7.8 and 7.9) when equal amounts of time are spent in each type of block, including rest or baseline periods as a type of block. That is, when two stimulation types, or one stimulation type and a rest period, are used, then 50% of the time should be spent applying each of these two conditions.

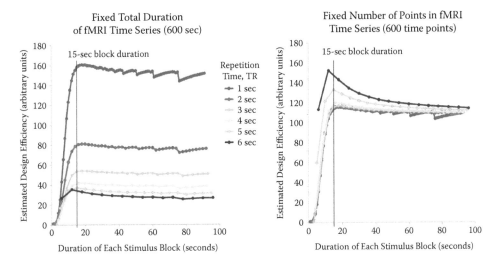

FIGURE 7.10 Estimated efficiencies of fMRI paradigm designs depending on the durations of stimulus/task blocks. Rest periods and stimulation/task periods have equal durations, to correspond with the highest efficiency designs shown in Figure 7.8. The left plot shows the efficiencies calculated for a fixed time-series duration, so that longer TR values have fewer sampled points and lower efficiency. The right plot shows efficiencies calculated for time series with a fixed number of points so that longer TR values have longer duration time series. The canonical hemodynamic response function (HRF) is used for these estimates. The results indicate that the efficiency is highest for blocks that are roughly 15 sec long or longer. The discontinuities in the plots with shorter TR values are expected to be an artifact of the calculation method.

For event-related and fast event-related designs, there are considerably more variables to consider. However, Murphy and Garavan (26) carried out an analysis to determine how many blocks should be used in event-related designs. The measure that was adopted to indicate the effectiveness of a design was the number of significantly active voxels. The authors assumed that more active voxels indicates higher sensitivity and therefore greater effectiveness of the study design. Because the number of active voxels detected can vary with the analysis method, as will be discussed in Chapter 8, they also applied three different analysis approaches to their data. The data were obtained in repeated studies from a single volunteer, and then they verified the sensitivity of the different analysis methods with data from 25 people. In each fMRI acquisition a single trial type was applied and was either a visual stimulus with a black-and-white flashing checkerboard or a combined task in which the person performed a bilateral finger-tapping task while the flashing checkerboard was presented. The results of their analysis based on a GLM indicate that stable fMRI results (i.e., consistent results across repeated acquisitions) can be achieved with approximately 25 events. The same authors have also pointed out that if more than one type of event is being studied, then 25 of each type will be needed in the study design (27).

Key Points

17. Each image or set of images is acquired over a period of a few seconds and represents one *time point* or *volume* in the time series.
18. Lower TR values produce fMRI results with higher statistical power, and optimally the TR should be relatively short at less than 1.5 sec.

19. Reducing the TR may require some reduction of spatial information, typically by reducing the number of slices.

20. Higher numbers of time points produce higher sensitivity for fMRI. However, increasing the number of time points has diminishing benefits as the time series gets longer.

21. The primary factor limiting the timing for fMRI studies is the speed of the BOLD response itself, because it takes ~5 sec to reach a peak and a total of ~25 sec to return to baseline.

22. The highest efficiency for block designs is achieved when the stimulation periods account for approximately 50% of the total duration for one trial type compared with some reference condition, and approximately equal duration in each condition when more conditions are applied.

23. With event-related designs the highest efficiency is achieved when events occur at a probability rate of approximately 50%, or occur in 50% of the time points.

24. Trials in event-related designs should be 25 sec apart to avoid overlapping BOLD responses.

25. Trials in fast event-related designs should be at least 4 sec apart so that responses to successive trials can be estimated as a linear combination (sums) of overlapping responses.

26. Nonstationary stochastic event-related designs (meaning the probabilities of events occurring are varied) are more efficient than stationary stochastic designs (fixed probabilities).

27. The efficiency of block designs increases with increasing block lengths, up to a length of around 15 sec.

28. The number of blocks for block design is determined by the duration (>15 sec) and having equal time spent in each condition (including rest conditions).

29. The number of trials for event-related designs is approximately 25 events of each trial type or more.

7.6 Summary of Factors Influencing fMRI Study Design

Optimal fMRI study design will include the greatest number of measurement points to describe the time series of signal changes, as long as it is practically achievable. The rate of sampling is determined by MRI system capabilities and the chosen imaging parameters, such as the volume that is imaged, and efforts should be made to sample as quickly as possible (shortest possible TR). The analysis of design efficiency indicates that having an equal number of time points during baseline and task/stimulation conditions is optimal for block designs, that block designs are more efficient than event-related designs, and event-related designs with variable ISI are more efficient than event-related designs with a single fixed ISI. The duration of the stimulation conditions and how often they are alternated with baseline periods depends primarily on the physiology, that is, the rate of BOLD signal change and how long the desired function can reasonably be sustained within the environment of the MRI system. However, as rough guidelines, the discussions above indicate that block designs appear to be optimal with blocks that are ~15 sec long or longer, trials in event-related designs should be ~25 sec apart, and trials in fast event-related designs should be ~4 sec apart or more.

Clearly, there is no single fMRI study design that is optimal for all situations, and the design must be adapted for each neuronal function of interest as well as to the types of questions that are being asked about this function. The questions can focus on sustained or transient features of neural functions, or interactions between functions, common features between different responses, and so on. The power of fMRI and the flexibility with which it can be applied is therefore a great advantage, but it can present a considerable challenge when trying to determine the best fMRI design for a specific application. However, by understanding the key elements that affect the sensitivity and effectiveness of fMRI results, and by having a well-defined question to investigate, many of the design choices become clear. One potential source of help with the design process is to use an fMRI simulator, such as is available at: http://www.cabiatl.com/CABI/resources/fmrisim/.

Key Points

The challenges presented for designing an effective fMRI study can be reduced by:

30. Having a well-defined question to investigate.
31. Understanding the key elements that affect the sensitivity and effectiveness of fMRI results.
32. Using an fMRI simulator: http://www.cabiatl.com/CABI/resources/fmrisim/.

References

1. Barry RL, Williams JM, Klassen LM, Gallivan JP, Culham JC, Menon RS. Evaluation of preprocessing steps to compensate for magnetic field distortions due to body movements in BOLD fMRI. *Magn Reson Imaging* 2009.
2. Leung HC, Skudlarski P, Gatenby JC, Peterson BS, Gore JC. An event-related functional MRI study of the stroop color word interference task. *Cereb Cortex* 2000;10(6):552–560.
3. Eatough EM, Shirtcliff EA, Hanson JL, Pollak SD. Hormonal reactivity to MRI scanning in adolescents. *Psychoneuroendocrinology* 2009;34(8):1242–1246.
4. Tessner KD, Walker EF, Hochman K, Hamann S. Cortisol responses of healthy volunteers undergoing magnetic resonance imaging. *Hum Brain Mapp* 2006;27(11):889–895.
5. Amaro E Jr, Barker GJ. Study design in fMRI: Basic principles. *Brain Cogn* 2006;60(3):220–232.
6. Friston KJ, Josephs O, Rees G, Turner R. Nonlinear event-related responses in fMRI. *Magn Reson Med* 1998;39(1):41–52.
7. Chee MW, Venkatraman V, Westphal C, Siong SC. Comparison of block and event-related fMRI designs in evaluating the word-frequency effect. *Hum Brain Mapp* 2003;18(3):186–193.
8. Bandettini PA, Cox RW. Event-related fMRI contrast when using constant interstimulus interval: Theory and experiment. *Magn Reson Med* 2000;43(4):540–548.
9. Glover GH. Deconvolution of impulse response in event-related BOLD fMRI. *NeuroImage* 1999; 9(4):416–429.
10. Boynton GM, Engel SA, Glover GH, Heeger DJ. Linear systems analysis of functional magnetic resonance imaging in human V1. *J Neurosci* 1996;16(13):4207–4221.
11. Donaldson DI, Petersen SE, Ollinger JM, Buckner RL. Dissociating state and item components of recognition memory using fMRI. *NeuroImage* 2001;13(1):129–142.
12. Visscher KM, Miezin FM, Kelly JE, Buckner RL, Donaldson DI, McAvoy MP, Bhalodia VM, Petersen SE. Mixed blocked/event-related designs separate transient and sustained activity in fMRI. *NeuroImage* 2003;19(4):1694–1708.
13. Madden DJ, Costello MC, Dennis NA, Davis SW, Shepler AM, Spaniol J, Bucur B, Cabeza R. Adult age differences in functional connectivity during executive control. *NeuroImage* 2010;52(2):643–657.
14. Menon RS, Ogawa S, Kim SG, Ellermann JM, Merkle H, Tank DW, Ugurbil K. Functional brain mapping using magnetic resonance imaging. Signal changes accompanying visual stimulation. *Invest Radiol* 1992;27 Suppl 2:S47–S53.
15. Ogawa S, Tank DW, Menon R, Ellermann JM, Kim SG, Merkle H, Ugurbil K. Intrinsic signal changes accompanying sensory stimulation: Functional brain mapping with magnetic resonance imaging. *Proc Natl Acad Sci USA* 1992;89(13):5951–5955.

16. Bandettini PA, Wong EC, Hinks RS, Tikofsky RS, Hyde JS. Time course EPI of human brain function during task activation. *Magn Reson Med* 1992;25(2):390–397.

17. Michielsen ME, Smits M, Ribbers GM, Stam HJ, van der Geest JN, Bussmann JB, Selles RW. The neuronal correlates of mirror therapy: An fMRI study on mirror induced visual illusions in patients with stroke. *J Neurol Neurosurg Psychiatry* 2010.

18. Westerhausen R, Moosmann M, Alho K, Belsby SO, Hamalainen H, Medvedev S, Specht K, Hugdahl K. Identification of attention and cognitive control networks in a parametric auditory fMRI study. *Neuropsychologia* 2010;48(7):2075–2081.

19. Mier D, Lis S, Neuthe K, Sauer C, Esslinger C, Gallhofer B, Kirsch P. The involvement of emotion recognition in affective theory of mind. *Psychophysiology* 2010.

20. Constable RT, Spencer DD. Repetition time in echo planar functional MRI. *Magn Reson Med* 2001; 46(4):748–755.

21. Worsley KJ, Friston KJ. Analysis of fMRI time-series revisited—Again. *NeuroImage* 1995;2(3): 173–181.

22. Mechelli A, Price CJ, Henson RN, Friston KJ. Estimating efficiency a priori: A comparison of blocked and randomized designs. *NeuroImage* 2003;18(3):798–805.

23. Dale AM. Optimal experimental design for event-related fMRI. *Hum Brain Mapp* 1999;8(2–3): 109–114.

24. Friston KJ, Zarahn E, Josephs O, Henson RN, Dale AM. Stochastic designs in event-related fMRI. *NeuroImage* 1999;10(5):607–619.

25. Murphy K, Bodurka J, Bandettini PA. How long to scan? The relationship between fMRI temporal signal to noise ratio and necessary scan duration. *NeuroImage* 2007;34(2):565–574.

26. Murphy K, Garavan H. Deriving the optimal number of events for an event-related fMRI study based on the spatial extent of activation. *NeuroImage* 2005;27(4):771–777.

27. Garavan H, Murphy K. Experimental Design. In: Filippi M, editor. *fMRI Techniques and Protocols*. New York: Humana Press; 2009. 133–149.

8

Functional MRI Data Analysis

The final step of a functional magnetic resonance imaging (fMRI) study is to analyze the data to reveal the anatomical locations and/or temporal properties of the neural function(s) of interest. Just as the study design has a wide range of flexibility, there are a variety of analysis methods to choose from. However, in many cases the study design and the analysis method must correspond, in order to provide the desired comparison or contrast of functions, reveal interactions, and so forth. Fortunately, there are a number of available sources of analysis software for fMRI, and most include all of the necessary features to prepare the data for analysis, apply the analysis, and then display the results. In addition, group analyses can be carried out, either based on the grouped time-series data or on the results of the individual analyses. The final outcome must then be interpreted to decide what conclusions can be drawn, and it is often very useful to again consider the original source of the fMRI data and how it links to changes in neural activity.

The ultimate goal of all fMRI data analysis methods is to detect magnetic resonance (MR) signal changes that are related to neural function, with the greatest possible reliability and sensitivity. This task would be trivial if the MRI signal intensity in each voxel in an image was perfectly constant, except for changes due to the blood oxygenation–level dependent (BOLD) effect or other neural activity–related contrast mechanisms. As described in previous chapters however, this is clearly not the case, and the signal intensity in each voxel can change due to random noise, physiological motion, changes in the person's position, and possibly changes in the MRI system hardware over time as well (such as with changes in temperature). The challenge for fMRI data analysis is therefore to determine which component of signal change, if any, can be accurately ascribed to changes in neural activity. The distinction between the signal changes of interest (related to neural activity) and confounding effects (motion, etc.) is made based on the different signal change patterns in time, and different spatial locations. As will be described below, the fMRI data can be analyzed by searching for particular patterns of signal change, or by detecting the spatial and/or temporal patterns of signal change in the data.

8.1 Hypothesis Testing

The purpose of fMRI data analysis is to determine whether or not a time series of MRI signal changes in a particular voxel or group of voxels shows a significant response to a stimulus or to the performance of a task. However, it is not enough to detect whether or not the MR signal intensity changed every time a certain task or stimulus was applied, but we need to know whether or not the change was consistent enough, and large enough, to be significant. If we were to repeat the fMRI experiment in the same person, or even a different person, and we evoked the same changes in neuronal activity, would we get the same result? Could the observed signal variation have been caused by random signal fluctuations (noise) or by movement (heartbeat, breathing, etc.)? These are the questions that we would like to have answered by the results of the analysis.

We can think of the analysis task as being to separate the measured signal into (1) the signal variations we are interested in (the signal), (2) the signal variations we can explain but are not interest in (the *confounds* or *covariates*), and (3) the signal variation that we cannot explain (the error). The signal intensity time-course from a single voxel can be a sum of all three of these sources of variation, or it might not contain the signal we are interested in. Prior to doing statistical testing, ideally we would want to remove the effects of confounds as much as possible, by means of the *preprocessing*, discussed below.

We want to know specifically whether or not the signal variations we are interested in show a significant response to a task or stimulus. We therefore need to ask a very specific question, and so we form a hypothesis about how the signal variations correspond to the predicted BOLD response for the neuronal activity of interest (Figure 8.1). However, to perform the statistical test, we use the *null hypothesis*—that the measured signal variations *do not* correspond with the BOLD response, or in other words, that the tasks or stimuli did not affect the measured MR signal. A statistical value is then calculated, such as the difference between the observed and predicted responses, divided by the standard deviation of the observed response, for example. This statistical value is used to determine the probability that the null hypothesis is true, based on knowing (or assuming) the distribution of the values that we could get, if the null hypothesis was indeed true. In other words, what is the probability of getting a value that is at least as extreme if the null hypothesis is true? If the probability is low enough (how low will be addressed next), then we reject the null hypothesis because it is unlikely that the calculated statistical value came from data for which the null hypothesis is true. This means that we reject the conclusion that the signal variations do not correspond with the expected BOLD response. While this is not the same as confirming that the signal changes do indeed correspond with the expected BOLD response, by rejecting the null hypothesis we accept

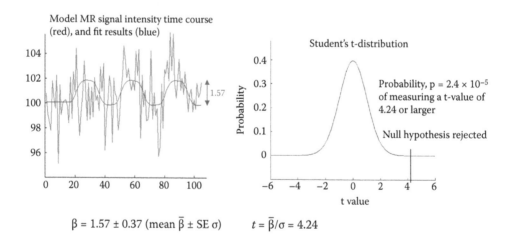

$$\beta = 1.57 \pm 0.37 \text{ (mean } \bar{\beta} \pm \text{SE } \sigma) \qquad t = \bar{\beta}/\sigma = 4.24$$

FIGURE 8.1 Example of a statistical test of the null hypothesis to determine the probability that the measured magnitude of a response, β, is equal to zero. Simulated data are used (shown in red in the plot on the left) for 2% BOLD response magnitude, in the presence of 2% noise (relative to the baseline intensity). The blue line in the plot on the left shows the fit response determined with a general linear model and the measured magnitude of the BOLD response, β. The plot on the right shows the probability distribution of t-values for data that agree with the null hypothesis. The results from the simulated data indicate that the probability is very low ($p < 10^{-5}$) that the null hypothesis is valid, and therefore the alternative hypothesis is accepted.

the alternative hypothesis, which is that the measured signal variations correspond with the BOLD response.

The limitations and inherent assumptions of any statistical test are important to understand to ensure that the results of the test are valid. For example, if we use a *t-test*, we must compare the result to *Student's t-distribution for the null hypothesis*; for an *F-test* we compare the result to an F-distribution; and if we use a *Z-test*, we need to compare the result to a normal, or Gaussian, distribution. In other words, it is important to know the expected distribution of values that could be measured when the null hypothesis is true for a given statistic. If the statistic we measure falls outside the expected distribution or rarely occurs, we may reject the null hypothesis. However, if the data do not meet the conditions that are assumed for the distribution, then the statistical test may not be valid. A common problem with fMRI data is the presence of slow or periodic trends that are not related to neuronal activity (called *autocorrelations* because the time-series data are correlated with themselves over some time lag). These trends increase the probability of obtaining a correspondence with the predicted BOLD time course, and a high statistical value, by chance. An important part of the analysis, as will be discussed in Section 8.3, is to prepare the data so that the statistical tests are valid, such as by removing or reducing auto-correlations. The choice of threshold for the probability that will result in a rejection of the null hypothesis is also important and will be discussed in detail in Section 8.5.

The results of testing the null hypothesis can produce four possible outcomes, based on whether or not the null hypothesis is true or false, and whether or not the statistical test indicates that the null hypothesis is true or false (Figure 8.2). If the null hypothesis is indeed true but the statistical test rejects the null hypothesis, then we have a Type I error, or a *false-positive* result, because the conclusion would be that the voxel being tested shows a significant BOLD response. If instead the null hypothesis is false but the statistical test fails to reject it, then we have a Type II error, or a *false-negative* result. One important goal of the analysis is to limit

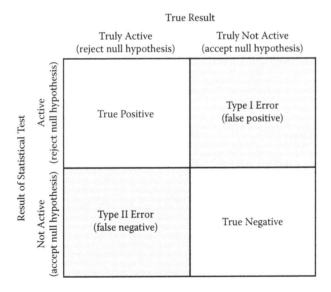

FIGURE 8.2 The possible outcomes of statistical tests of the null hypothesis for BOLD fMRI data. If the null hypothesis is rejected, we conclude that the signal intensity time course being tested shows a significant BOLD response, reflecting a change in neuronal activity in the voxel, and the voxel is therefore active. If this conclusion is incorrect, it is a Type I error, or false-positive result. If the null hypothesis is not rejected then we conclude that the signal intensity time course does not have a significant BOLD response and the voxel is not active. If this conclusion is incorrect, it is a Type II error, or a false-negative result.

the number of errors (Section 8.5), but it is necessary to recognize that the error rate is never expected to be zero.

8.2 fMRI Analysis Software

A number of software packages are available that perform all of the necessary steps for fMRI analysis. Still, to use these software packages it is necessary to understand the analysis steps because you will need to specify how you want the analysis to be done. For example, if a general linear model (GLM) is used, then it will be necessary to specify the timing of the stimulation paradigm that was used when your fMRI data were collected. That is, you need to tell the software what you are looking for in terms of the timing of the MR signal changes. It will be necessary to choose how the data should be prepared for analysis, and possibly also how you want the results to be displayed and so on. With an understanding of the basic principles underlying the analysis and the reasons for doing each of these steps, it is much easier to learn how to use any of these packages.

The list of software packages in Table 8.1 is not likely complete and is expected to change over time. However, the purpose of this list is to provide new practitioners of fMRI with the knowledge that such software is readily available, and with starting information to guide their research into which software package would suit their needs. Some packages are specifically for fMRI analysis and some include, or are entirely, other tools to help with analysis, processing, data format conversions, display, and so forth and can be very useful resources.

Table 8.1 List of Software Resources That Support fMRI Analysis		
Name (in alphabetical order)	Web Site	Basic Method(s) Used for fMRI Analysis (most also include other tools)
Analysis of Functional NeuroImages (AFNI)	http://afni.nimh.nih.gov/afni/	GLM
BrainVoyager	http://www.brainvoyager.com/	GLM and ICA
Fiasco/FIAT	http://www.stat.cmu.edu/~fiasco/	GLM
FMRIB Software Library (FSL)	http://www.fmrib.ox.ac.uk/fsl/	GLM and ICA
FMRLAB	http://sccn.ucsd.edu/fmrlab/	ICA
FreeSurfer	http://surfer.nmr.mgh.harvard.edu/	Morphometric analysis
MEDx	http://www.medicalnumerics.com/products/medx/index.html	GLM and ICA
NIfTI	http://nifti.nimh.nih.gov/	Data format conversion
NITRC	http://www.nitrc.org/	A large source of neuroimaging tools
Statistical parametric mapping (SPM)	http://www.fil.ion.ucl.ac.uk/spm/	GLM
VoxBo	http://www.voxbo.org/index.php/Main_Page	GLM

Note: GLM: General linear model; ICA: independent components analysis.

Some of these analysis packages that are distributed for free are made to function within the programming environment MATLAB™, which must be purchased. This means that you first start MATLAB running, and then within MATLAB you enter a command to run the fMRI analysis software. Programming environments are used to simplify the development of analysis software, and to help with sharing of the software across many different types of computers.

8.3 Preprocessing

The term *preprocessing* refers to the steps that are taken prior to analyzing fMRI data to remove sources of signal variation within the data that are not of interest (such as movement, etc.) and to prepare the data for statistical analysis. The order in which these steps are done is important and generally consists of (1) global normalization, meaning that the entire data set is corrected to remove variations in time or unwanted spatial features, (2) motion correction, (3) slice timing correction, (4) distortion correction or spatial normalization, (5) spatial smoothing, and then (6) temporal filtering. How (or if) these steps are applied can depend on the analysis software that is used.

8.3.1 Global Normalization

One very important goal of preprocessing steps is to remove slow or periodic trends, that is, *autocorrelations* as mentioned in Section 8.1, because these can increase the rate of false-positive results. *Global normalization*, or correction of the image intensity at each point of the time series, is intended to remove any trends that occur across the entire image volume.

Images acquired with a repetition time, TR, that is less than about three or four times the T_1-value of the tissues in the brain (see Chapter 4) will be T_1-weighted to some degree. However, a constant level of weighting is only reached after the first few time points are acquired. The number of time points needed to reach a steady state depends on the TR and the flip angle selected for the RF excitation pulse. For most fMRI acquisitions, with a TR of 1500 msec or more, and if the flip angle is set at the Ernst angle or lower, then by the third time point the MR signal intensity has reached a steady state. One of the first steps of preprocessing is to delete the first two time points (or more if necessary) to avoid the initial global reduction of signal intensity from entirely nonweighted to T_1-weighted.

Other sources of global signal intensity change can be reduced by means of the global normalization mentioned above. The general process is to multiply the entire three-dimensional (3D) image volume at each time point by a scaling factor to shift the average signal intensity to a constant value (1). Other mathematical processes for normalizing the signal intensity have also been proposed, but the end goal is the same. It is assumed that voxels with significant neuronal-activity–related signal changes are a small proportion of the entire volume that was imaged, and so variations of the average volume signal intensity are related to movement, MR hardware instability, and so forth and are not related to the signal changes of interest. By scaling the entire volume, the signal changes of interest will not be masked out. As an alternative, expected sources of global signal change, or low frequency trends, can be included as components in the GLM to account for these variations without scaling all of the image data (2,3). However, global normalization also has the potential to introduce error into the results and may not be suitable for every situation, particularly if activity related to a task or stimulus is correlated with global signal changes or if active regions are widespread (4).

8.3.2 Motion Correction

One common source of confounding effects is that of bulk movement of the person being imaged. This typically occurs when the person becomes uncomfortable or restless or when the task or stimulus being used involves a motor component. Subtle movement can also occur slowly over time as padding under the head or body becomes compacted over the course of an fMRI session. Such movement has the potential to create two effects: (1) the tissue represented by the signal within a given voxel is not constant throughout the fMRI time series, and (2) the time interval between two successive RF excitation pulses may vary if the tissues move between slice locations, potentially causing variations in T_1-weighting. The ideal solution is to ensure that the person being imaged remains motionless, by means of comfortable and supportive padding and keeping the duration of each fMRI time series at a reasonable minimum. In practice, a small amount of movement is almost inevitable, and it may be necessary to correct for movement that occurred between successive images by applying slight translations, rotations, and in some cases even changes in curvature to the images (such as with the spinal cord and brainstem).

The most commonly used method for aligning two MR images that are quite similar is the *affine transformation*. This is a method for mapping each voxel in an image to a slightly modified location in another image, where the amount of shift in voxel position may vary across the image. For motion correction of fMRI data, the relative positioning of each voxel and its immediate neighbors is typically changed by only small amounts, if at all, so that shapes of fine features in the image are not grossly distorted but can nonetheless be rotated, translated, and scaled in size. The shift in voxel position can be described mathematically as:

$$\vec{x} = A\vec{x} + \vec{b}$$

Here, $\vec{x}$ is the new two-dimensional (2D) vector of coordinates for the voxel in the shifted image, and $\vec{x}$ is the original vector of coordinates. The matrix A describes rotation and scaling, whereas $\vec{b}$ describes the linear translation of the voxel. For example:

$$\begin{array}{c} x \\ y \end{array} = \begin{array}{cc} \cos\theta & \sin\theta \\ -\sin\theta & \cos\theta \end{array} \begin{array}{c} x \\ y \end{array} + \begin{array}{c} x_0 \\ y_0 \end{array}$$

describes rotation through the angle θ around $(x, y) = (0, 0)$ and a shift by the amount (x_0, y_0). If θ is set to 10° and $(x_0, y_0) = (2, -2)$ then the point at $(x, y) = (5, 5)$ will be mapped to the point $(x', y') = (7.79, 2.06)$. This illustrates a key feature of the effects of applying spatial image transformations, that the voxels at regular 2D grid coordinates in the transformed image may need to be interpolated based on all of the voxels that are mapped to nearby locations. Alternatively, the *nearest neighbor* approach may be used and the voxel at the nearest grid point at $(x', y') = (8, 2)$ may be set to the value of the voxel that was mapped to (7.79, 2.06). Whichever approach is used, a small amount of spatial distortion and effective smoothing of the image data is expected to occur.

For the purposes of fMRI analysis, image registration methods are incorporated into available analysis packages, and these methods will determine the optimal spatial transformations to correct for shifts in position of the course of the time-series data. The transformations that are applied may be more complex than the example above, as they can vary with position, and allow for nonrigid distortions. In every case, however, the image registration will reduce motion effects, but will not eliminate them entirely because the corrections are imperfect, and because of variations in T_1-weighting. An important additional piece of information that is provided by the registration procedure, therefore, is a record of the timing and magnitude of motion that was detected over the fMRI time series. For GLM analysis, this motion record can be included in the set of basis functions to model any residual effects of the motion that may appear as variations in the MR signal. It is important to keep in mind that the greater the amount of displacement, the less the effectiveness of the image registration procedure and the greater the effects on the MR signal due to the change in position. Movements greater than 2 or 3 voxel dimensions may not be sufficiently corrected and the data may therefore be unreliable for detecting neural activity.

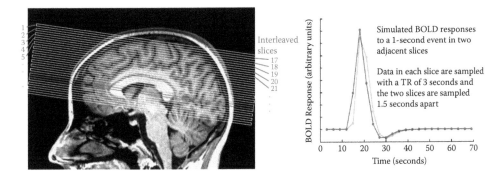

FIGURE 8.3 Slice timing correction demonstrated by means of simulated BOLD responses (without noise) to a 1 sec duration event, starting 10 sec after the sampling begins. Sampling is simulated every 3 sec, but at different times for the two slices. The figure on the left indicates the positions of 32 interleaved slices, and a simulated region with a BOLD response indicated in red, which spans several slices. Because the slices are interleaved and acquired in a 3 sec span (TR), adjacent slices are sampled 1.5 sec apart. If the sampling is not corrected for the different slice acquisition times and are assumed to be sampled simultaneously, then the two slices appear to show different shapes for the BOLD response time course.

8.3.3 Slice Timing Correction

Another preprocessing step that is commonly necessary is to account for the fact that all of the 2D images within an imaging volume at one time point of the fMRI series are not acquired at precisely the same time. If a single-shot imaging method is used, such as gradient-echo EPI, then the slices are imaged one at a time, most often in an interleaved order. Slices 1, 3, 5 … may be acquired first, followed by slices 2, 4, 6 … (as described in Chapter 5). The image acquisition will typically take up the entire TR period, with each 2D image acquisition spread out uniformly in time. The time interval between the acquisitions of slice 1 and slice 2 will therefore be approximately TR/2, and these two slices therefore represent different time points in the BOLD response to a change in neural activity (Figure 8.3). However, the slice timing depends on the slice order chosen, and possibly on the number of slices (odd or even number) as well. The actual slice order must be determined from the MRI system documentation or image header information, and must be defined for the analysis software to correct for the slice timing differences. To compensate for the differences in slice timing, one method is to interpolate the images that were acquired for each slice to estimate the data that would have been acquired if all of the image slices could be acquired simultaneously. This may not be necessary if the data are acquired very quickly or if long blocks are used. If long TR values are used, then the interpolation may not be very accurate. It is therefore necessary to decide whether or not slice timing correction by means of interpolation is necessary, or suitable, for a particular fMRI study. Another approach is instead to shift the predicted signal intensity response patterns to model the expected BOLD response at the times that each slice was acquired. However, this approach may not be practical for most fMRI studies, which may include a large number of slices. Another alternative, when using a GLM analysis, is to define more than one timing pattern to fit to the data to accommodate slight shifts in the recorded BOLD response.

8.3.4 Spatial Normalization

Spatial normalization refers to transforming images so that the anatomy shown for all people is the same shape and size. The goal of the normalization procedure is to have any selected voxel contain the MR signal from the same anatomical location in every person. The fMRI time-series data for that selected voxel can therefore be compared across a group of people to detect

Superior margin of   Posterior margin of   Vertical axis through mid-line
Anterior Commissure (AC) Posterior Commissure (PC)   plane orthogonal to AC-PC line

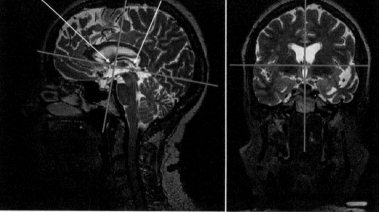

Horizontal axis
(R/L) orthogonal
to both vertical
axis and AC-PC
line

FIGURE 8.4 Definition of the axes and center of the Talairach coordinate system for representing positions of anatomical regions in the brain and for spatially normalizing brain image data.

similarities or differences. Spatial normalization of adult human brain images was originally based on the work of Talairach and Tournoux (5), who observed that the brain has consistent spatial proportions. They defined a coordinate system (Figure 8.4) with the horizontal axis along the line joining the superior margin of the anterior commissure (AC) and the posterior margin of the posterior commissure (PC) (the *AC-PC line*), and a vertical axis parallel to the mid-sagittal plane and orthogonal to this AC-PC line. The third axis is right–left and is orthogonal to both the horizontal and vertical axes. The origin of the coordinate system (0, 0, 0) is at the AC point. This standardized coordinate system for the human brain is called the *Talairach* coordinate system, and Talairach and Tournoux created a reference brain atlas based on a single brain from an elderly woman. This idea has since been further refined, and a new atlas and set of reference MRI brain images has been constructed from images of 305 human brains by researchers at the Montreal Neurological Institute (6). A slightly modified coordinate system has been defined and is termed the *MNI* coordinate system, and this system and related template images are used for most fMRI analysis.

The spatial normalization procedure is similar in principle to that used for image registration, but must allow for much larger scale image transformations given that brains from different people must be coaligned. One method used for spatial normalization is *ANIMAL* (automatic nonlinear image matching and anatomical labeling) developed by Collins and Evans (7), and other widely used methods are included in fMRI analysis software packages such as SPM and AFNI (listed in Table 8.1). The ANIMAL method is an iterative approach based on an initial course alignment of 3D image data with a reference data set, by means of three rotation, three translation, and three scaling parameters. In each subsequent iterative step of the normalization process, finer scale alignments are applied within increasingly localized regions. The result is the complete deformation field required to map the image data to the reference data set. Another widely used method for image registration is a 12-parameter affine transformation (8). The transformation needed to map one 3D image data set to another (such as a 3D reference image) is determined by means of an iterative process to minimize the differences between the two. With both of these methods, the affine transformation or the deformation field can also be applied in reverse to map a predefined region of interest masks, region labels, and so forth from a 3D reference image set to the original (i.e., nontransformed) image data.

The image normalization process can be applied to images at each time point of the fMRI time series prior to applying one of the analysis methods, or can be applied only to the results of the analysis. The advantage of normalizing only the analysis results is that any spatial smoothing imposed by the normalization process and any imperfections in the normalization will not affect the analysis itself. However, normalizing all of the data prior to analysis is useful for group analysis methods (discussed in Section 8.6), and normalizing the data at every time point serves the purpose of image registration as well.

In addition to spatial normalization for comparing fMRI results or group analyses, the deformation field and its inverse can be used for characterizing differences in brain structures and for image segmentation and labeling. Tissue probability maps (9) have been constructed from a large number of data sets, have been accurately labeled, and can be mapped back onto image data from an individual to identify probable areas of white matter, gray matter, and cerebrospinal fluid (CSF) (10). Structural characterization, such as measures of changes in cortical thickness as a function of age, has also been demonstrated as a very useful application of normalization methods (11–14).

8.3.5 Spatial Smoothing

Many of the mathematical processes used to extract the neuronal-activity–related signal change patterns from among the various sources of noise and confounds are based on the assumptions that the noise in the data is white noise with a Gaussian amplitude distribution, with a constant variance across all voxels (15–18). In practice, however, fMRI data include both random noise, which fits these assumptions, and also physiological noise, which is often related to cardiac- and respiratory-driven motion or physiological fluctuations. These structured noise sources do not match the underlying assumptions, and can influence the fMRI analysis by altering the estimates of the variance of the fit parameters. That is, even if the fit parameters (such as β-values) themselves are not affected by the structured noise, changes in the estimated variance of these values alter the statistical tests that are used to determine whether or not the amplitude of a particular response pattern is significantly different than zero. Recall that the T-statistic used to determine the significance of a component in the GLM is given by $T = \beta/\sqrt{\mathrm{Var}(\beta)}$. Therefore if the estimate of $\mathrm{Var}(\beta)$ is artificially low, the T-value is artificially high and may cause a false-positive result to be inferred.

To reduce structured noise, which may bias the results one way or another, a number of approaches have been proposed. One of the simpler approaches is to reduce the noise by spatially smoothing the fMRI data (Figure 8.5) with a Gaussian smoothing kernel with a full width at half maximum (FWHM) that is twice the voxel dimensions (16,19). *Smoothing* means reassigning each voxel in the image a value that is a weighted average of the original voxel value and the values of its neighboring voxels. Ideally, the width of spatial smoothing should match the expected extent of the active regions, but the size of active regions can be expected to vary across parts of

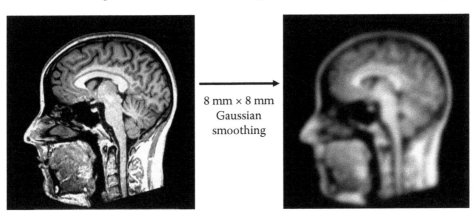

8 mm × 8 mm Gaussian smoothing

FIGURE 8.5 Effects of smoothing (i.e., spatial filtering) with an 8 mm Gaussian smoothing kernel.

the brain, brainstem, or spinal cord. Smoothing across too large an extent reduces efficiency and spatial accuracy and can eliminate the detection of smaller active regions. Spatial smoothing is not expected to eliminate structure in the noise but can reduce it nonetheless. Importantly, spatial smoothing has been shown to improve the estimates of noise structure, thereby improving the performance of methods to reduce the bias it can impose on statistical tests (16).

In addition to improving the signal-to-noise ratio of fMRI data, spatial smoothing can improve the correspondence of spatial locations of active regions across spatially normalized data sets, and can therefore improve the sensitivity of group analyses. It also reduces the number of independent tests that are done, and therefore impacts on the problem of multiple comparisons as discussed below. However, these improvements are at the expense of spatial resolution as mentioned above, and so the benefits and drawbacks of smoothing must be balanced, and an appropriate amount of smoothing should be used.

8.3.6 Temporal Filtering

Temporal filtering of the time-series data in each voxel is another way of reducing structured noise, such as high-frequency fluctuations or slow trends that cannot be attributed to the BOLD response (or other neural activity–related contrast). Applying a transformation to *whiten* the data (make the noise more random), as has been shown to be effective for independent components analysis (ICA) and principal components analysis (PCA) methods, requires a suitable estimate of the noise structure and has been suggested to be inappropriate for GLM methods as it can lead to invalid statistical tests (16). The optimal filtering strategy has been shown to be a *band-pass* filter (Figure 8.6), meaning that slow trends below a certain frequency (producing autocorrelations), and also rapid signal fluctuations above a certain frequency, are removed by the filtering process (16). The choice of the high- and low-frequency limits depends on the study design that is used, because relevant signal fluctuations that could be part of the BOLD signal response must not be removed. The low-frequency limit used in a number of published examples is at 0.008 Hz, which corresponds to a fluctuation with a period around 120 seconds. The appropriate high-frequency limit, for the low-pass part of the band-pass filter, depends on the timing between stimulus blocks or events and on the repetition time, TR.

Key Points

1. *Preprocessing*—Refers to the steps that are taken prior to analyzing fMRI data.
2. *Global normalization*—Refers to adjusting the average intensity of each image volume to some constant value in order to eliminate signal intensity variations in time that are consistent across all voxels.
3. *Motion correction*—Refers to aligning, or *registering*, images at all time points to correct for bulk body movement during the fMRI time series.
4. *Spatial normalization*—Is the process of rotating and spatially scaling image data to match the size, shape, and orientation of another image data set or a standardized set of reference data.
5. *Slice timing correction*—Is used to account for the fact that fMRI data from different slice locations is not acquired at the same time and therefore may not reflect the same timing of the signal intensity response.
6. *Smoothing*—Refers to spatial smoothing and means reassigning each voxel in the image a value that is a weighted average of the voxel and its neighbors.
7. *Filtering*—Refers to temporal filtering and is the removal of selected frequency components from the time-series data.
8. *Whitening data*—Means removing structure in the noise and making it closer to a purely random process.

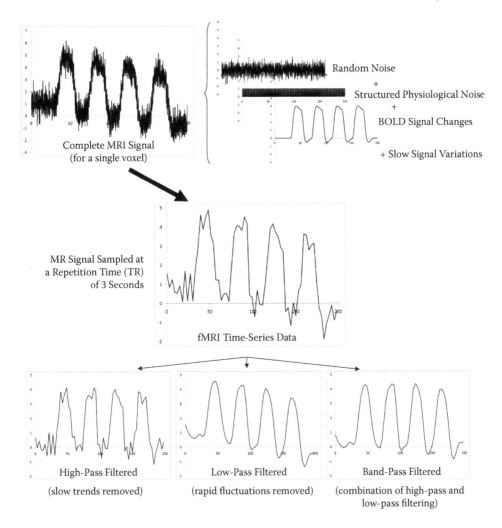

FIGURE 8.6 Effect of temporal filtering on simulated fMRI data for a single voxel.

8.4 Data Analysis Methods

A variety of methods can be used to analyze fMRI data (after preprocessing) to detect changes in neuronal activity and to compare or contrast responses. The methods can be categorized as either *model driven* or *data driven*. With model-driven fMRI analysis, one or more specific time-course responses are predicted, and a statistical test is used to determine which voxels, if any, correspond with these responses. In data-driven analysis, significant signal variations that are consistent across sets of voxels are detected to reveal both anatomical locations and temporal patterns that may be of interest. Model-driven methods used for fMRI are generally *univariate*, meaning that each voxel is tested independently. Data-driven methods, on the other hand, are *multivariate* because data from multiple voxels are analyzed together and compared to detect consistent components of signal variation. The choice of approach to use depends on the questions that are being asked about the data and whether or not a specific response pattern can be reliably predicted. The most common form of analysis used for fMRI is the GLM, which is a model-driven, univariate approach.

8.4.1.1 Correlation

One of the earliest analysis methods used for fMRI (20), which is still used to some extent, is based on the correlation between a predicted response and the measured signal intensity time course for each voxel. The degree of correlation is quantified with the R-value, which ranges over all values from –1 to 1, with a value of 1 indicating a perfect correlation and –1 meaning a perfect inverse correlation (the signal variations match perfectly but in the opposite direction), whereas a value of zero indicates no correlation whatsoever. The correlation value, R, is calculated with:

$$R = \frac{\sum (x - \bar{x})(y - \bar{y})}{\sqrt{\sum (x - \bar{x})^2 \sum (y - \bar{y})^2}}$$

where x is the predicted response, $\bar{x}$ is the average value of x, and y is the measured signal intensity time course, with an average value of $\bar{y}$, and Σ indicates the sum of all values. A lower limit threshold for R must be chosen, and any voxels with a higher R-value are then concluded to be significantly correlated with the predicted signal intensity time course. The correlation method is relatively easy to use and is sensitive to the shape of the predicted response and can be very specific. However, correlation values are reduced in the presence of confounding signal changes from other sources or if the predicted response is not accurate.

8.4.1.2 General Linear Model

The general linear model (GLM), as mentioned in previous chapters (Chapters 2 and 7), is the most widely used analysis method for fMRI because it is flexible and can reliably detect signal intensity variations of interest, even in the presence of confounding effects. The GLM is based on decomposing the measured signal intensity time series into a weighted sum of model time courses, called *regressors*. The result of the GLM for each voxel provides the weightings, or β-values, for each regressor, as well as the uncertainty (specifically the variance) of each weighting. The β-value for the regressor of interest, and its uncertainty, are used to test the null hypothesis that the voxel time-course data do not have a component of signal change that matches the predicted pattern, that is, β = 0. With the GLM, more than one pattern can be modeled for different possible responses to a sequence of stimulation or performance of a task, and patterns of signal change arising from physiological motion, or any other predictable confounding effects, can also be modeled and used as regressors (Figure 8.7). The GLM is described in detail in Section 2.3, and for the specific case of fMRI data analysis it can be written as:

$$X = H \cdot \eta + D \cdot \gamma + e$$

Here the notation used by Friston et al. (15) is adopted, with characters in bold indicating matrices, and X describes the time-course data from one voxel in a set of time-series fMRI data. H and D are matrices containing the expected patterns of signal intensity changes arising from neuronal activity in H and from confounding effects such as physiological motion in D. The values of η and γ are the parameters to be determined with the GLM, which indicate the magnitude of each of the modeled regressors to fit the time-series data, X. The expected patterns of signal intensity changes, H, are determined by the timing we choose for when a task is performed or a stimulus is applied, and are therefore expected to result in higher MR signal during stimulation conditions compared with during baseline periods, in active regions of the brain, brainstem, or spinal cord (as described in Chapter 6). The signal intensity changes in time arising from confounding effects, D, if they can be predicted, may be estimated from recordings of heartbeat and breathing during the fMRI time series or from motion correction steps, for example. The two matrices H and D are typically combined in the GLM, and can be written in the form used in previous chapters:

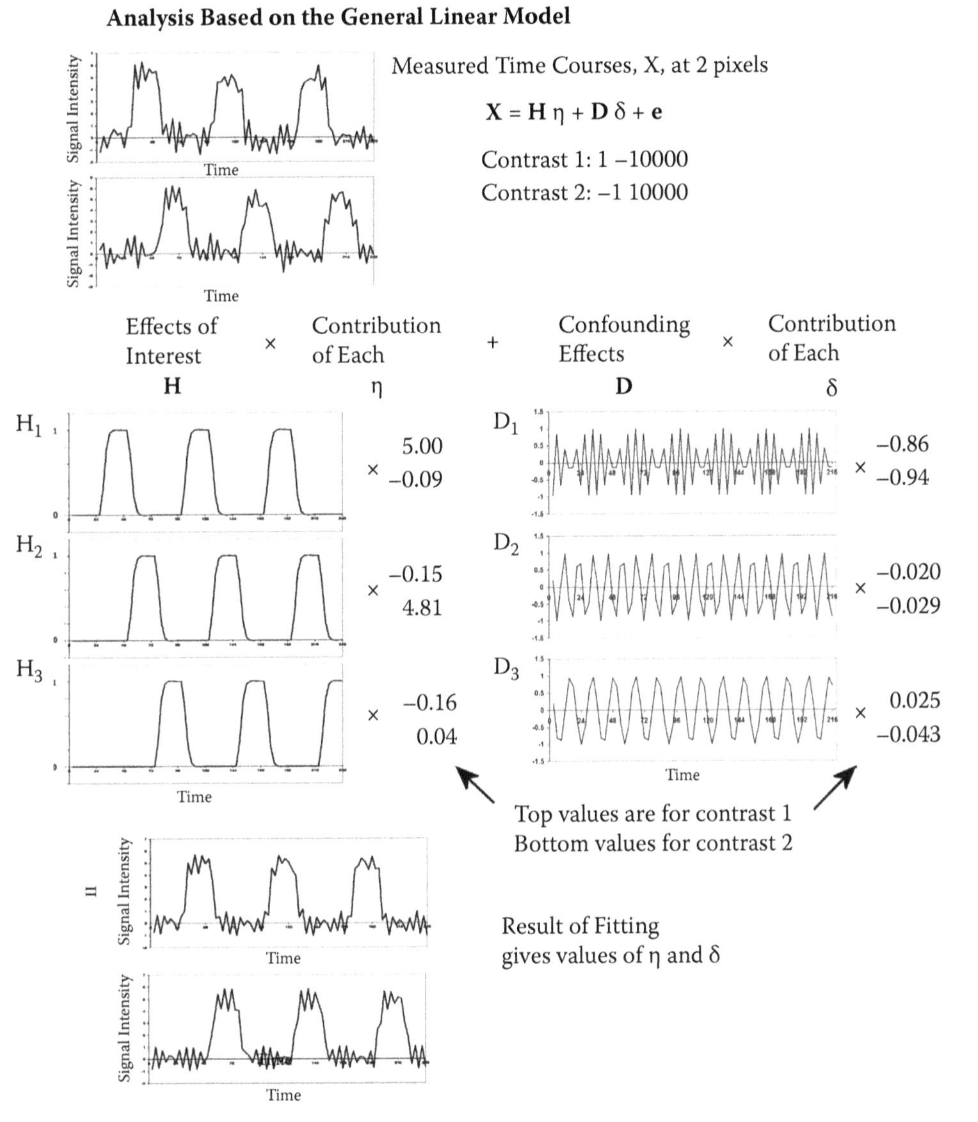

Analysis Based on the General Linear Model

Measured Time Courses, X, at 2 pixels

$$X = H \eta + D \delta + e$$

Contrast 1: 1 −1 0 0 0 0

Contrast 2: −1 1 0 0 0 0

Effects of Interest **H**	×	Contribution of Each η	+	Confounding Effects **D**	×	Contribution of Each δ

H_1 ⟶ × 5.00, −0.09

D_1 ⟶ × −0.86, −0.94

H_2 ⟶ × −0.15, 4.81

D_2 ⟶ × −0.020, −0.029

H_3 ⟶ × −0.16, 0.04

D_3 ⟶ × 0.025, −0.043

Top values are for contrast 1
Bottom values for contrast 2

Result of Fitting
gives values of η and δ

FIGURE 8.7 Example of GLM analysis of two modeled fMRI time series.

$$X = G \cdot \beta + e$$

Here the matrix G is equal to the matrices H and D concatenated together to make an $n \times t$ matrix, where n is the number of regressors and t is the number of time points in each regressor. The term β (beta) represents all of the parameter values, one for each of the patterns in G. This is the reason why the magnitudes of the fMRI responses are commonly referred to as β-values. These responses can be determined with the equation (15,21):

$$\beta = (G^T G)^{-1} G^T X$$

The uncertainty (specifically, the variance) of each β-value computed with this equation is given by:

$$\mathrm{Var}\beta = c \, \sigma^2 (G^T G)^{-1} c^T$$

The value of σ^2 is the sum of the squares of the values in e, divided by the number of degrees of freedom (i.e., approximately the number of values in X minus the number of β-values being determined, n). The matrix c is a one-dimensional (1D) list of n-values, being either 0s and 1s, to indicate which β-values are to be included. For example, if only the first β-value is being considered, then only the first time-course pattern in G is used in the calculation, and only the first value in c is set equal to 1 and all the other values are set to zero. A particular β-value can also be specified from the matrix of β-values indicated by β, with $β = c \cdot β$. For example, $β_1 = [1 \ 0 \ 0...] \cdot β$. For fMRI analysis we can use these values to compute the significance of a particular β-value, or combination of β-values, in terms of the probability that the result could be equal to zero (i.e., the null hypothesis), with the t-statistic from Student's t-test given by:

$$t = β/s_{\bar{x}}(β)$$

Here, $s_{\bar{x}}$ is used to indicate the sample standard deviation, or standard error, given by $\sqrt{\mathrm{Var}(β)}$. Alternatively, the significance may be expressed as the *standard score* or Z-score with $Z = β/s_x(β)$, where s_X is the population standard deviation. With a large number of samples, and therefore large degrees of freedom, the sample standard deviation approaches the same value as the population standard deviation, and t-values and Z-scores are very similar in this situation. Otherwise, Z-scores tend to be lower than t-values for the same level of significance.

The advantages of the model-driven approach with a GLM are that specific signal change patterns in time can be detected, even in the presence of confounding effects that produce very different signal fluctuations in time, and the results are quantitative. The disadvantages of this approach are that the expected pattern(s) of signal change arising from the neural activity of interest must be predicted, and a statistical threshold must be chosen to determine when a β parameter is *significantly* different from zero. The choice of the statistical threshold to infer significance is not trivial and is discussed in more detail below (Section 8.5).

8.4.2 Data-Driven, Multivariate Analysis Methods

With data-driven methods, as indicated by the name, the patterns of signal intensity change that occur consistently across voxels in the fMRI data set are detected and models of the expected responses are not needed (Figure 8.8). Typically the data preprocessing includes subtracting the average value of each voxel in the set, so that only the variation of the pattern in time or space affects the separation into components. Data-driven methods include principal components analysis (PCA) (18), independent components analysis (ICA) (22), and fuzzy clustering analysis (FCA) (23). One advantage of data-driven methods is that unknown or unexpected patterns of signal change in time in response to a sequence of tasks or stimuli can be determined. Another advantage is that these methods, being multivariate, can be applied to search for either the dominant temporal or spatial patterns of signal change, meaning that both when and where signal changes occur consistently can be detected in the fMRI data. A disadvantage of data-driven methods, however, is that the dominant pattern(s) of signal change within a set of time-series fMRI data may not be due to neuronal activity, but may be due to body movement or physiological motion such as related to the heartbeat and breathing. It is therefore necessary to assess the patterns that are detected to determine which may be attributed to neuronal activity and which may be ascribed to motion or other effects (18). This assessment may be based on involvement of specific regions of interest, or on expected properties of temporal responses, depending on the neural function of interest.

The results of data-driven methods (when used for temporal analysis) reveal the dominant patterns of signal change in time, within the fMRI data set, whereas for model-driven methods (i.e., the GLM) these patterns in time are predicted, or modeled, by the person doing the analysis. This means that if all of the signal change patterns are as expected, then the data-driven method results will be very similar to the predicted patterns used in the GLM. Data-driven methods

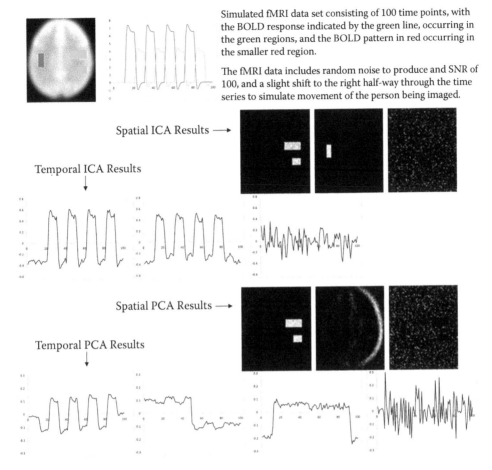

Simulated fMRI data set consisting of 100 time points, with the BOLD response indicated by the green line, occurring in the green regions, and the BOLD pattern in red occurring in the smaller red region.

The fMRI data includes random noise to produce and SNR of 100, and a slight shift to the right half-way through the time series to simulate movement of the person being imaged.

Spatial ICA Results ⟶

Temporal ICA Results

Spatial PCA Results ⟶

Temporal PCA Results

FIGURE 8.8 Examples of the application of principal components analysis (PCA) and independent components analysis (ICA) for fMRI.

are therefore typically used when the expected responses are unknown or cannot be accurately described for all regions of interest (24,25). However, even if the signal change patterns in time can be predicted with a good degree of accuracy, data-driven methods can provide additional valuable information about spatial patterns of responses, and therefore identify regions that can be inferred to be functionally connected.

PCA reduces the data set into uncorrelated components, with the first component describing the largest amount of variance in the data, and each successive component describing the largest amount of remaining variance after the previous components have been removed. As is implied by the name, ICA finds components that are independent and does not assume any dependency between successive time points in the temporal patterns of signal change (26,27). Here an important distinction is that *independent* means that any one of the time-course components cannot be written as a linear combination (i.e., weighted sum) of the other components, whereas *uncorrelated* means that none of the time-course components are correlated with any of the others. Independent time courses are uncorrelated, but uncorrelated time courses are not necessarily independent. Independence is a property of the set of components, not of any particular component or pair of components. FCA is different from both ICA and PCA because voxels are identified in groups, or *clusters*, based on their temporal signal change patterns (23,28,29). However, each voxel in the image data can be assigned to more than one cluster, and the probability of a voxel being a member in each cluster is identified. Hence, the term *fuzzy clustering* discriminates this method from clustering methods that assign each voxel to a single cluster.

The math behind the ICA, PCA, and FCA methods is unique for each method and can be complex; it is only introduced briefly here to explain the differences between the methods. In all data-driven methods, the data are first reorganized into a 2D matrix with the rows containing the time-series data for each voxel, and each column represents the voxel data at a particular time point. If there are n voxels and t time points, then the data matrix is $n \times t$ (i.e., a two-dimensional grid with n rows and t columns). Even though the original fMRI data are typically four-dimensional (4D) $(N_x \times N_y \times N_z \times t)$, with 3D image data acquired at each time point, the spatial information can be retained in the 2D matrix that is used for the data-driven analysis. The voxel data are listed in columns in order of their spatial positions in the image data, and consistent spatial patterns are therefore changed in shape. The new spatial patterns nonetheless remain consistent and can be identified, and can later be mapped back to their original shapes. The number of voxels, n, is therefore the product of N_x, N_y, and N_z. As mentioned previously, the data typically also need to be preprocessed by subtracting the mean values of each time series so that the variances, as opposed to the differences in average voxel intensities, can dominate the separation into components. The data are also typically *whitened* as described in Section 8.3.

In brief terms, PCA analysis proceeds by first subtracting the mean value of each voxel time series from each row of the $(n \times t)$ matrix described above (call this matrix X). A *covariance* matrix is constructed by multiplying this matrix by the transpose of itself, and dividing by the number of voxels minus one (i.e., $A = 1/(n - 1) X^T X$). This covariance matrix, A, is square $(t \times t)$, making it possible to calculate its *eigenvenvalues* and *eigenvectors* (*eigen* is a German word meaning inherent or characteristic). The matrix A has eigenvectors, **x**, and eigenvalues λ, that fulfill the equation:

$$Ax = \lambda x$$

The eigenvector *x* is one of the components that we are trying to find, and λ is a scalar value that indicates the relative amount of the variance in the data that is accounted for by *x*. The equation above means that *x* is not changed by multiplying it by A, but it is only scaled by a constant factor, λ. The set of all eigenvectors (up to t separate vectors, for a $t \times t$ matrix), which are in fact the principal components of the data, are therefore calculated. The resulting components are sorted in descending order of their eigenvalues, so that the component that accounts for the most variance is listed first, and so on.

Determination of independent components for ICA is even more complex than with PCA, and several variations of methods can be used, depending on the assumptions of noise in the data, the number of components to be determined, and so forth (17). The details of the math will therefore not be described here. In very brief terms, an iterative process is used to estimate, and then fine-tune, patterns in the data that are independent and can be summed linearly to equal each of the patterns in the original data. The ICA method can be used without knowing the underlying math and it is included in several fMRI analysis packages (mentioned in Section 8.2), or the ICA can be computed with software packages that are available such as *FastICA* from the Helsinki University of Technology, Laboratory of Computer and Information Science.

FCA is also done by means of an iterative process and treats each time series, consisting of t time points, as a point in *t*-dimensional space (23). The process can be visualized as though the *t*-dimensional point defined by the time series for each voxel is plotted, and then all of the points are clustered based on their position in this plot. Clusters of points with similar time series are thus identified, and the most appropriate number of clusters is determined. Some points may not clearly belong to one particular cluster and are labeled according to the probability of fitting within each of the clusters. Again, the math is too complex to describe in detail here, but the fuzzy clustering method can be used for fMRI analysis in software packages that are either shared or commercially available.

Theoretically, or perhaps ideally, all of the data-driven methods such as ICA, PCA, and FCA and the model-driven methods based on the GLM and correlation should yield similar results if the analysis is done to answer the same question or test the same hypothesis. That is, the results should accurately reveal the locations and time-series patterns of signal change that have occurred as a result of neuronal activity that was elicited by the pattern of tasks or stimulation. The main differences between the results would be the form in which they are revealed and how spatial or temporal patterns are grouped to show consistent features in the results. However, the choice of analysis method is often made based on the question being asked or the hypothesis being tested, and different methods are often applied to answer different questions and to yield different results.

8.4.3 Data Analysis for Resting-State Studies

Functional MRI studies of the resting state, that is, in which the person being studied performs no voluntary tasks at all, is an important emerging area that presents unique requirements for data analysis. While many resting state studies use ICA or PCA, as described in the preceding sections, other specialized analysis methods have emerged (30). While the PCA method has been used successfully for fMRI analysis, it provides the best results when the signals of interest have a large magnitude compared with other signal components, such as those from physiological motion and random noise, and when the sampling rate is fast relative to the BOLD responses. ICA appears, from the numbers of published papers on the topic, to be used more often than PCA for fMRI analysis. With any of these methods, the signal intensity time courses that are of interest must be selected from among the results. This selection is commonly done by rejecting components that fall outside certain frequency ranges of fluctuation or that are correlated with physiological noise, and components are selected that occur within areas known to match the default-mode network (31). Specific software packages, such as MATLAB (The MathWorks, Inc., Natick, MA), are also available for connectivity analysis with resting-state or stimulation fMRI data (32). Another alternative for the analysis of resting-state data is the *temporal clustering algorithm*, or TCA (33), and the two-dimensional variant 2dTCA (34), which can detect multiple different timing patterns within a data set.

The original TCA is applied to fMRI data on a voxel-by-voxel basis (33). An intensity threshold is first applied to exclude voxels outside of the head, and each voxel time series is then scaled so that values represent the percent signal change from the baseline intensity. A histogram is then constructed by counting the number of voxels that are at the peak value of their time series, at each time point of the data set. For example, if the fMRI data consists of N time points, then the histogram also contains N time points. The value of the histogram at time point t is then the number of voxels that were at their peak value at that time. The voxels containing random noise are expected to have their peaks distributed evenly across all time points, whereas the voxels with temporally related neuronal events (and related BOLD signal changes) are expected to have peaks at consistent time points. The peaks in the histogram are characterized by fitting a Gaussian function to determine the peak time and the width. The peak positions reflect the consistent resting-state function.

The 2D variant, 2dTCA (34), is also applied on a voxel-by-voxel basis, after smoothing with a 3-point averaging filter and normalizing values to percent signal changes from the baseline intensity. Voxels that are not expected to contain BOLD responses are determined as those whose peak percent signal change is outside the expected range of 0.5% to 8%, or with peak values less than two standard deviations above the average value over time (to find transient responses). These voxels that are then assumed to reflect signal changes that are not of interest are averaged to determine a global signal time course. The remainder of the voxels are expected to contain BOLD responses, and the global signal changes are subtracted from each voxel time course, and slowly varying trends are also removed.

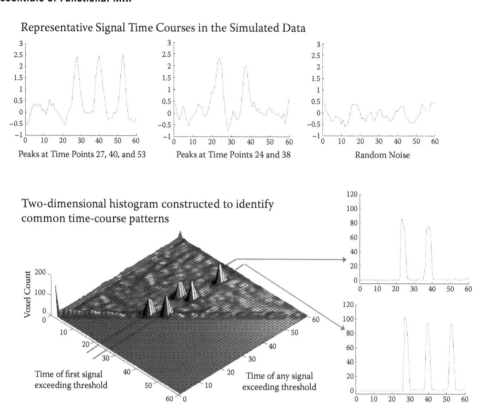

FIGURE 8.9 Description of the temporal clustering algorithm (TCA) for a simulated data set consisting of 1000 voxels: 10% of the voxels are set with transient responses peaking at time points 27, 40, and 53; and another 10% have peak responses at time points 24 and 38, plus random noise; whereas the remaining voxels contain only random noise.

The resulting signal intensity time courses are therefore expected to reflect the true BOLD responses of interest, and a two-dimensional histogram is constructed from these data. For time-course data consisting of N time points, a two-dimensional histogram of size $N \times N$, is created, called hist2d(x,y), as depicted in Figure 8.9. A threshold value is set for each voxel to be 1.5 times the standard deviation of the voxel value over time above the mean voxel value. The x-axis of the 2D histogram indicates the number of voxels that had their first significant signal increase over the threshold value at each time point, and the y-axis indicates the number of voxels for which the time point, y, is any significant value over the threshold, not just the first occurrence. That is, the value of hist2d(x,y) indicates the number of voxels for which the xth time point was their first significant value over the threshold and the yth time point exceeded the threshold. This is different from the one-dimensional (1D) histogram used in the original TCA method, which indicated only the number of voxels that were at their peak value at each time point.

Next, columns of hist2d that reflect the transient BOLD responses of interest are determined by looking at the diagonal elements of hist2d. These are the values where $x = y$, and points on the diagonal with large values reflect time points at which a large number of voxels had their first significant signal increase. The corresponding columns of hist2d are the resulting signal components or reference time courses. These signal components are then normalized by subtracting the mean and dividing by the standard deviation and used as basic functions in the GLM.

The main advantage of the 2dTCA over TCA is that voxels with similar time courses are grouped together, but those with dissimilar time courses are kept separate. The main disadvantage is that that reference time courses from transient physiological motion or other effects must be considered until other means can be found to rule them out. Another weakness of the method is that dissimilar time courses are not separated if they happen to have their first significant signal increase, even though subsequent signal increases do not coincide.

Key Points

9. *Model-driven* analysis methods are based on predicted models of responses, and parameters are determined that describe the fit to the measured fMRI data.
10. The *general linear model* is the basic model-driven method.
11. *Data-driven* methods do not require a response to be predicted, and predominant patterns of signal change are detected in the fMRI data.
12. Data-driven methods commonly used for fMRI include independent components analysis (ICA), principal components analysis (PCA), and fuzzy clustering analysis (FCA).

8.5 Statistical Threshold, and Correction for Multiple Comparisons

After the analysis methods have been applied, a critical step in the determination of the fMRI results remains: that is, to decide how well the responses that were detected must match the signal change patterns in time and/or anatomical locations that were expected for the fMRI response, for us to infer that the responses detected do indeed arise from changes in neural activity (i.e., we reject the null hypothesis). The results of the GLM provide a statistical measure that the responses matching predicted patterns of signal change have magnitudes that are significantly different from zero. The results of data-driven methods demonstrate the temporal MR signal patterns and spatial locations of the dominant sources of signal variance in the data set. These results must still be compared with expected patterns of signal change or expected locations of activity to determine whether or not they can be attributed to neural activity. As a result, the final analysis outcome can depend on a single subjective choice of the statistical threshold that is used to infer that a response is significant. However, with a well-designed study it can be possible that the responses detected are specific enough that varying the statistical threshold over a reasonable range will not significantly alter the final interpretation of the results. In most cases the choice of statistical threshold should be made with a careful balance of sensitivity and specificity while limiting the rate of occurrence of false-positive results. An excellent detailed review of this problem is provided by Bennett et al. (35). Methods for reducing the occurrence of false-positive (Type I errors) or false-negative results (Type II errors) include limiting the false-discovery rate (FDR) or using one of the methods for controlling the family-wise error (FWE) rate, as described below. The family-wise error rate refers to the probability of there being a false-positive result in the whole volume of results, such as across the entire brain. The false-discovery rate, on the other hand, is considered less restrictive and refers to the proportion of voxels in the data set that are expected to be false-positive results. Methods for controlling the error rate are included in all fMRI analysis software packages and in many cases require only that the desired level and type of control be indicated by the user.

Any of the statistics that are typically used, such as t-value, Z-value, and so forth, can be converted into a p-value, which reflects the probability of measuring another value that is at least as different from the reference value, given the known variance in the values. The p-value is useful because the expected number of false positives, as a result of noise in data, can be estimated. If the p-value threshold is chosen to be 0.001, for example, then the expected number of false positives is 0.1% of the number of voxels spanned by the data set. A data set spanning 100,000 voxels can therefore be expected to contain 100 voxels that appear to have a significant BOLD response, just by random chance. Controlling the number of false-positive results that are expected as a result of the large number of voxels that are tested therefore becomes very important. This is referred to as the *multiple comparison problem* or *multiple testing problem* 36). The *Bonferroni correction* is a widely used method for limiting the family-wise error rate by proportionally reducing the p-value threshold. For example, if the Bonferroni-corrected p-value threshold is desired to be at the level p < 0.05, and 10,000 voxels are analyzed, then the p-value threshold to use is $0.05/10,000 = 5 \times 10^{-6}$. This correction is considered conservative because although it will reduce the number of false-positive results, it may increase the number of false-negative results.

Another correction method is to control the false-discovery rate, which is the estimated proportion of false-positive results (37). In a set of results from V voxels, there are four different possible outcomes based on whether the voxels are actually active or inactive and whether they are declared as being active or inactive, as discussed in Section 8.1. The value V_{ia} describes the number of voxels that are truly inactive but were declared as being active, and V_{aa} is the number of voxels that were declared as being active and were truly active. The false-discovery rate is given by:

$$FDR = \frac{V_{ia}}{V_{ia} + V_{aa}}$$

But, only the total number of voxels that are declared as being active or inactive are known; whether or not they are truly active or inactive is not known. The expected false-discovery rate is estimated with:

$$E(FDR) \leq \frac{V_{ia}}{V_{ia} + V_{aa}} \quad q \leq q$$

To control the FDR one has to select a value for q (between 0 and 1) to indicate the maximum false-discovery rate that can be tolerated.

The p-values for all of the voxels must then be ordered from smallest to largest:

$$p_1 \leq p_2 \leq p_3 \leq \ldots \leq p_i \leq \ldots \leq p_V$$

Now find the largest value of i for which

$$p_i \text{ " } \frac{i}{V} \frac{q}{c(V)}$$

where V is the total number of voxels being test, and $c(V)$ is a constant that depends on the distribution of p-values in the set of voxels, with two possibilities:

1. $c(V) = 1$, or
2. $c(V) = \Sigma_{i=1}^{V}(1/t) \approx \ln(V) + 0.5772$

The first is valid when p-values across voxels are independent and the noise in the voxels is Gaussian with nonnegative correlation across voxels. The second is valid under less restrictive conditions and applies for any joint distribution of p-values across voxels. The second choice also provides a larger constant that leads to a lower p-value threshold.

In the first case above, with $c(V) = 1$, if our data set contains 10,000 voxels and we want a false-discovery rate q of 0.05 (analogous to the p-value used in the Bonferroni correction above), then the lowest p-value in the set of results must meet the condition:

$$p_1 \text{ " } \frac{1}{V} 0.05$$

and therefore $p_1 \leq 5 \times 10^{-6}$, just as with the Bonferroni correction. However, the next highest p-value in the set of results needs to meet the slightly less restrictive condition:

$$p_2 \text{ "} \frac{2}{V} \, 0.05$$

At some value of i the value of p_i no longer meets this condition, and so there are i–1 voxels detected as being significant.

Imposing lower limits on the spatial extent of active regions can also increase the significance of the results, because if false-positive results are caused by random signal variations, they are less likely to occur by chance in adjacent voxels (38). The errors (false positive or false negative) are expected to be distributed randomly in space across the image data. However, errors occurring due to physiological motion are more likely to occur near large blood vessels, cerebrospinal fluid, and edges of features where the image signal intensity changes abruptly. As a result, clusters of active voxels are not guaranteed to be accurate results, but false-positive results caused by physiological motion may be recognized based on their anatomical location. The process of determining the cluster size (i.e., the number of adjacent active voxels) at a given T-value threshold for each voxel that is needed to reach a desired significance level is somewhat complex. Fortunately, software methods to estimate the appropriate cluster size are freely available via the Internet, such as fmristat (39–41), and are included in some fMRI analysis packages. The important features to understand are that the cluster size is greater with more stringent desired significance levels, with greater extent of spatial smoothing applied to the data, and with lower numbers of degrees of freedom in the data.

Key Points

13. A statistical threshold must be chosen for the inference that the observed signal change patterns in time, and/or anatomical locations, arise from changes in neural activity.
14. The p-value statistical threshold reflects the probability of a false-positive result occurring by chance because of random noise.
15. Corrections for multiple comparisons are required because an fMRI data set consists of a large number of voxels, and therefore some false-positive results may occur by chance.
16. The *Bonferroni correction* accounts for multiple comparisons by dividing the desired p-value by the number of voxels to define a new, corrected p-value, and is one method for limiting the rate of family-wise errors (FWEs).
17. Another method of correcting for multiple comparisons is to control the *false discovery rate* (FDR) by adjusting the p-value.
18. By requiring that active regions occupy a certain minimum spatial volume (the cluster size threshold), the occurrence of false positive results is reduced, with a given T-value threshold.

8.6 Group Analysis

While fMRI results from an individual are needed for clinical assessments of function or to detect the effects of disease or trauma, research to characterize these effects in general and reference information from typical healthy people are also needed for comparison with the individual results. The general characterizations and reference information require group analysis of fMRI results to demonstrate the consistent features of the areas of activity and the magnitude of the signal change responses, as well as to detect the normal range of variation across people. All group analysis methods have the common features of combining results of individual analysis or the original fMRI data across multiple data sets on a voxel-by-voxel basis. The method of combination may be a simple average, the difference between two or more groups, or more complex comparisons with other measured properties such as the speed of performance of a task, rate of correct answers, scores on questionnaires, and so forth. An essential task before any group analysis can be done is therefore to normalize the results or data so that the anatomy of interest is the same shape, size, and position in the images. This is the same normalization process that is described in Section 8.3.4. Any given voxel is then expected to give information about the same anatomical location in all of the people/studies that are being combined.

8.6.1 Fixed-Effects, Random-Effects, and Conjunction Analyses

A number of methods have been developed for group analysis, and the choice of method depends on the reason for the group analysis or the question that is being asked. Three methods that have some common features are the *fixed-effects*, *random-effects*, and *conjunction* analyses. A *first-level* analysis is done first on the fMRI data from each individual using a model-driven method as described above, in order to determine the significance of the response to the stimulus or task, at each voxel. The resulting maps of statistical values from the individuals are then spatially normalized and combined with the group analysis, or *second-level* analysis.

The fixed-effects and random-effects analysis methods are applied to the estimates of the magnitude of the responses in each voxel. These are the β-values described above that are determined with the GLM. The fixed-effects analysis shows the significance of the average response across the group of subjects relative to the uncertainty of the response in each person, and it is assumed that the variance of each of the β-values is the same. The group response therefore describes the consistent features across the specific group of people that was studied, and it may not be valid to assume that this result can be generalized to all people. Random-effects analysis, on the other hand, shows the significance of the average response across the group of subjects relative to the variation of that response across the people that are studied. It therefore takes into account differences between people and can be used to make generalized conclusions about all people. Because the uncertainty in the response across people (more specifically, the standard error of the mean) has fewer degrees of freedom than the response measured in each individual, the fixed-effects analysis tends to give higher significance values and is considered to be more sensitive (42). Specifically, these values are calculated for each voxel with:

$$\text{Fixed effects:} \qquad T = \bar{\beta} \big/ \sqrt{\text{Var}(\beta)}$$

$$\text{Random effects:} \qquad T = \bar{\beta} \big/ \sqrt{\text{Var}(\bar{\beta})}$$

where $\bar{\beta}$ indicates the average of the β-values across the data sets (people, studies, etc.) being combined in the group analysis, $\text{Var}(\beta)$ is the variance of each individual β-value as described above (it is assumed to be consistent across data sets), and $\text{Var}(\bar{\beta})$ is the variance of the β-values across the data sets being combined.

An alternative approach is the conjunction analysis, which is intended to provide the sensitivity of fixed-effects analysis with the ability to make inferences about the population as a whole (42). This analysis method uses the T-value maps produced by the first-level analysis for each individual. The minimum T-value across all of the people or studies being combined is then taken for each voxel. The results therefore indicate that people in general can be expected to have a response that is at least as significant as the conjunction analysis value. Another conjunction approach to characterize the response within a predefined region of interest (such as some anatomy of interest) is to take the minimum T-value out of the maximum T-values for each voxel within the region.

8.6.2 General Linear Model for Group Analysis

A very different form of group analysis from those described above is to use a general linear model to detect all regions that responded significantly in a way that corresponds with some predicted response or with some measured parameters (43). The predicted response could simply indicate an expected difference between a patient group and a control group, could be based on measured parameters such as performance scores on a task (speed, reaction time, rate of correct answers, etc.), or could be responses on questionnaires that are completed either before or after the fMRI session (43). The exact response does not need to be predicted; only the relative expected differences between groups or individuals need to be indicated. For example, in a given voxel some participants in a study may show little or no signal change response and may rate a thermal sensation as causing no discomfort at all, while other participants rate the same thermal sensation and being uncomfortable and have a significant signal change response. The measured responses are therefore related to the rating of the thermal sensation, and the GLM would show a significant response at that voxel.

8.6.3 Partial Least Squares Analysis

A data-driven approach for group analysis is provided by the *partial least squares* (PLS) method (44,45). This method is applied to the image data from all participants, after the images have been spatially normalized. It is distinct in this way from the methods described above, because the PLS method is not applied to the results of a previous, first-level analysis, but rather to the image data itself. The data for each stimulation block, or each event (in an event-related study), and the subsequent rest or baseline period are typically averaged to produce a mean response consisting of t time points (Figure 8.10). The responses from all m voxels of interest in a data set are then represented by concatenating the data into a long vector of $m \times t$ data points. A matrix of all responses from n studies (which could be some number of fMRI data sets with different stimuli or tasks from each of a number of people studied) is then created. This matrix containing all of the data is therefore n rows by $m \times t$ columns, and for this desription will be called D.

The responses across different studies are compared by defining sets of contrasts, with one value for of the n studies. Mutiple independent contrasts can be defined to investigate multiple effects. For example, the n studies may consist of four different fMRI studies in each of n/4 people, and the first n1 people may be younger participants and the remaining n–n1 participants are older. A contrast to detect the differences between older and younger participants across all four study types would then be a list with the first $4 \times n1$ values set at 1, and the remaining $4 \times (n-n1)$ values set at –1. Another contrast, for example, could be set to detect only the differences in responses between study 1 and study 2 in each person, regardless of their age. This would be a list consisting of 1 –1 0 0 …, repeating for total of $4 \times n$ values. The set of all k different contrasts is formed into a matrix of k rows by n columns.

The matrix product of the contrasts, C, and the data, D (i.e., $C \times D$), is a matrix of size k rows by $m \times t$ columns, and represents the average differential responses according to the contrasts that were defined. The principal components of this matrix are then calculated, as described above in Section 8.4.2 on PCA. The principal components are the voxel *saliences* as described by McIntosh and Lobaugh (44) and represent the weight of the covariance contribution of each voxel. Since each principal component is also of length $m \times t$, the results can be redistributed into saliences of length t time points for each of the m voxels. Positive weights imply increasing signal intensity changes with increasing subject expression of a covariance pattern, while negative weights imply simultaneous decreasing signal intensity changes with increasing subject expression. As a result, both spatial and temporal differences in responses are determined.

The significance of the voxel saliences are determined with a bootstrap method in which the original data are resampled by substitution, without changing the order of the experiments for

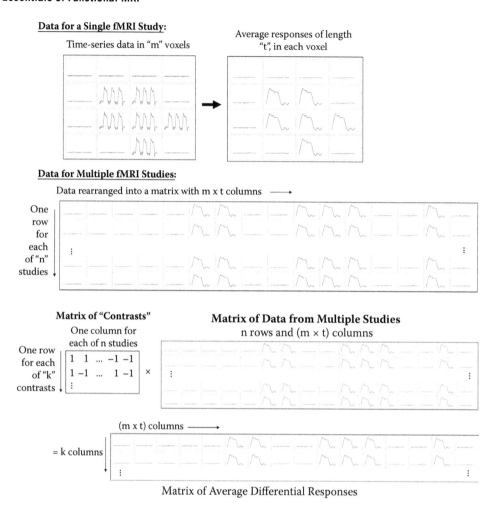

FIGURE 8.10 Graphical description of the partial least squares (PLS) method. The principal components of the matrix of average differential responses are then calculated to determine the salient features of the signal intensity responses.

each volunteer. For example, the data for volunteer 2 would be replaced with that from volunteer 1 in one permutation. A large number (~100–200) such permutations are analyzed and the standard error of the saliences is thus determined. The ratio of the salience to its standard error is termed the *bootstrap ratio* and provides an estimate of its significance. This analysis constitutes a single statistical test and so corrections for multiple comparisons are not required (44).

8.6.4 Functional and Effective Connectivity, and Dynamic Causal Modeling

Connectivity in the context of fMRI refers to detection of regions of the central nervous system (CNS) that are interconnected and function as a network. However, important distinctions are made between *functional* connectivity, which simply means that the fMRI responses detected in two regions are correlated, and *effective* connectivity, which is the influence one region has over another (46). Effective connectivity is therefore more informative than functional connectivity because it demonstrates how two regions interact, but functional connectivity is much easier to determine. Although a complete discussion of the methods used to determine connectivity

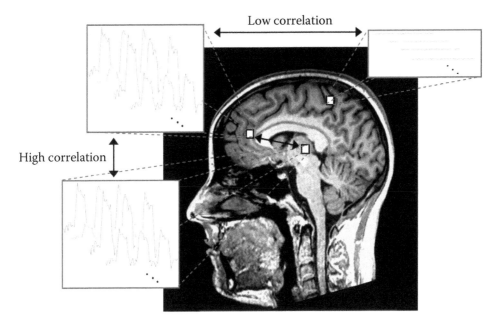

FIGURE 8.11 Illustration of functional connectivity based on the correlation between signal intensity time courses in each voxel, across multiple data sets, such as multiple people studied or repeated experiments.

would be useful, it is beyond the scope of this book, and so the detailed mathematical methods that underlie connectivity analyses are only introduced briefly here.

Functional connectivity can be detected by means of the correlation between time-series responses across all voxels in an fMRI data set (Figure 8.11). It is similarly demonstrated by the results of PCA and ICA as described in Section 8.4.2. These two methods demonstrate the voxels that have components of signal change in common and therefore can be inferred to be functionally connected. Of course, two regions can have highly correlated signal change responses because they receive common input from a third region, and the correlation does not imply a direct anatomical or effective connection. Connectivity inferred from a correlation between time-series data does not reveal the direction of connectivity either.

One method for estimating connectivity that is able to provide information about directionality, and therefore effective connectivity, is based on the principle of *Granger causality* (47). The basic principle proposed by Granger is that if we compare the time series of two voxels, X and Y (that is, applying Granger's idea to fMRI), then the processes in Y cause the processes in X if the inclusion of past observations of Y helps to improve the predictability of the next observations of X. The method requires high enough temporal resolution to include some information about causality in the data, and it has been applied to fMRI data with repetition times (TR) of 2 seconds (48) and 2.44 seconds (49), as examples. The application of this method for fMRI has been termed *Granger causality mapping* (GCM) (50,51). The distinction between GCM and connectivity based on correlation between two time-series responses is that the GCM method specifically addresses the question as to whether one area can be inferred to have caused the effect seen in the other. If subtle differences in the timing of responses are obscured by long repetition times or temporal filtering or noise, then the GCM method may show that we cannot infer causality, even if a correlation-based method shows that we can infer functional connectivity. This is an important distinction when looking at the results or when deciding which method or combination of methods to use for a particular analysis. The underlying math is relatively complex and

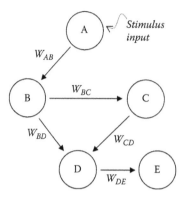

FIGURE 8.12 Example of structural equation modeling (SEM). The model is based on known or expected anatomical connections between regions of the CNS, which are here represented as *A*, *B*, *C*, *D*, and *E*. The responses in each region are modeled as a linear combination of inputs from other regions. The output of the SEM demonstrates the magnitude and significance of the weightings, *W*, of the inputs from one region to another.

so will not be described here, but a MATLAB toolbox for Granger causality mapping is freely available (50).

Effective connectivity can also be estimated by means of a method called *structural equation modeling*, or SEM (46,52) (Figure 8.12). This method can be used to quantify the interactions between a number of brain areas at the same time by making use of the correlations between the responses in different brain regions and the known anatomical connections between these regions. The observed fMRI time-series responses in each region are modeled as being linear combinations of inputs from the other regions, with weighting factors on the inputs defined for each connecting pathway. That is, some inputs may produce strong influence over another region, or the influence could be very weak or even zero. The weighting of an input can be either positive or negative as well. The result therefore demonstrates the relative direction and magnitude of the influence one region has over another, specific to the task that is used to generate the fMRI data, and within the limits of the known anatomical connections that are included in the model.

The SEM approach is similar to *dynamic causal modeling* (DCM) (53,54) (Figure 8.13) because both methods use anatomical information to define realistic limits on the possible

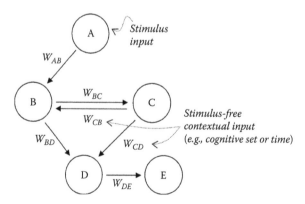

FIGURE 8.13 Example of dynamic causal modeling (DCM). (Adapted from Friston, K.J., et al., *NeuroImage* 19, 1273–1302, 2003 [53].) As with SEM, the model is based on known anatomical connections, and the weights of inputs are determined. However, the DCM allows for nonlinear interactions, as well as modulation of the weights.

connections between regions and attempt to explain the observed signal change responses in each region based on the influences of other regions. However, the DCM method allows for non-linear interactions, and one region can influence another region directly or can modulate the coupling between two other anatomical regions. Again, the output is a model of the anatomical network that is specific to the task used for the fMRI study and includes the effective connectivity between the anatomical regions of interest.

Regardless of the method used to determine connectivity and whether functional or effective connectivity is determined, the information provided can be very helpful for the interpretation of fMRI results. Simply knowing where the brain, brainstem, or spinal cord responds to a stimulus or is involved with the performance of a task may not provide an unambiguous demonstration of how the CNS functions or the effects of injury or disease. The additional information provided by the relationships between active regions can reduce the number of possible interpretations, or at least help to exclude errors caused by physiological motion. The interpretation of fMRI results is discussed further in the following section.

Key Points

19. A statistical threshold must be chosen for the inference that the observed signal change patterns in time, and/or anatomical locations, arise from changes in neural activity.
20. *Group analysis* methods are used to combine results of individual analyses, or the original fMRI data, across multiple data sets on a voxel-by-voxel basis.
21. *Fixed-effects* group analysis shows the significance of the average response across the group of subjects relative to the variation of the response in each person.
22. *Random-effects* group analysis shows the significance of the average response across the group of subjects relative to the variation of the response across the group.
23. *Conjunction* analysis takes the minimum significance value (T-value) across the group at each voxel.
24. *General linear model* (GLM) group analysis detects regions that responded significantly in a way that corresponds with some predicted response or measured parameters, such as task performance, questionnaire scores, and so forth.
25. *Partial least squares* (PLS) group analysis is a data-driven approach that is applied to the fMRI time-series data to detect consistent features of the responses across people or studies.
26. *Connectivity* refers to detection of regions of the CNS that are interconnected and function as a network.

8.7 Interpretation of fMRI Results—What Do They Really Mean?

As discussed in Chapter 6, the key physiological process underlying MRI signal changes that are related to neural activity is the change in energy consumption (metabolism) by both neurons and astrocytes (55,56). The function and energy consumption of the astrocytes and neurons are closely linked, and together they produce the net neural signaling (57–60). The energy consumption is determined by the amount of work done to release and recycle neurotransmitters, reestablish and maintain membrane potentials and intracellular and extracellular ion concentrations, and the like. As a result, the energy consumption is more closely linked to synaptic input than to the neuronal output spiking rate. To reiterate from Chapter 6, this relationship has been demonstrated by the observation that hemodynamic changes correlate better with local field potentials (LFPs) than either single-unit or multiunit recordings

(55,61–65). BOLD contrast, signal enhancement by extravascular water protons (SEEP), and diffusion-tensor imaging (DTI) signal changes arising from cellular swelling, and vascular changes shown by vascular space occupancy (VASO) and perfusion-weighted imaging (PWI) therefore reflect net changes in presynaptic input, whether excitatory or inhibitory. This relationship is also demonstrated by the dependence of relaxation times on oxygen metabolism in the brain (from Chapter 6):

$$R_2 \text{ and } R_2^* \propto V \, \frac{CMRO_2}{4 \, S_A \, CBF}^{\beta}$$

A strong coupling between LFPs and changes in tissue oxygen concentration has even been demonstrated in the absence of spiking output, supporting the conclusion that hemodynamic changes reflect synaptic more than spiking activity (63,64).

The best evidence to date therefore indicates that fMRI results, regardless of the contrast mechanism used, are most closely related to synaptic input than to the local neuronal spiking rate. Increases in either inhibitory or excitatory input can produce increased MRI signals, and decreases in either type of input can produce decreased MRI signals. For example, a region may receive tonic inhibitory input during the condition that is considered to be the baseline, or resting state, condition. During a task or stimulus this inhibitory input may be diminished to produce a response. The BOLD signal change would nonetheless be expected to be in the negative direction, because of the reduced synaptic input. Although this is a negative BOLD response, it would be incorrect to interpret it as either a deactivation or local inhibition. Similarly, a local positive BOLD response implies increased synaptic input, but this could be increased inhibitory input, resulting in a local reduction of neuronal activity. Functional MRI results can therefore be ambiguous, particularly if only a localized region of the CNS is studied. One possible way of improving the accuracy of the interpretation is to look at the fMRI responses in anatomically connected regions, as in the connectivity analyses described in Section 8.6.4. The local output of a region may be indicated by those regions that receive this output as their inputs, within the limitations that any region may receive multiple inputs.

Key Points

27. The key physiological process underlying MRI signal changes that are related to neural activity is the change in energy consumption (metabolism) by both neurons and astrocytes.

28. The best evidence to date therefore indicates that fMRI results, regardless of the contrast mechanism used, are most closely related to synaptic input than to the local neuronal spiking rate.

References

1. Gavrilescu M, Shaw ME, Stuart GW, Eckersley P, Svalbe ID, Egan GF. Simulation of the effects of global normalization procedures in functional MRI. *NeuroImage* 2002;17(2):532–542.
2. Friston KJ, Jezzard P, Turner R. Analysis of functional MRI time series. *Hum Brain Mapp* 1994;1:153–171.
3. Friston KJ, Holmes AP, Worsley KJ, Poline JB, Frith CD, Frackowiak RS. Statistical parametric maps in functional imaging: A general linear approach. *Hum Brain Mapp* 1995;2:189–210.
4. Aguirre GK, Zarahn E, D'Esposito M. The inferential impact of global signal covariates in functional neuroimaging analyses. *NeuroImage* 1998;8(3):302–306.

5. Talairach J, Tournoux P. *Co-Planar Stereotaxic Atlas of the Human Brain*. New York: Thieme Medical; 1988.

6. Evans AC, Collins DL, Mills SR, Brown ED, Kelly RL, Peters TM. 3D statistical neuroanatomical models from 305 MRI volumes. Proc IEEE Nucl Sci, Syposium and Medical Imaging Conference 1993:1813–1817.

7. Collins DL, Evans AC. Animal: Validation and applications of nonlinear registration-based segmentation. *Int J Patt Recog Artif Intel* 1997;11(8):1271–1294.

8. Ashburner J, Friston K. Multimodal image coregistration and partitioning—A unified framework. *NeuroImage* 1997;6(3):209–217.

9. Mazziotta JC, Toga AW, Evans A, Fox P, Lancaster J. A probabilistic atlas of the human brain: Theory and rationale for its development. The International Consortium for Brain Mapping (ICBM). *NeuroImage* 1995;2(2):89–101.

10. Hellier P, Barillot C, Corouge I, Gibaud B, Le GG, Collins DL, Evans A, Malandain G, Ayache N, Christensen GE, Johnson HJ. Retrospective evaluation of intersubject brain registration. *IEEE Trans Med Imaging* 2003;22(9):1120–1130.

11. Cocosco CA, Zijdenbos AP, Evans AC. A fully automatic and robust brain MRI tissue classification method. *Med Image Anal* 2003;7(4):513–527.

12. Evans AC. The NIH MRI study of normal brain development. *NeuroImage* 2006;30(1):184–202.

13. Im K, Lee JM, Lyttelton O, Kim SH, Evans AC, Kim SI. Brain size and cortical structure in the adult human brain. *Cereb Cortex* 2008;18(9):2181–2191.

14. Tisserand DJ, van Boxtel MP, Pruessner JC, Hofman P, Evans AC, Jolles J. A voxel-based morphometric study to determine individual differences in gray matter density associated with age and cognitive change over time. *Cereb Cortex* 2004;14(9):966–973.

15. Friston KJ, Holmes AP, Poline JB, Grasby PJ, Williams SC, Frackowiak RS, Turner R. Analysis of fMRI time-series revisited. *NeuroImage* 1995;2(1):45–53.

16. Friston KJ, Josephs O, Zarahn E, Holmes AP, Rouquette S, Poline J. To smooth or not to smooth? Bias and efficiency in fMRI time-series analysis. *NeuroImage* 2000;12(2):196–208.

17. Hyvarinen A. Fast and robust fixed-point algorithms for independent component analysis. *IEEE Trans Neural Netw* 1999;10(3):626–634.

18. Andersen AH, Gash DM, Avison MJ. Principal component analysis of the dynamic response measured by fMRI: A generalized linear systems framework. *Magn Reson Imaging* 1999;17(6):795–815.

19. Worsley KJ, Friston KJ. Analysis of fMRI time-series revisited—Again. *NeuroImage* 1995;2(3):173–181.

20. Bandettini PA, Wong EC, Hinks RS, Tikofsky RS, Hyde JS. Time-course EPI of human brain function during task activation. *Magn Reson Med* 1992;25(2):390–397.

21. Watson GS. Serial correlation in regression analysis. 1. *Biometrika* 1955;42(3–4):327–341.

22. McKeown MJ, Makeig S, Brown GG, Jung TP, Kindermann SS, Bell AJ, Sejnowski TJ. Analysis of fMRI data by blind separation into independent spatial components. *Hum Brain Mapp* 1998;6(3):160–188.

23. Baumgartner R, Ryner L, Richter W, Summers R, Jarmasz M, Somorjai R. Comparison of two exploratory data analysis methods for fMRI: Fuzzy clustering vs. principal component analysis. *Magn Reson Imaging* 2000;18(1):89–94.

24. Valsasina P, Agosta F, Caputo D, Stroman PW, Filippi M. Spinal fMRI during proprioceptive and tactile tasks in healthy subjects: Activity detected using cross-correlation, general linear model and independent component analysis. *Neuroradiology* 2008;50:895–902.

25. Esposito F, Bertolino A, Scarabino T, Latorre V, Blasi G, Popolizio T, Tedeschi G, Cirillo S, Goebel R, Di SF. Independent component model of the default-mode brain function: Assessing the impact of active thinking. *Brain Res Bull* 2006;70(4–6):263–269.

26. McKeown MJ, Sejnowski TJ. Independent component analysis of fMRI data: Examining the assumptions. *Hum Brain Mapp* 1998;6(5–6):368–372.

27. Viviani R, Gron G, Spitzer M. Functional principal component analysis of fMRI data. *Hum Brain Mapp* 2005;24(2):109–129.

28. Moser E, Diemling M, Baumgartner R. Fuzzy clustering of gradient-echo functional MRI in the human visual cortex. Part II: Quantification. *J Magn Reson Imaging* 1997;7(6):1102–1108.

29. Baumgartner R, Scarth G, Teichtmeister C, Somorjai R, Moser E. Fuzzy clustering of gradient-echo functional MRI in the human visual cortex. Part I: Reproducibility. *J Magn Reson Imaging* 1997;7(6):1094–1101.

30. Greicius MD, Srivastava G, Reiss AL, Menon V. Default-mode network activity distinguishes Alzheimer's disease from healthy aging: Evidence from functional MRI. *Proc Natl Acad Sci USA* 2004;101(13):4637–4642.

31. Cauda F, Sacco K, Duca S, Cocito D, D'Agata F, Geminiani GC, Canavero S. Altered resting state in diabetic neuropathic pain. *PLoS One* 2009;4(2):e4542.

32. Zhou D, Thompson WK, Siegle G. MATLAB toolbox for functional connectivity. *NeuroImage* 2009;47(4):1590–1607.

33. Liu Y, Gao JH, Liu HL, Fox PT. The temporal response of the brain after eating revealed by functional MRI. *Nature* 2000;405(6790):1058–1062.

34. Morgan VL, Li Y, bou-Khalil B, Gore JC. Development of 2dTCA for the detection of irregular, transient BOLD activity. *Hum Brain Mapp* 2008;29(1):57–69.

35. Bennett CM, Wolford GL, Miller MB. The principled control of false positives in neuroimaging. *Soc Cogn Affect Neurosci* 2009;4(4):417–422.

36. Nichols T, Hayasaka S. Controlling the family-wise error rate in functional neuroimaging: A comparative review. *Stat Methods Med Res* 2003;12(5):419–446.

37. Genovese CR, Lazar NA, Nichols T. Thresholding of statistical maps in functional neuroimaging using the false discovery rate. *NeuroImage* 2002;15(4):870–878.

38. Lieberman MD, Cunningham WA. Type I and Type II error concerns in fMRI research: Rebalancing the scale. *Soc Cogn Affect Neurosci* 2009;4(4):423–428.

39. Cao J, Worsley K. The geometry of correlation fields with an application to functional connectivity of the brain. *Ann Appl Prob* 1999;9:1021–1057.

40. Worsley KJ. An improved theoretical P-value for SPMs based on discrete local maxima. *NeuroImage* 2005;28(4):1056–1062.

41. Worsley KJ. Spatial smoothing of autocorrelations to control the degrees of freedom in fMRI analysis. *NeuroImage* 2005;26(2):635–641.

42. Friston KJ, Holmes AP, Price CJ, Buchel C, Worsley KJ. Multisubject fMRI studies and conjunction analyses. *NeuroImage* 1999;10(4):385–396.

43. Beckmann CF, Jenkinson M, Smith SM. General multilevel linear modeling for group analysis in FMRI. *NeuroImage* 2003;20(2):1052–1063.

44. McIntosh AR, Lobaugh NJ. Partial least squares analysis of neuroimaging data: Applications and advances. *NeuroImage* 2004;23 Suppl 1:S250–S263.

45. McIntosh AR, Bookstein FL, Haxby JV, Grady CL. Spatial pattern analysis of functional brain images using partial least squares. *NeuroImage* 1996;3(3 Pt 1):143–157.

46. Buchel C, Friston KJ. Modulation of connectivity in visual pathways by attention: Cortical interactions evaluated with structural equation modelling and fMRI. *Cereb Cortex* 1997;7(8):768–778.

47. Granger CWJ. Investigating causal relations by econometric models and cross-spectral methods. *Econometrica* 1969;37(3):414–&.

48. Deshpande G, LaConte S, James GA, Peltier S, Hu X. Multivariate Granger causality analysis of fMRI data. *Hum Brain Mapp* 2009;30(4):1361–1373.

49. Abler B, Roebroeck A, Goebel R, Hose A, Schonfeldt-Lecuona C, Hole G, Walter H. Investigating directed influences between activated brain areas in a motor-response task using fMRI. *Magn Reson Imaging* 2006;24(2):181–185.

50. Seth AK. A MATLAB toolbox for Granger causal connectivity analysis. *J Neurosci Methods* 2010; 186(2):262–273.

51. Roebroeck A, Formisano E, Goebel R. Mapping directed influence over the brain using Granger causality and fMRI. *NeuroImage* 2005;25(1):230–242.

52. McIntosh AR, Grady CL, Ungerleider LG, Haxby JV, Rapoport SI, Horwitz B. Network analysis of cortical visual pathways mapped with PET. *J Neurosci* 1994;14(2):655–666.

53. Friston KJ, Harrison L, Penny W. Dynamic causal modelling. *NeuroImage* 2003;19(4):1273–1302.

54. Kasess CH, Stephan KE, Weissenbacher A, Pezawas L, Moser E, Windischberger C. Multi-subject analyses with dynamic causal modeling. *NeuroImage* 2010;49(4):3065–3074.

55. Logothetis NK, Pauls J, Augath M, Trinath T, Oeltermann A. Neurophysiological investigation of the basis of the fMRI signal. *Nature* 2001;412(6843):150–157.

56. Figley CR, Leitch JK, Stroman PW. In contrast to BOLD: Signal enhancement by extravascular water protons as an alternative mechanism of endogenous fMRI signal change. *Magn Reson Imaging* 2010.

57. Fox PT, Raichle ME. Focal physiological uncoupling of cerebral blood flow and oxidative metabolism during somatosensory stimulation in human subjects. *Proc Natl Acad Sci USA* 1986;83(4):1140–1144.

58. Fox PT, Raichle ME, Mintun MA, Dence C. Nonoxidative glucose consumption during focal physiologic neural activity. *Science* 1988;241(4864):462–464.

59. Gordon GR, Choi HB, Rungta RL, Ellis-Davies GC, MacVicar BA. Brain metabolism dictates the polarity of astrocyte control over arterioles. *Nature* 2008;456(7223):745–749.

60. Pellerin L, Magistretti PJ. Glutamate uptake into astrocytes stimulates aerobic glycolysis: A mechanism coupling neuronal activity to glucose utilization. *Proc Natl Acad Sci USA* 1994;91(22):10625–10629.

61. Logothetis NK. The underpinnings of the BOLD functional magnetic resonance imaging signal. *J Neurosci* 2003;23(10):3963–3971.

62. Logothetis NK, Pfeuffer J. On the nature of the BOLD fMRI contrast mechanism. *Magn Reson Imaging* 2004;22(10):1517–1531.

63. Viswanathan A, Freeman RD. Neurometabolic coupling in cerebral cortex reflects synaptic more than spiking activity. *Nat Neurosci* 2007;10(10):1308–1312.

64. Logothetis NK. The ins and outs of fMRI signals. *Nat Neurosci* 2007;10(10):1230–1232.

65. Logothetis NK. What we can do and what we cannot do with fMRI. *Nature* 2008;453(7197):869–878.

9

Clinical Applications of Functional MRI

If functional magnetic resonance imaging (fMRI) is to be used routinely as a tool for clinical practice, it must be shown to produce reliable and robust results that can be accurately interpreted. It is also important that the method is practical to use in MRI units in hospitals. In this chapter a range of examples are presented showing how fMRI has been used to date for clinical practice, and in many more cases, how it has been used for research of specific patient groups and how it has been shown to have potential for clinical assessments. Toward this end, this chapter presents examples of how fMRI studies reported in the literature were designed and how the data were acquired, but it does not discuss the findings of the studies or the significance of the results, except to show the potential clinical value. Such discussions are better suited to specialized books or papers on each neurological disorder, as opposed to here where the focus is on the theory and practice of fMRI. The examples that are discussed are organized according to (1) the method used: resting-state studies and studies using saccadic eye movements, and (2) the neurological disorder that was studied: multiple sclerosis, stroke, consciousness disorders, traumatic brain injury, and spinal cord injury. As a result, there is some overlap between sections. This choice of organization is necessary, though, to illustrate the fact that these two specific methods can be used to investigate many different types of neurological disorders, with very similar study designs and experimental setups for different patients. These examples may therefore represent a practical approach to fMRI for clinical practice. The examples discussed in groups by type of disorder are to demonstrate the current capabilities of fMRI for clinical assessments. The span of the studies that are described here is intended to give an overview of the current state of *clinical* fMRI, and to direct the reader to specific papers on each application or neurological disorder.

The stated goals or expected future outcomes of many published scientific papers on fMRI are clinical applications to provide improved diagnosis of diseases or effects of trauma and improved treatment strategies. However, relatively few clinical applications of fMRI have been put into practice. Important clinical applications of fMRI are expected in the not-too-distant future, though, because of the clinically relevant results that have been reported in scientific papers. After roughly 20 years of development of fMRI, and the demonstrated importance of the information that can be obtained with maps of neural function in the brain, the question then naturally arises as to why this method is not yet in widespread use in hospitals. The main challenges that are believed to have contributed to slowing the development of fMRI for diagnosis of disease and for treatment planning and monitoring of results (1–3) include:

1. Lack of reference data

2. No standardization of methods (acquisition, analysis, and modes of stimulation)

3. Challenges with interpretation of results (4)

4. The highly multidisciplinary nature of fMRI studies

The first of these points refers to the fact that fMRI results can often only be interpreted by observing how function has been altered by disease, injury, and so forth. Therefore, data from age- and gender-matched healthy control subjects may be needed with the same tasks and acquisition parameters. It is not practical for each hospital to acquire its own set of reference data to aid with interpretation of results. There are also important questions as to how the presence of a tumor, edema, tissue damage, and the like, and also drug treatments, may alter the blood oxygenation–level dependent (BOLD) response (5), or how the emotional aspects of having suffered an injury or dealing with a neurological condition may affect task performance (6). Any of these factors could directly affect fMRI results and potentially render reference data from healthy control subjects useless. A more practical alternative is to use suitable reference results from other sites. This alternative therefore relates to points 2 and 3 above. Since fMRI results depend on the modes of stimulation or tasks and the baseline conditions to which they are compared, comparisons are only meaningful if the tasks and so forth are the same. The sensitivity and specificity of fMRI results are also affected by the acquisition parameters (timing, BOLD sensitivity, spatial and temporal resolution, etc.) and analysis methods. Although these are expected to have less of an impact than the choice of stimulation and baseline conditions, they can nonetheless affect the interpretation of results. The specific examples that are discussed below demonstrate that there is a good degree of consistency in many of the acquisition parameters that are used across studies, although the temporal and spatial resolution varies between studies. The methods of stimulation or types of tasks studied and the corresponding baseline conditions are not standardized and appear to be different between almost every published study. This presents a challenge for the creation of a useful fMRI reference database to aid with interpretation of clinical fMRI results. Point 3 above also notes that, for some applications, standardization may not be possible because fMRI studies may need to be designed for each specific patient, depending on the location and extent of neurological deficits. Finally, the fourth point above refers to the fact that effective clinical use of fMRI may require a team approach that includes the necessary mix of neurosurgeons, neurologists, neuroradiologists, psychologists, rehabilitation therapists, MR physicists, and others to develop effective fMRI study designs and acquisition parameters for each patient and to analyze the data and correctly interpret the results. Again, the alternative may be the future development of standardized fMRI acquisition methods and user-friendly analysis software, so that clinical use of fMRI can be time- and cost-effective.

9.1 Examples of Current Clinical Applications of fMRI

The dominant clinical uses of fMRI that have been demonstrated to date and used for clinical practice are presurgical mapping of essential speech and motor regions of the brain, and epilepsy seizure-focus localization, for the purposes of treatment planning with minimal postoperative neurological deficit (2). The range of accuracy that has been reported varies between studies and has been tested by comparison with intraoperative electrocortical stimulation (ECS) for motor areas determined with fMRI, and language areas detected with fMRI have been tested with the results of the intracarotid Amytal procedure, that is, the *Wada* test (7).

Yetkin et al. (8) studied 28 patients with fMRI using motor tasks involving the fingers, lips, and tongue, and also language tasks involving silent word-generation and counting. They reported agreement between sites of activity detected with fMRI and corresponding areas detected by means of intraoperative electrical stimulation, of 100% within 20 mm, and 87% within 10 mm. Petrella et al. (9) evaluated the impact of preoperative fMRI localization of language and motor areas. Out of 39 patients with brain tumors, the fMRI results altered the therapeutic plan in 49% and enabled a more aggressive approach in 46%. Out of this group, 30 patients underwent surgery, and fMRI helped to shorten the surgical time in 60% of the cases. Fitzgerald et al. (10) assessed the sensitivity of fMRI compared with ECS for mapping of language areas and determined it to be 81% when the areas were required to be in contact, and 92% for the areas being within 2 cm of each other. In another example, Majos et al. (11) reported that the agreement in the location defined by fMRI and intraoperative mapping techniques was 84% for motor areas, 83% for sensory areas, and 98% when both kinds of activity are taken into account. Similarly, Arora et al. (12) state that in patients with epilepsy they found excellent agreement between language lateralization determined with fMRI and with the Wada test. The agreement was 84% of patients (n = 40) with a reading task, 83% with an auditory task, 77% for a verbal fluency task, and 91% for the combination of tasks. However, they caution that it is important "to understand the neurophysiology of language processing, both in control subjects and in epilepsy patients, in order to better interpret the results from fMRI" (p. 2239). Bartos et al. (13) report that the fMRI sensitivity for localizing the primary motor cortex detected by ECS was only 71%, and therefore concluded that fMRI was not very reliable. They caution that fMRI should be used in combination with ECS, and this same recommendation is made by Xie et al. (14). Overall, it therefore appears from the scientific papers that have been published to date that the use of fMRI for preoperative mapping shows great promise and has demonstrated benefits, yet should be used with caution and with a very good understanding of both the underlying physiology and the fMRI method.

The specific fMRI methods for preoperative mapping that have been reported depend on the magnetic field strength of the MRI system and the specific application, and have evolved somewhat as MRI systems have improved. Nonetheless, there are consistent aspects of the methods, as all so far have employed the BOLD contrast mechanism and are based on T_2^*-weighted imaging with a single-shot gradient-echo (GE) method with echo-planar imaging (EPI) spatial encoding (with rare exceptions), as described in Chapter 6.

The choices of imaging parameters illustrated in Table 9.1 are presumably selected to optimize the sensitivity to BOLD contrast, while balancing the needs of imaging time and spatial resolution, and spanning all of the anatomy of interest, as described in Chapter 6. The choice of EPI encoding scheme provides high-speed imaging with a tolerable reduction in image quality. The choice of gradient-echo method with the echo time (TE) set approximately equal to the value of T_2^* of the tissues in the brain has been shown to provide T_2^*-weighted images with maximum BOLD sensitivity. At 1.5 tesla (T) the value of T_2^* in the brain is expected to be around 60 msec, whereas at 3 T this value is around 30 msec. This difference is reflected in the choices of TE values at the two different MR field strengths listed in Table 9.1, although some of these examples use a relatively short TE at 1.5 T and may not have optimal BOLD contrast as a result. The choice of repetition time (TR) is typically made to allow the minimum amount of time required to acquire the number of slices needed to span the anatomy of interest. In the examples in Table 9.1, the TR ranges from 1 sec to 6 sec, likely because the examples listed do not all have the same requirements for timing or coverage of the anatomy. The flip angle of the radio-frequency (RF) excitation pulse must then be selected to balance the need for high signal while avoiding T_1-weighting of the data, which could cause significant in-flow artifacts in blood vessels. With a TR of 1000 msec at 1.5 T, the optimal flip angle, given by the Ernst angle (see Chapter 4), is approximately 68°, whereas with a TR of 2000 msec it is 82°, and almost 90° with a TR of 3300 msec (assuming a T_1-value for gray matter of 1000 msec at 1.5 T, as described in Chapter 3). The variation of flip angle corresponding to the choice of repetition time can also be seen in most of the examples listed in Table 9.1. Together, these reasons explain the choices of imaging method (gradient echo); the spatial encoding scheme (EPI); and the key imaging parameters, TE, TR, and flip angle, in the examples listed, although not all of these examples appear to be optimal.

Table 9.1 Examples of Imaging Parameters Used for Preoperative Mapping of Eloquent Cortical Areas Based on fMRI: Selected Examples Were Taken from a Cross-Section of Published Scientific Papers, and All Use a Gradient-Echo Imaging Method with an EPI Encoding Scheme

Reference	Field Strength (Tesla)	TE (msec)	TR (msec)	Flip Angle (deg)	Resolution (mm)	Time Points	Slices
Voyvodic et al. 2009	3	35	1500	90	3.75 × 3.75 × 5	256	22 slices, axial
Berntsen et al. 2008	3	35	3000	90	1.64 × 1.64 × 1.64 SENSE factor of 2	81	33 slices, coronal
Bartos et al. 2009	1.5	54	6000	90	Unspecified in-plane 4 mm slices	64	28 slices, axial
Arora et al. 2009	1.5	50	1800	80	3.13 × 3.13 × 6	134, 138, or 160	18 slices, axial
Petrella et al. 2006	1.5	40	2000	Unspecified	3.75 × 3.75 × 5	Unspecified	Unspecified number, axial
Majos et al. 2005	1.5	64	1680	90	3.28 × 1.64 × 3	80	21 slices, axial
Gasser et al. 2005	1.5 Intraoperative	60	3300	90	3.44 × 3.44 × 3	80	30 slices, axial
Nelson et al. 2002	1.5	40	2000	85	3.75 × 3.75 × 6	114 or 134	20 slices, coronal
Yetkin et al. 1997	1.5	40	1000	Unspecified	3.75 × 3.75 × 10	140	7 slices, sagittal

Sources: Arora, J. et al., *Epilepsia* 50, 10, 2225–2241, 2009 (12); Bartos, R. et al., *Acta Neurochir* (Wien) 151, 9, 1071–1080, 2009 (13); Berntsen, E.M. et al., *Neurol Res* 30, 9, 968–973, 2008 (15); Gasser, T. et al., *NeuroImage* 26, 3, 685–693, 2005 (16); Majos, A. et al., *Eur Radiol* 15, 6, 1148–1158, 2005 (11); Nelson, L. et al., *J Neurosurg* 97, 5, 1108–1114, 2002 (17); Petrella, J.R. et al., *Radiology* 240, 3, 793–802, 2006 (9); Voyvodic, J.T. et al., 29, 4, 751–759, 2009 (5); Yetkin, F.Z. et al., *Neuroradiol* 18, 7, 1311–1315, 1997 (8).

Another key factor affecting the quality of fMRI results that are obtained is the image resolution. As described in Chapter 5, the choice of imaging field of view and sampling matrix size, which determine the resolution, can also influence the limits of imaging speed and TE and TR that can be achieved. In the examples in Table 9.1, the chosen image resolution reflects, in part, a trend toward finer resolution with the more recent studies. This is somewhat speculative because the rationale for choice of imaging parameters is not always provided in scientific papers. Nonetheless, this trend may reflect the improvements in imaging methods and MRI system hardware that have been realized over the past decade, and the fact that low-resolution images need no longer be tolerated for the sake of achieving an adequate signal-to-noise ratio. One common reason for using a relatively thick slice, however, is to span the anatomy of interest with a smaller number of slices, thereby achieving greater imaging speed and better temporal resolution in the fMRI data. The examples in Table 9.1 therefore provide a good cross-section of fMRI methods suitable for preoperative mapping, and they depict the range of options that may be employed to meet the needs of a particular study in terms of balancing speed, resolution, and image quality.

The tasks or stimuli that have been used for preoperative mapping with fMRI are designed to elicit activity in motor or language areas, since these have been the main areas of interest. All of the studies listed in Table 9.1 have used block designs, with periods of task performance interleaved with approximately equal duration periods of rest. The corresponding methods of stimulation or the tasks that were used to elicit activity in language or motor areas for preoperative mapping are listed in Table 9.2. In each of these examples, roughly equal periods of rest and stimulation were used, giving optimal design efficiency as described in Chapter 7. The shortest stimulation period listed, among these selected examples, is 9 sec, and the longest period is 48 sec. The shortest period is therefore sufficiently long for the BOLD effect to reach its peak (by ~6 sec after the onset of stimulation), and in this example 18 sec is allowed between repeated motor task blocks with the same hand, allowing time for the BOLD effect to return to baseline values. In this study a repetition time of 1.5 sec is used, and so 6 time points are acquired within the stimulation period and 12 within the rest period to describe the time course of BOLD signal changes. The example with the longest stimulation periods of 48 sec has 8 time points acquired within each stimulation block and in each rest block. These two examples therefore have roughly equal numbers of sampled time points, although in the total span of each study (coincidentally, 384 sec for both) more points would be acquired with the faster sampling rate of 1.5 sec. The study with the faster sampling would therefore provide greater sensitivity in the same time to acquire the data, as discussed in Chapter 7.

Another important practical consideration that is illustrated by the examples listed in Table 9.2 is that of how to communicate with the patient in the magnet during an fMRI study, in order to apply stimuli or to change tasks. In most of the published papers for the examples listed, the authors specify that prior to having the patient enter the MRI system, the tasks were carefully explained and the patients practiced the tasks for several minutes. This is an important step, because with careful explanations and practice of the task, the fMRI practitioner can verify that the patient understands the task and that the patient will understand the brief instructions given during the fMRI study when tasks are to be initiated or stopped. The mode of communication during the fMRI study is also very important because the instructions must be brief and clear, given the inherent acoustic noise that is produced during imaging, and the communication must not incite the patient to move, even slightly. Common communication methods are to use the two-way intercom system that is integrated into every clinical MRI system, and also to use an MRI-compatible liquid crystal display (LCD) screen or a rear-projection screen viewed via a mirror with a projector positioned outside the high magnetic field region around the MRI system. At the same time, it is important to have some means of monitoring the performance of the tasks to ensure that the desired neural function was engaged as designed. This requirement is particularly important for the current examples of preoperative mapping because all of the functions require active tasks (motor and language) to be performed.

Table 9.2 Examples of fMRI Study Designs and Methods of Stimulation or Tasks to Elicit Alternated Changes in Activity in Motor or Language Areas for Preoperative Mapping

Reference	Area of Interest	Task	Paradigm	Communication
Voyvodic et al. 2009	Motor Cortex	Opening and closing of hand	Repeated cycles of 9 sec rest, 9 sec left hand motion, 9 sec rest, 9 sec right hand motion—total of 384 sec	Prompted by visual cues
Berntsen et al. 2008	Motor Cortex	Six motor tasks: finger, tongue, lip, and toe movements, and isometric contraction of upper arm and thigh	Four 27 sec periods of motor activity (one type of movement for each fMRI session) interleaved with five equally long rest periods	The task to was described via intercom, and an LCD screen was used to signal rest or movement periods
Bartos et al. 2009	Motor Cortex	Alternated touching of the tip of the thumb to the other fingers (or opening and clenching of hand for patients who were too weak)	Four alternating periods of active and rest phases, each covered by 8 dynamic scans—total of 384 sec	*Unspecified*
Arora et al. 2009	Language Areas	Reading, auditory task, verbal fluency task	Alternated blocks of stim and baseline—4 stim + 5 rest blocks Reading: 37.3 sec for stim blocks and 20 sec for rest blocks Auditory: 33 sec for stim and 31.5 sec for rest Verbal fluency: 24 sec for stim and 22.5 sec for rest	*Unspecified*
Petrella et al. 2006	Motor and Language Areas	Motor: Alternating hand squeezing Language: Silently fill in the blanks in written sentences with words missing, compared WITH baseline condition with scrambled letters parsed into groups to simulate text	*Unspecified*	*Unspecified*

Majos et al. 2005	Sensory and Motor Areas	Sensory: Stroking palm and digits of one hand with a brush—hand opposite the affected hemisphere, or right hand in controls Motor: Simple motor task	Periods of rest or stimulation each lasted 13.44 sec, and periods of rest and task conditions were repeated 5 times, each beginning with a rest period—total of 134.4 sec	*Unspecified*
Gasser et al. 2005	Motor areas	Electrical stimulation of median and tibial nerves (patients were anesthetized for surgery)	*Unspecified*	None
Nelson et al. 2002	Language and Motor Areas	Language: Antonym word-generation task or a letter word-generation task or both Motor: Finger tapping or foot movement or both—alternating blocks of right or left limb activity interspersed with periods of rest	Five periods of task, interspersed with 6 periods of rest, each of 20 sec duration	Letters were provided at the beginning of each task period, or words for antonym generation were projected on a screen every other second Auditory cues were provided for the initial and end of each motor task cycle
Yetkin et al. 1997	140	Language: Silent word generation, counting by twos, word-generation task Motor: Pursing of the lips, movement of the tongue, finger tapping task by opposing thumb and fingers	20 sec periods of rest and activation were alternated	*Unspecified*

Sources: Arora, J. et al., *Epilepsia* 50, 10, 2225–2241, 2009 (12); Bartos, R. et al., *Acta Neurochir (Wien)* 151, 9, 1071–1080, 2009 (13); Berntsen, E.M. et al., *Neurol Res* 30, 9, 968–973, 2008 (15); Gasser, T. et al., *NeuroImage* 26, 3, 685–693, 2005 (16); Majos, A. et al., *Eur Radiol* 15, 6, 1148–1158, 2005 (11) Nelson, L. et al., *J Neurosurg* 97, 5, 1108–1114, 2002 (17); Petrella, J.R. et al., *Radiology* 240, 3, 793–802, 2006 (9); Voyvodic, J.T. et al., 29, 4, 751–759, 2009 (5); Yetkin, F.Z. et al., *Neuroradiol* 18, 7, 1311–1315, 1997 (8).

9.2 Examples of Forthcoming Clinical Applications

As mentioned earlier, quite a number of published scientific papers report fMRI studies with results that provide diagnostic information or characterize diseases, and show that these methods could be used for clinical assessments. While these applications are clinically relevant, they have not yet been used for clinical practice, such as to diagnose disease or to guide interventions or treatments. The purpose here is to show how the fMRI studies were designed and how the data were acquired, and not to discuss the findings of the studies or the significance of the results, except to point out how they might be useful for future clinical applications.

9.2.1 Resting-State Studies, Default-Mode Network Studies

Resting-state fMRI studies, which can be used to investigate the *default-mode network* (DMN), have been used to investigate a number of neurological conditions (such as multiple sclerosis in Section 9.2.2). The DMN has been reported to directly support internally directed thought (such as daydreaming, envisioning the future, retrieving memories) when focused attention is relaxed (18). Activity in the DMN is negatively correlated with areas of the brain involved with focus on external visual signals. The network includes part of the medial temporal lobe for memory, part of the medial prefrontal cortex for mental simulations, and the posterior cingulate cortex for integration, and also involves the precuneus and the medial, lateral, and inferior parietal cortex. Specifically, key areas are the ventral medial prefrontal cortex (vMPFC), posterior cingulate/retosplenial cortex (PCC/Rsp), inferior parietal lobule (IPL), lateral temporal cortex (LTC), dorsal medial prefrontal cortex (dMPFC), and hippocampal formation (HF). The activity in this network is detected in fMRI because it produces low-frequency (<0.1 Hz) BOLD fluctuations that are coordinated across regions.

The resting-state fMRI method is therefore particularly well suited for the study of diseases that may not have specific localized foci that can be detected, but instead affect the coordination between regions or alter complete networks (not necessarily just the DMN but networks for motor, attention, and so forth, as well). However, resting-state fMRI studies and the methods for analysis are still emerging and evolving, and there are concerns about the validity of the results, whether a true "resting state" can be ensured during an fMRI study. For example, it has been proposed that methods used to subtract global signal fluctuations (as described in Chapter 8) may actually cause apparent concurrent signal changes in spatially separated regions, giving the impression of functional connectivity (19). There is debate as to whether participants should have their eyes open or closed during resting-state studies (20). It has also been suggested that apparent coordinated signal changes across brain regions may reflect changes in blood CO_2 levels, and at the very least must be accounted for in the analysis of resting-state data, as a confounding effect (21). A study characterizing the test–retest repeatability of resting-state fMRI indicated that global connectivity patterns determined with resting-state fMRI have good reproducibility nonetheless (22). It is therefore important to keep these questions and controversies in mind, yet it is expected that they can be resolved, and that methods for data acquisition and analysis may still change and improve, if needed. The considerable number of published papers to date that report the use of resting-state fMRI methods indicate that this method has a great deal of potential for future clinical applications, with a vast range of neurological conditions.

Examples of these applications include mild cognitive impairment (23), Alzheimer's disease (24,25), temporal lobe epilepsy (26), diabetic neuropathic pain (27), Parkinson's disease (28), and schizophrenia (29). The imaging parameters used to acquire the fMRI data are listed in Table 9.3.

One study compared resting-state activity in 14 amnestic mild cognitive impairment (aMCI) patients and 14 healthy elderly mild cognitive impairment (MCI) patients (23) and used an independent components analysis (ICA) approach to investigate DMN activity. Compared with the aMCI patients, the healthy elderly exhibited increased functional activity in the DMN regions, including the bilateral precuneus/posterior cingulated cortex, right inferior parietal lobule, and

Table 9.3 Examples of Imaging Parameters Used for Resting-State fMRI Studies; Selected Examples Were Taken from a Cross-Section of Published Scientific Papers, and All Use a Gradient-Echo Imaging Method with an EPI Encoding Scheme, Unless Otherwise Indicated

Reference	Field Strength (Tesla)	TE (msec)	TR (msec)	Flip Angle (deg)	Resolution (mm)	Time Points	Slices
Bluhm et al. 2007 (GE—spiral sequence, 2-shot)	4	15	1500	60	4 × 4 × 4	110	Unspecified
Qi et al. 2010	3	40	2000	90	4 × 4 × 4	239	28 slices, axial
Morgan et al. 2010	3	35	2000	Unspecified	3.75 ×3.75 × 4.5 0.5 mm gap	200	30 slices, axial
Greicius et al. 2004 (GE—spiral sequence)	3	30	2000	80	Unspecified	Unspecified	Unspecified
Helmich 2009	3	30	1450	Unspecified	3.5 × 3.5 × 5 1.5 mm gap	265	21 slices, axial
Greicius et al. 2004 Asymmetric SE-EPI	1.5	50	2680	90	3.75 × 3.75 × Unspecified	Unspecified	Unspecified
Cauda et al. 2009	1.5	50	2000	90	3.125 × 3.125 × 4.5	200	19 slices, axial
Wang et al. 2007	1.5	60	2000	90	3.75 × 3.75 × 5 2 mm gap	180	20 slices, axial

Sources: Bluhm, R.L. et al., *Schizophr Bull* 33, 4, 1004–1012, 2007 (29); Cauda, F. et al., *PLoS One* 4, 2, e4542, 27, 2009; Greicius, M.D. et al., *Proc Natl Acad Sci USA* 101, 13, 4637–4642, 2004 (24); Helmich, R.C. et al., *Cereb Cortex* 20, 5, 1175–1186, 2010 (28); Morgan, V.L. et al., *Epilepsy Res* 88, 2–3, 168–178, 2010 (26); Qi, Z. et al., *NeuroImage* 50, 1, 48–55, 2010 (23); Wang, K. et al., *Hum Brain Mapp* 28, 10, 967–978, 2007 (25).

left fusiform gyrus, as well as a trend toward increased right medial temporal lobe activity. The aMCI patients exhibited increased activity in the left prefrontal cortex, inferior parietal lobule, and middle temporal gyrus, compared with the healthy elderly. Increased frontal–parietal activity may indicate compensatory processes in the aMCI patients. The authors proposed that abnormal DMN activity could be useful as an imaging-based biomarker for the diagnosis and monitoring of aMCI patients.

MCI is the transitional stage in the progression from full health to Alzheimer's disease. Studies of mild Alzheimer's patients compared with age-matched elderly control participants have been carried out to investigate the DMN as well. Grecius et al. (24) showed prominent activity in the hippocampus in both groups, suggesting that the DMN is closely involved with episodic memory processing. The Alzheimer's disease group showed decreased resting-state activity in the posterior cingulated cortex and hippocampus, suggesting that disrupted connectivity between these two regions accounts for the posterior cingulated hypometabolism commonly detected in positron emission tomography (PET) studies of early Alzheimer's disease. Results from Wang et al. (25) showed that Alzheimer's patients had decreased positive correlations between prefrontal and parietal regions and some increased positive correlations either within the prefrontal regions or between the prefrontal regions and other brain regions.

A resting-state study of diabetic neuropathic pain (27) demonstrated a cortical network consisting of two anticorrelated patterns. These are summarized as one that had activity correlated negatively with pain and positively with the controls and included the bilateral dorsal anterior cingulate cortex, the bilateral pre- and postcentral gyrus, and areas involved with vision and object recognition. The other had activity correlated positively with pain and negatively with the controls and included areas involved with episodic memory, visuospatial processing, reflections upon self, and aspects of consciousness (the precuneus), and areas of the dorsolateral prefrontal cortex, frontopolar cortex, frontal gyrus, thalamus, insula, and parietal lobes. The authors concluded that, in this study, pain was the result of aberrant default-mode functional connectivity.

Yet another study used resting-state fMRI to investigate Parkinson's disease (PD) (28). The authors compared the functional connectivity profile of the posterior putamen, the anterior putamen, and the caudate nucleus between 41 PD patients and 36 matched controls. The conclusions from this study were that dopamine depletion in Parkinson's disease leads to a re-mapping of cerebral connectivity that reduces the spatial segregation between different cortico-striatal loops. PD patients had decreased connectivity between the posterior putamen and the cortex (bilateral primary and secondary somatosensory cortex, inferior parietal cortex [IPC], insula, and cingulate motor area [CMA]). It was also proposed that the observed alterations of network properties may underlie abnormal sensorimotor integration in Parkinson's disease.

A group of 17 patients with schizophrenia were studied by Bluhm et al. (29) and compared with healthy volunteers. They observed correlations between spontaneous signal fluctuations in the posterior cingulate and fluctuations in the lateral parietal, medial prefrontal, and cerebellar regions in control subjects. In comparison, patients with schizophrenia had significantly less correlation between activity in the posterior cingulate with that in the lateral parietal, medial prefrontal, and cerebellar regions.

In spite of the wide range of neurological conditions that were investigated in this small sampling of resting-state fMRI studies described in the literature, the acquisition methods can be seen to be quite consistent. As with the applications discussed earlier, most of these examples employed gradient-echo methods to acquire T_2^*-weighted data to be sensitive to the BOLD effect, as described in Chapter 6. The echo times were selected for optimal BOLD sensitivity and are in the ranges of 15 msec, 30 to 40 msec, and 50 to 60 msec, at field strengths of 4 T, 3 T, and 1.5 T, respectively. Again, this choice of TE quite closely matches the expected values of T_2^* of the tissues in the brain at these field strengths. Most of these examples used EPI or spiral spatial-encoding schemes to obtain high acquisition speed and relatively short repetition times (TR), again for high speed, with the flip angles reduced to avoid significant T_1-weighting. A notable difference between the fMRI data acquisition methods used for resting-state studies and the examples of presurgical mapping above is that the resting-state studies make use of faster acquisitions and a higher number of sampled data points to improve the ability to detect the spontaneous low-frequency BOLD signal fluctuations.

These examples demonstrate that the various neurological conditions cause different alterations of the correlations between activity in regions of the DMN. Craddock et al. (30) have demonstrated the use of pattern recognition algorithms to automatically identify conditions of depression by means of the alterations in DMN connectivity, and also point out the applicability of this method to other neurological conditions as well. However, this study illustrates the complexity of this method and supports the conclusion that a considerable amount of research may be needed before such methods can be used to diagnose individual patients in a clinical setting (31–33). The fact that the same acquisition parameters can be used to interrogate a wide range of neurological conditions is a highly attractive feature for a clinical diagnostic method. Since only the resting-state condition is investigated, this method does not require additional stimulation or monitoring equipment to be set up in clinical MRI facilities. If the challenges of obtaining reliable and specific diagnoses in individuals can be overcome, resting-state fMRI may one day prove to be a very valuable, and practical, clinical application of fMRI.

9.2.2 Multiple Sclerosis

Multiple sclerosis (MS) is a disease that is being studied quite extensively with fMRI, in both the brain and spinal cord (34). Because of its transient and yet slowly progressive nature, MS can take years to diagnose definitively. Diagnosis and research into the disease can be further confounded by the fact that there are four subtypes: relapsing-remitting (RR), secondary progressive (SP), primary progressive (PP), and progressive relapsing (PR), each with different characteristics of the timing of progression of the disease. As a result, new ways of obtaining more information about the disease in patients, such as with MRI and fMRI, are being investigated (Figure 9.1).

A key challenge that has been identified for fMRI assessments of MS is that interpretation may be affected by disease-driven differences in task performance (34). To overcome this challenge, studies have been carried out with larger populations and use of several motor, visual, and cognitive tasks in groups representing all major clinical phenotypes of the disease. However, this solution will not be effective for individual patient assessments. One main conclusion drawn from fMRI studies to date is that cortical re-organization occurs in patients affected by MS. As a result, patients adapt and their abilities do not necessarily match the axonal/neuronal loss seen with imaging until the disease burden becomes too great and they are no longer able to adapt further.

A number of published examples of fMRI studies of MS to date use a range of approaches. In one recent example, changes in cognitive functions were assessed by using a working memory task with fMRI and looked specifically at patients with primary progressive MS (PPMS) (35). Working memory was investigated by using an n-back task, which consists of viewing a series of sequentially presented items, and deciding whether or not the current item is identical to the one seen n items back. For this study n = 2 was chosen, so that the task was challenging enough to probe cognitive function and working memory, without being so challenging that people with some cognitive decline would not be able to perform it. For the fMRI study, individual consonants were presented on a visual display for 1.5 sec followed by a black screen for 0.5 sec. Subjects responded with a button press with their right hand to indicate their response as to whether the character displayed matched the one displayed two previously. During rest periods subjects fixated on a stationary cross in the center of the screen. Prior to each study, the subject underwent training to become familiar with the task. A total of eight task periods were alternated with eight rest periods, and a total of 200 time points were acquired. The results of the study showed that the regions involved with the performance of the 2-back task differed between PPMS patients who had preserved cognitive function from those who were cognitively impaired.

Motor task performed with the right hand (dominant hand)

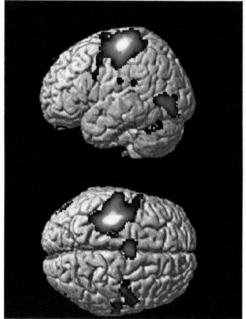 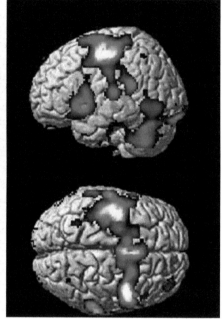

Healthy Volunteer　　　　Patient with Relapsing-Remitting
Multiple Sclerosis (RRMS)

FIGURE 9.1 An example of fMRI results obtained from a healthy volunteer (left) and from a patient with relapsing-remitting multiple-sclerosis (RRMS) (right). A more widespread and bilateral recruitment of the sensorimotor network is clearly visible in the patient. (Figure courtesy of Dr. Massimo Filippi, Neuroimaging Research Unit, Scientific Institute and University San Raffaele, Milan, Italy.)

Functional MRI data were acquired at 3 tesla and spanned the entire brain using a single-shot gradient-echo sequence with EPI encoding (GE-EPI). The data were T_2^*-weighted by setting the repetition time, TR, and echo time, TE, to 2000 msec and 30 msec, respectively, with an 85° flip angle. The acquisition matrix was 128 × 128 and the field of view (FOV) was 240 mm × 240 mm, resulting in an in-plane resolution of 1.875 mm × 1.875 mm. The imaging plane was axial, set parallel to the AC-PC line (the line between the anterior commissure and the posterior commissure), and 30 slices were imaged with a thickness of 4 mm. For brevity, these imaging parameters can be summarized using notation as follows: GE-EPI, TR/TE/flip = 2000/30/85°, resolution = 1.875 × 1.875 × 4 mm³, 128 × 128 matrix, 30 axial slices. These parameters provide optimal BOLD contrast, given that the chosen TE is roughly equal to the T_2^*-value of brain tissues at 3 tesla (30 msec). The combination of TR and flip angle provides minimal T_1-weighting, relatively high temporal resolution (2 sec per volume), and the highest signal-to-noise ratio possible with the chosen TR. This optimum is provided by setting the flip angle very close to the Ernst angle. The imaging parameters give a relatively high in-plane resolution of 1.875 mm × 1.875 mm, which is expected to yield high spatial precision in the results. The choice of slice thickness of 4 mm is relatively common and was presumably made to enable whole-brain coverage with the 30 slices that could be acquired within the given TR. For fMRI this combination of parameters enabled acquisition of 200 time points in 400 sec (6 min 40 sec). Again, this is within the relatively common range of timing and fMRI study duration and is expected to provide high sensitivity for detecting BOLD responses.

In this study anatomical reference images were acquired with a dual echo turbo spin-echo sequence. This is the same as a fast spin echo sequence described in Chapter 6 but provides images at two different echo times within the same acquisition. To provide both proton-density-weighted and T_2-weighted anatomical images, the parameters were set as follows: TR/TE = 3500/24 and 120, echo train-length (ETL) = 5, flip angle for refocusing pulses was set to 150° to reduce the energy deposited in the tissues (specific absorption ratio), and 44 contiguous 3 mm axial sections were acquired. The image acquisition matrix was 256 × 256, with a 240 mm × 240 mm FOV, providing 0.94 mm × 0.94 mm in-plane resolution. A separate set of anatomical images was also acquired with a three-dimensional T_1-weighted fast GRE with TR/TE/flip = 25/4.6/30°, 220 contiguous axial sections, resolution = 0.89 × 0.89 × 1 mm, matrix = 256 × 256, and FOV = 230 × 230 mm. In any fMRI study, anatomical reference images are valuable for displaying the results and accurate anatomical localization of areas of activity, given that images acquired with EPI may be spatially distorted.

In a related multisite study of MS (36), 56 patients and 60 age-matched healthy control volunteers were studied to characterize the relationships between clinical expression of MS and brain function. The fMRI results showed greater task-related activity bilaterally in a number of brain areas, and the key conclusions from this study were that there are motor system responses common to patients across a wide range of degrees of pathological progression of the disease, and plasticity and progression with aging may play a role in disease progression. However, there were no consistent relationships found between fMRI results and clinical behavior, "suggesting that fMRI is unlikely to be used as a simple 'surrogate' for clinical response across a heterogeneous population" (p. 121). The findings of this study therefore do not provide support for the use of fMRI for clinical assessments of individual MS patients. However, this study investigated only right-hand motor function and may therefore have been somewhat limited in the range of functional changes that could be detected. This study raised another interesting point by assessing whether or not there were systematic differences in results across data acquired at eight different sites. The results showed no significant differences between sites for the main effects of the motor tasks, either for patients or for healthy volunteers. This finding is important for the prospect of the future use of fMRI for clinical assessments of MS (or other diseases) because consistency, regardless of where the assessment is done, is essential.

The fMRI methods used for this study incorporated a block design, with six periods of hand movement for 30 sec, interleaved with rest periods of equal duration. The hand movement consisted of flexing and extending the fingers of the right hand (right-handed participants only) paced by flashes of light from a visual cue. The entire sequence was repeated four times in each scanning session with short breaks in between. Functional MRI data were acquired at 1.5 tesla, and again employed GE-EPI, with TR/TE/flip = 3000/60/unspecified, which is expected to provide high BOLD sensitivity. The spatial parameters were 240 × 240 mm FOV, 64 × 64 matrix, with 21 contiguous 6 mm axial slices parallel to AC-PC line. Data were acquired for 6 minutes to acquire 120 volumes per fMRI time series. The resulting image resolution is 3.75 mm × 3.75 mm × 6 mm.

Another example of the use of fMRI to study MS employed fMRI of synchronous spontaneous low-frequency BOLD fluctuations (LFBFs), or resting-state activity, as described above, to determine functional connectivity between regions, combined with diffusion tensor imaging (DTI) to demonstrate anatomical connectivity (37). For this study, conventional block-design fMRI data were acquired at 3 tesla during a unilateral finger opposition task to identify motor areas. A typical GE-EPI method was used with the following imaging parameters: TR/TE/flip = 2800/29/80°, matrix = 128 × 128, 256 mm × 256 mm FOV, 31 axial slices, each 4 mm thick. In a separate acquisition, resting-state data were acquired, meaning that the participants did not perform any tasks; they were instructed to refrain from any voluntary motion and simply remained at rest with eyes closed. With the same imaging parameters as for the motor fMRI study, resting-state fMRI data were acquired spanning 132 time points.

In the same imaging session, DTI data were acquired with diffusion weighting in 71 different directions with b = 1000 mm²/s, as well as eight images with b = 0. The imaging parameters were TE/TR = 102/7700 ms, 128 × 128 matrix, 256 mm × 256 mm FOV, 5/8 partial k-space sampling, 48 axial slices, each 2 mm thick, and data were acquired four times and averaged.

Regions of interest in the motor cortex were determined using the fMRI results from the motor task data described above. The region of interest for each hemisphere was defined as the one containing the highest T-value within the motor areas on the right and left sides. Reference time-series data were then determined for each voxel within these regions from the resting-state data, and the cross-correlation with the resting-state time series with each other voxel in the brain was calculated to determine the functional connectivity. The DTI data were used to determine anatomical pathways connecting these regions, and anatomical and functional connectivity were compared.

9.2.3 Stroke

Studies of the effects of stroke, and changes over time with recovery, have been a topic of considerable interest for study with fMRI. Specific challenges are presented by stroke for fMRI studies, however. These include the very changes in blood flow that may occur in and around affected regions that resulted in the stroke, which can also cause changes in the neurovascular coupling and the resulting BOLD response, and the functional deficits that may prevent a patient from performing a task for fMRI. Nonetheless, a considerable number of published fMRI studies have investigated the effects of stroke and show promise for potential future clinical applications for assessment and monitoring. A small cross-section of recent examples is discussed here to illustrate the potential clinical applications.

One example of the potential clinical use of fMRI for the assessment of stroke and monitoring the effects of rehabilitation efforts is described by Page et al. (38). In this study, the effects of 8 weeks of training with a neuroprosthesis were investigated in a group of eight chronic stroke patients. Functional MRI was only one of several tests done to monitor the outcomes of the therapy, in addition to the Action Research Arm Test (ARAT), the upper extremity section of the Fugl-Meyer assessment (FM), and the amount-of-use scale of the Motor Activity Log (MAL). Comparisons of pre- and posttherapy showed increases in ARAT and FM scores and increased affected arm use, and fMRI showed significant increases in cortical activity.

The fMRI results did not show any changes between pre- and posttherapy when the task was performed with the unaffected hand. For the affected hand (right hand in seven participants, and results were mirrored for the one participant with the left hand affected), however, significant increases in the brain volume with BOLD signal changes were seen (pre- versus posttherapy) in the right precentral and postcentral gyri, right inferior parietal lobule, right middle frontal gyrus, and right precuneus. Additional increases in activation were present in the basal ganglia and thalamus, bilaterally. The authors concluded that these fMRI results are consistent with other studies showing that the contralesional sensorimotor cortex can help to compensate for stroke-induced motor impairments. The functional changes that the authors were able to demonstrate as a result of therapy after stroke provide a good example for the potential use of fMRI for monitoring the effects of other rehabilitation efforts in patients following stroke.

Functional MRI data were acquired at 4 tesla, while participants performed four runs of an attempted wrist flexion task (two left side, and two right side). Participants practiced the task prior to entering the MRI system. During the fMRI acquisition, directions were presented via goggles and via an auditory reminder prior to each movement or rest block. The task consisted of alternating periods of 30 sec of rest and 30 sec of movement, and each fMRI acquisition lasted 5 minutes. The experimenters monitored the participants visually to ensure that they performed the tasks correctly. The specific fMRI acquisition parameters are summarized in Table 9.4, for comparison with other fMRI studies of stroke.

Table 9.4   Examples of Imaging Parameters Used for fMRI Studies of Stroke; Selected Examples Were Taken from a Cross-Section of Published Scientific Papers, and All Use a Gradient-Echo Imaging Method with an EPI Encoding Scheme

Reference	Field Strength (Tesla)	TE (msec)	TR (msec)	Flip Angle (deg)	Resolution (mm)	Time Points	Slices
Page et al. 2010	4	25	3000	*Unclear*	4 × 4 × 5	96	30 slices, coronal
Menke et al. 2009	3	40	3000	90	3.6 × 3.6 × 3.6	510	36 slices, axial
Riecker et al. 2010	1.5	39	3000	90	3 × 3 × 4 1 mm gap	185	28 slices, axial

Sources: Menke, R. et al., *BMC Neurosci* 10, 118, 2009 (40); Page, S.J. et al., *Neurorehabil Neural Repair* 24, 2, 195–203, 2010 (38); Riecker, A. et al., *Hum Brain Mapp* 2010 (39).

In another study of stroke employing fMRI, Riecker et al. (39) investigated the specific contribution of the ipsilateral (nonaffected) hemisphere to recovery of motor function following stroke. In this study, fMRI was applied while participants performed a repetitive movement task with the right index finger, at a range of different frequencies of movement. The participants studied were eight well-recovered patients, each with a chronic striatocapsular infarction of the left hemisphere, and eight age-matched healthy volunteers for comparison. The fMRI results revealed BOLD responses that increased linearly in magnitude with the frequency of the finger motor task, in the left supplementary motor cortex (SMA) and the left primary sensorimotor cortex (SMC) in both the patient and control groups. However, a similar graded response was only seen in the patient group in the right premotor cortex (PMC) and the right SMC. The authors concluded that "these results support the model of an enhanced bihemispheric recruitment of preexisting motor representations in patients after subcortical stroke" (p. 1027).

Again, the specific fMRI acquisition parameters are listed in Table 9.4 for comparison with other studies. The task paradigm, consisting of a tapping task with the right index finger, was carried out by means of audio tones to indicate to the participants when to move and at what pace, and when to rest. Audio clicks were presented to the participants at six frequencies: 2.0, 2.5, 3.0, 4.0, 5.0, and 6.0 Hz for pacing of the movement. Each series of clicks was presented for 6 sec, and the series was presented 15 times over the course of each fMRI acquisition. The timing of each block was pseudo-randomized with onset-to-onset intervals ranging from 12 to 24 sec. When the participants moved their finger, they pressed a button, which enabled monitoring of their performance. Prior to the fMRI experiment, participants practiced the motor task several times together with measures of electromyography, and in this way the experimenters were able to ensure that the task was performed correctly and without any mirror movements of the opposite hand.

Whereas the previous two studies focused on motor deficits after stroke, Menke et al. (40) carried out a study of language deficits (aphasia) following stroke. In this example, eight patients were studied with chronic aphasia and moderate to severe word-finding difficulties (anomia) as a result of a single left-hemisphere stroke. A group of nine healthy participants was also studied for comparison. Participants were treated for anomia by means of 3 hours of computer-assisted naming therapy per day over a period of 2 weeks.

Functional MRI studies were carried out to determine the effects of this therapy by having the participants perform an object-naming task, similar to the tasks employed for therapy. During each fMRI study, photos of concrete objects were presented on a screen and viewed via a mirror, and the participants were instructed to overtly name each object. Pictures of each of 90 different objects were presented for 5 sec, followed by variable interstimulus intervals of 8 to 16 sec (average 12 sec) in an event-related fMRI study design. The task performance was monitored during each fMRI session by means of a specialized dual-channel communication system for MRI.

Performance of the naming task was reported to have improved in general during training, from no participants correct in naming responses at the baseline assessment to a mean of 64% correct immediately after training. Functional MRI results showed a relationship between short-term treatment success and increased activity bilaterally in the parahippocampi. The authors also reported increases in activity in the right parietal cortex (precuneus), the right cingulate gyrus, and bilaterally in the occipital lobes that were correlated with behavioral improvement. Long-term success for trained object names was associated with the degree of activity increase in the right-sided Wernicke's homologue. In healthy control subjects, the comparison between two fMRI sessions did not reveal any increases in brain activity. The authors proposed that different brain regions are required for initial learning versus long-term consolidation. Activity increases in bilateral brain regions involved in memory, attention, and multimodal integration predicted treatment success during initial learning. In contrast, long-term success appeared to be mediated by activity increases in the right-sided Wernicke's homologue and left temporal language areas, "suggesting bilateral recruitment of 'normal' task-related areas during the consolidation process" (p. 11).

These three examples were selected from among the available studies because they are very recent and provide a sampling of the cross-section of applications and imaging parameters. Together, these studies illustrate the capabilities of fMRI for assessing the effects of stroke and changes over time as a result of therapeutic efforts. However, as with other examples discussed earlier in this chapter, these studies do not demonstrate the ability to assess individual patients.

The imaging parameters used for fMRI in these three examples are summarized in Table 9.4, and span the commonly used field strengths of 1.5 T, 3 T, and 4 T. Again, as with the previous examples in this chapter, these parameters are approximately optimal for BOLD sensitivity. The echo times, TE, in most cases are set approximately equal to the expected values of T_2^* of the tissues in the brain, being around 15–20 msec, 30–40 msec, and 50–60 msec at 4 T, 3 T, and 1.5 T, respectively. The cubic voxels acquired by Menke et al. (40) with sides of 3.6 mm are expected to be optimal with high signal-to-noise ratio and ease of reformatting to other orientations, spatial normalization, and minimal effects of spatially dependent partial volume effects (as described in Chapter 6).

9.2.4 Consciousness Disorders/Coma

Disorders of consciousness include coma, in which a person is not awake or aware, the vegetative state, in which a person lacks only awareness, and the minimally conscious state, in which a person can be awake and aware but awareness is impaired (41). It has been shown that PET and fMRI of unresponsive patients may be able to complement neurological examinations, electroencephalogram (EEG), and so forth for the assessment of unresponsive patients and that fMRI may be able to provide evidence of conscious awareness when other neurological examinations cannot.

One example of fMRI of a patient in a vegetative state that attracted a considerable amount of attention was reported by Owen et al. (42). The patient had been in a vegetative state for approximately 5 months as a result of a traumatic brain injury, and fMRI was carried out to detect differences in responses to hearing spoken sentences in comparison with acoustically matched noise sequences. The results showed bilateral activity in the middle and superior temporal gyri, equivalent to that observed in healthy volunteers. The authors also tested the responses to sentences containing ambiguous words, and these produced an additional response in the left inferior frontal region, again similar to that seen in healthy volunteers. To further probe whether or not the patient was demonstrating conscious awareness, the authors performed fMRI studies while the patient was given spoken instructions to perform two different mental imagery tasks. One task was to imagine playing a game of tennis, and the other was to imagine visiting all of the rooms in their house. During studies in which the patient imagined playing tennis, fMRI results showed activity in the supplementary motor area, and when imagining walking through her house, activity was detected in the parahippocampal gyrus, the posterior parietal cortex, and the lateral premotor cortex. As with the other tasks, the observed responses corresponded with those seen in healthy control volunteers while performing the same tasks. The authors therefore concluded that the patient retained the ability to understand spoken commands and to intentionally cooperate and perform the tasks, indicating that she was consciously aware of herself and her surroundings.

The fMRI studies of responses to spoken sentences or noise were carried out by playing single sentences at a time, within silent periods of 7.4 sec, prior to a single MRI scan lasting 1.6 sec. A total of 118 spoken sentences were presented, as well as 59 trials where noise was presented instead of a coherent sentence, and 60 silent trials. A total of 237 scans were therefore acquired over a total span of 2133 sec, or roughly 35.6 minutes. However, the fMRI acquisition was divided into three sessions of 79 trials each (237 total). The onset of the scan timing was jittered (systematically shifted in time to be slightly earlier or later) for each scan in order to detect the BOLD responses at slightly different points in time relative to the stimulus. The responses to mental imagery tasks were detected in block-design fMRI studies with alternating periods of task and rest. The patient was presented with prerecorded instructions of what to imagine, or to rest, at the start of each block. Each task and rest condition was repeated 10 times. Unfortunately, the authors do not provide any details of how the fMRI data were acquired.

In another example of the use of fMRI to assess patients in vegetative or minimally conscious states, Coleman et al. (43,44) carried out an fMRI investigation of the speech-processing abilities of 41 patients with varying degrees of impaired consciousness. Functional MRI was performed while the patients were presented with three different auditory stimuli including sentences with high and low ambiguity, or unintelligible noise, or during silent periods. The timing of the presentation of sentences, noise, or silent periods was very similar to that of the study described above, with each being presented within a period of 7.4 sec prior to a single image acquisition lasting 1.6 sec. The timing was selected so that the mid-point of each stimulus occurred 5 sec before the mid-point of the subsequent image acquisition. This was in an effort to detect the peak of the BOLD response, based on the assumption that this response would lag the change in neural activity that was elicited by the stimulus, by 5 sec.

The fMRI data in this study were acquired at 3 tesla, presumably by means of a gradient-echo imaging method with an EPI encoding scheme, although the authors do not identify the imaging method. The imaging parameters (summarized in Table 9.5) include a 9 sec repletion time and a 27 msec echo time, and the flip angle is not specified although it is presumed to be 90° with such a long repetition time. The long repetition time provides the 7.4 sec without any sounds produced by the MRI system, so that the auditory stimuli can be presented without any interference. The image data spanning the entire brain is acquired within the remaining 1.6 sec period, as indicated above. The spatial parameters include a 25 cm × 25 cm field of view, 128 × 128 acquisition matrix, and 4 mm slices with a 1 mm gap between slices. This produces a resolution of 1.95 mm × 1.95 mm × 4 mm. It is presumed that the slice thickness and inclusion of a gap were chosen to enable imaging of the entire brain, within the 1.6 sec acquisition. The authors also specify that the slices were in an axial-oblique orientation, tilted away from the eyes, to avoid motion artifacts produced by eye movements.

The results showed a range of responses across the group of patients, with 6 responding to sound only, 19 responding to both sound and speech, and 16 showed no significant activity in response to auditory stimuli. Assessments done 6 months after the fMRI investigation was carried out showed a strong association between the level of auditory processing shown on the fMRI results and the 6-month Coma Recovery Scale (CRS) score. Two patients who were diagnosed as being in a vegetative state were particularly interesting because they showed different

Table 9.5   Examples of Imaging Parameters Used for fMRI Studies of Patients with Consciousness Disorders; Selected Examples Were Taken from a Cross-Section of Published Scientific Papers, and All Use a Gradient-Echo Imaging Method with an EPI Encoding Scheme

Reference	Field Strength (Tesla)	TE (msec)	TR (msec)	Flip Angle (deg)	Resolution (mm)	Time Points	Slices
Coleman et al. 2009, 2007	3	27	9000	*Unspecified*	1.95 × 1.95 × 4 1 mm gap	237 in total	*Unspecified no.,* axial-oblique
Eickhoff et al. 2008	3	30	2200	*Unspecified*	3.1 × 3.1 × 3	393 (estimated)	*Unspecified*

Sources: Coleman, M.R. et al., *Brain* 130, 10, 2494–2507, 2007 (44); Coleman, M.R. et al., *Brain* 132, 9, 2541–2552, 2009 (43); Eickhoff, S.B. et al., *Exp Neurol* 214, 2, 240–246, 2008 (45).

responses to sentences with high and low ambiguity, indicating some aspects of speech comprehension. These patients had progressed to a minimally conscious state by 6 months after the fMRI data were acquired. Out of 8 patients in a vegetative state in the study who subsequently progressed to a minimally conscious state, 7 had shown a speech-specific or semantic response to sentence stimuli during the fMRI studies. These results therefore provide evidence for the fMRI results having prognostic value. The authors propose that these results show that fMRI offers "valuable additional information to inform the diagnostic decision-making process and importantly provides accurate prognostic information" (p. 2542). However, the authors also stress that this method is not without limitations, that only positive findings can be interpreted, and that this method should be used with caution. In spite of the limitations and the fact that a considerable amount of research and development may need to be undertaken before this method is ready for routine clinical use, this study appears to represent a very important step toward this ultimate goal.

In a related study, Eickhoff et al. (45) carried out a detailed fMRI study of a single female patient who had been in a comatose state for approximately 3 years. She had never regained consciousness or opened her eyes and showed only stereotypical movements of her left arm. Functional MRI studies were carried out with a block design, with 24 cycles of 18 sec of stimulation followed by 18 sec of rest. In a first fMRI examination, stimuli that were used in separate acquisitions included visual stimulation with a full-field flickering light, auditory stimuli with random nonemotional words, and brushing of the right and left forearms (in separate acquisitions) with a sponge. In a second examination, responses to different speech stimuli were tested. These consisted of recordings of speech from the patient's two female children, two close female friends, and another female who was unknown to the patient.

Functional MRI data were acquired at 3 tesla with a gradient-echo EPI sequence, as summarized in Table 9.5. The imaging parameters were TR/TE/flip = 2200/30/unspecified, with a spatial resolution of 3.1 mm × 3.1 mm × 3 mm. Presuming that a 90° RF excitation pulse was used, this provided approximately optimal BOLD sensitivity and spatial precision. In contrast with the study described above, the authors did not present the auditory stimuli only within periods in which the MRI system was silent, but rather presented them over the background sounds.

The results showed activity in the right somatosensory areas and the left cerebellum with tactile stimulation of the left forearm, and stimulation of the right side produced small bilateral activity in secondary somatosensory areas. Visual stimulation produced activity only in the left visual cortex. Auditory stimulation with words produced bilateral activity in primary auditory cortex and auditory association areas on the superior temporal gyrus, as well as in the left inferior frontal gyrus. Spoken phrases directed toward the patient resulted in activity in the left amygdala and right anterior superior temporal sulcus, and the magnitude of the responses depended on the voice that the patient heard. The strongest responses were reported when the patient heard her children's voices, followed by those of her friends, and the weakest responses when the speaker was unknown to the patient. These results therefore show evidence of cognitive and emotional capabilities that could not be detected with conventional assessment methods.

The imaging parameters that were selected for fMRI of patients in vegetative or comatose states in these two examples are again quite similar to those used to investigate other neurological conditions, as described earlier in this chapter. The echo time was chosen to be approximately equal to the value of T_2^* of the tissues in the brain at 3 tesla. The repetition time is chosen to be adequately long to enable imaging of the whole brain, and in the example from Coleman et al., to allow silent periods for presentation of the auditory stimuli as well.

9.2.5 Traumatic Brain Injury

Traumatic brain injury (TBI), like many neurological conditions, produces a wide range of deficits and levels of severity depending on the mechanism and location of the trauma, and the outward signs of neurological deficits change over time following the initial injury and with recovery. Functional MRI may provide a very useful supplement to conventional neurological tests for assessing the effects of TBI and monitoring of recovery. In one recent example of the use of fMRI to study the effects of mild TBI, Slobounov et al. (46) studied 15 neurologically normal student-athletes with no history of TBI, and 15 student-athletes who had recently suffered sport-related mild TBI. (The participants did not suffer a loss of consciousness but had complaints including loss of concentration, dizziness, fatigue, headache, irritability, visual disturbances, and light sensitivity).

Functional MRI studies were carried out by having the participants view a virtual reality display of a series of intersecting corridors, complete with a selection of doors and rooms. The display was projected onto a screen and viewed via a mirror. While lying inside the MRI system, the participant could control the view that was displayed and virtually move through the scene by means of a joystick. In different parts of the study the participants were (a) shown the navigation route to follow, (b) navigated randomly, or (c) navigated with the goal of reaching a specific room. The purpose was to impose different degrees of memory load on the participants, in different parts of the fMRI acquisition. These conditions were also compared with a baseline condition in which the participant simply tracked a cursor on the screen by moving the joystick.

Functional MRI data were acquired at 3 tesla with a gradient echo with EPI spatial encoding, an echo time of 25 msec, repetition time of 2000 msec, and a flip angle of 79°, as summarized in Table 9.6. The selected spatial parameters produced an image resolution of 3.1 mm × 3.1 mm × 5 mm, with a 64 × 64 matrix. The field of view was 220 mm × 220 mm, and 30 slices were acquired to span the entire brain.

The results of this study showed different brain areas involved with learning the navigation route compared with navigating by memory, and also some differences with people after mild TBI compared with healthy control subjects. The areas showing different activity between learning and navigating by memory included the lateral extrastriate visual cortex extending into the visual area V2, bilateral parietal cortex, bilateral precuneus, and right dorsolateral prefrontal cortex (DL-PFC). Overall increased activity in several brain regions during learning the route, compared with navigating by memory, was reported to be similar for both normal controls and people after mild TBI. The authors proposed that the common regions of activity during learning a route and retrieval of spatial information may indicate a consistent brain network, including visual cortex, parietal cortex, DL-PFC, and hippocampal cortices. Structural and

Table 9.6 Examples of Imaging Parameters Used for fMRI Studies of Patients with Traumatic Brain Injury; Selected Examples Were Taken from a Cross-Section of Published Scientific Papers, and All Use a Gradient-Echo Imaging Method with an EPI Encoding Scheme

Reference	Field Strength (Tesla)	TE (msec)	TR (msec)	Flip Angle (deg)	Resolution (mm)	Time Points	Slices
Slobounov et al. 2010	3	25	2000	79	3.1 × 3.1 × 5	Unspecified	30, axial slices
Kohl et al. 2009	3	60	2000	90	3.75 × 3.75 × 5	192 across 3 runs	32, axial slices
Nakamura et al. 2009	3	30	2000	89	Unspecified	196	Unspecified

Sources: Kohl, A.D. et al., *Brain Inj* 23, 5, 420–432, 2009 (47); Nakamura, T. et al., *PLoS One* 4, 12, e8220, 2009 (48); Slobounov, S.M. et al., *Exp Brain Res* 202, 2, 341–354, 2010 (46).

functional disruption of connectivity between these regions or deficits within some regions may be caused by a concussive blow.

In another example of fMRI of traumatic brain injury, Kohl et al. (47) assessed cognitive fatigue in TBI patients, compared with a matched group of healthy control subjects. The authors investigated the concept that the increased cognitive effort required by patients with TBI to meet the demands of daily life leads to fatigue. This idea was tested by having patients perform a modified version of the Symbol Digit Modalities Test (mSDMT) during fMRI studies. The mSDMT consists of viewing a panel of nine symbols paired with nine numbers, and a "probe" box containing one number and one symbol. The participants had to respond as quickly and as accurately as possible, by means of a button press, to indicate whether or not the number and symbol in the probe box was a correct match, as indicated in the panel of all matching symbols and numbers. The panel of exemplar stimuli and the probe box were presented simultaneously for 6 sec. A total of 192 trials were presented over the course of three fMRI acquisitions, with each acquisition lasting 7 minutes. The interval between successive stimuli was varied randomly between 0, 4, 8, or 12 sec.

Functional MRI data were acquired at 3 tesla with a gradient-echo sequence with EPI spatial encoding. The imaging parameters included an echo time of 60 msec, repetition time of 2000 msec, and a flip angle of 90°. Spatial parameters resulted in a resolution of 3.75 mm × 3.75 mm × 5 mm. The field of view was 24 cm × 24 cm, with a 64 × 64 matrix, and 32 axial slices to span the entire brain. These parameters are also summarized in Table 9.6.

The results of this study show that patients with TBI had significantly slower response times on the tasks, and both groups showed improvements over time in their task performance. The patients with TBI showed increased BOLD responses in the left middle frontal gyrus within fMRI sessions, whereas the healthy control subjects showed constant values or a slight decrease. This was interpreted as demonstrating growing cognitive fatigue. Activity (in terms of magnitude of BOLD response) in the right superior parietal lobule and left basal ganglia (caudate/putamen) was also shown to decrease during fMRI sessions in healthy control subjects, but was constant or increased in patients with TBI. Changes over time across repeated tasks were also detected in the anterior cingulated and superior parietal cortex. The results are therefore intriguing, as they reveal effects of traumatic brain injury on cognitive function, including as fatigue develops, but the authors identify that this study requires more investigation. Nonetheless, this study may represent a very important first step in the development and evolution of clinical methods for assessing TBI.

In a related study, Nakamura et al. (48) carried out fMRI studies of six people with severe traumatic brain injury (Glasgow Coma Scale 3–8) at 3 months and at 6 months after the injury occurred. They used a working memory task with a delayed response during fMRI studies to probe the changes that had occurred as a result of injury, and recovery over time. The task consisted of viewing one, two, or four images of male and female faces in black-and-white pictures for 3 sec. During a subsequent delay period of 3 sec the participants fixed their gaze on a target at the center of the viewing screen, and then were presented with a picture of a single face and had to indicate a yes/no answer as to whether the face and location matched the previously displayed images. The participant was given another 3 sec to give their response. However, this study is unique from the previous examples because the data were analyzed to determine the task-related BOLD signal changes, and the data were then processed to remove these effects, with the intention of leaving only underlying "resting-state" effects. As a result, this example would also appear to fit in the section on resting-state studies, above. The processed data were then analyzed to characterize connectivity between brain regions, by means of *graph theory* (49). Briefly, this analysis consists of identifying anatomically relevant regions (called *network nodes*), and determining the effective connectivity between each pair of regions based on correlations between the fMRI time-series data. A threshold must be chosen to differentiate correlations that are sufficiently strong to be of interest from weakly or noncorrelated regions. Network parameters can then be quantified, depending on the properties that are of interest. In the study presented by Nakamura et al. (48), the "global efficiency" of the network was quantified as an average of the 1/(distance) between all pairs of significantly connected regions, or with a "local efficiency" determined by the average of 1/(distance) of regions that were significantly connected with one chosen region.

The results of this study demonstrated changes in networks of connected regions in the brain between 3 and 6 months following severe TBI in the six patients that were studied. Both global and local efficiency decreased significantly toward healthy control values in the 3 months between the repeated studies, and the mean path length in the network increased significantly toward healthy control values. This study therefore presents very interesting examples of how fMRI can be used to monitor the progression of changes in diffuse effects such as may be caused by TBI, and in the analysis methods that may be used to characterize these complex networks.

The fMRI data used for this study were acquired at 3 tesla, again with a gradient echo with EPI encoding method as in most other studies. The acquisition parameters included an echo time of 30 msec, repetition time of 2000 msec, and flip angle of 89°, for optimal BOLD sensitivity. The spatial parameters that were used are not clearly specified.

The fMRI acquisition parameters listed in Table 9.6 for the selected examples of studies of traumatic brain injury can be seen to be quite similar to those used in other fMRI studies. The echo time should be around 30 msec for optimal BOLD contrast-to-noise ratio at 3 tesla. With the repetition time used for each of these studies, 2000 msec, the optimal flip angle (Ernst angle) is 80° (assuming a T_1-value for gray matter of approximately 1150 msec, as indicated in Chapter 3).

9.2.6 fMRI Studies Using Saccadic Eye Movements

A current area of research that is particularly interesting from the point of view of future clinical applications is the use of fMRI during saccadic eye movements. Saccadic eye movements are rapid movements of both eyes in the same direction, and in many instances in everyday life occur as a reflex movement in response to a change in the visual field, such as a moving object, change in reflected light, and so forth. These reflexive movements can also be intentionally suppressed, but this requires attention and impulse control. As a result, the study of saccadic eye movements has been shown to reveal information about cognitive and attention functions and how they are altered by neurological conditions.

One of the earlier fMRI studies of eye movements in healthy volunteers helps to illustrate the method and its potential for obtaining clinically relevant information. Connolly et al. (50) investigated the cognitive functions involved with *preparatory set*, which is the readiness to respond to a sensory stimulus and the intention to perform a predetermined action depending on the stimulus. This function is important for investigations of neurological diseases or effects of injury and so forth, because it involves high-level cognitive function, attention, and in some cases the ability of overcome reflex responses. In their study, Connolly et al. used a visual display in which either a fixation point was displayed or the fixation point disappeared and another point, displaced to the right or left, was instead shown. The reflex response is to look toward a suddenly flashed visual stimulus, and this is termed a *prosaccade*. In some trials the participants were instructed to instead move their gaze in the opposite direction, away from the flashed stimulus, and this is termed an *antisaccade*. A prosaccade involves an automatic visuomotor response, whereas an antisaccade requires the automatic response to be suppressed and a voluntary response to be performed.

Functional MRI studies were carried out using an event-related design, and stimuli were presented on a screen viewed via a mirror. Eye movements were recorded by means of an MRI-compatible eye-tracking system. In each trial, a white fixation point was displayed at the center of the display for 3 sec, and then it changed color to either green or red for 3 sec to indicate to the participant to prepare to do a prosaccade or an antisaccade, respectively. The fixation point then disappeared and the display was blank for variable durations of 0, 2, or 4 sec, before a target point was flashed for 100 msec at 9° to 15° visual angle to either the right or left of center. Participants were instructed to maintain their gaze at the new target point until the fixation point was again displayed at the center of the display, 2 sec later.

The fMRI data were acquired at 4 tesla, using a gradient-echo sequence with EPI sampling scheme. An initial study was carried out using a block design to identify the location of the frontal eye field (FEF), and the data were acquired with an echo time of 28 msec, repetition time of 2000 msec, and a flip angle of 60°. These data were analyzed while the participant remained in the magnet, and the FEF was identified and used for subsequent slice positioning. For the event-related studies described above, the echo time remained at 28 msec, but the repetition time was reduced to 500 msec, and the flip angle reduced accordingly to 30°. A set of 6 axial slices were selected that included the FEF and entire parietal lobe. The field of view was 192 mm × 192 mm, with a 64 × 64 sampling matrix, resulting in an image resolution of 3 mm × 3 mm × 6 mm. Each experimental run was 528 sec in duration and included 12 prosaccade and 12 antisaccade trials.

The results showed that in the FEF and intraparietal sulcus (IPS) the peak of the BOLD response occurred at approximately the same time after the appearance of the target. The authors ascribe this activity to the combined visual and motor activation of FEF and IPS neurons. In the FEF a pretarget rise in BOLD signal was also observed, which depended on the duration of the gap between fixation point disappearance and target appearance, and the magnitude of the BOLD responses in the FEF also increased with gap duration. The authors were able to confirm, via additional tests, that the pretarget BOLD increase specifically in the FEF was indeed related to preparatory set.

In a related study, the regions of the brain that are involved with switching a planned response were investigated (51). As in the previous study described above, participants viewed a fixation point which changed color for 3 sec to indicate whether to perform a pro- or antisaccade. While the fixation point remained visible, a target circle appeared 5.5° of visual angle to the right or left. However, in 50% of the trials the fixation point changed color, thereby changing the task instruction, 100 msec or 200 msec after the target appeared. The participants therefore had to change their actions after they had already prepared, and started initiating, a response. After

3 sec the target disappeared, the participant returned his or her gaze to the fixation point, and 12 sec later another trial was initiated. The reaction times and error rates were recorded via an MRI-compatible eye-tracking system. In a second set of experiments the brain functions that are involved were further investigated by stepping the switching times by plus or minus 50 msec, depending on how well each participant performed the task. The purpose was to adjust the level of difficulty based on performance.

Functional MRI data were acquired at 3 tesla, again with a gradient echo with EPI encoding scheme. In this case the echo time was set to 30 msec, the repetition time was chosen to be 750 msec to obtain high temporal resolution, and the flip angle was set accordingly to 56°. A set of 11 axial slices, 3.3 mm thick, was selected to target specifically the caudate nucleus. The field of view was set to 211 mm × 211 mm, and the acquisition matrix was 64 × 64, resulting in an image resolution of 3.3 mm × 3.3 mm × 3.3 mm. A spatial saturation pulse was also applied to eliminate/reduce signal from the region of the eyes to reduce motion artifacts in the images arising from the eye movements. For the second set of experiments, the frontal eye fields were imaged, as well as the caudate nucleus. In this case the number of slices needed to be increased to 16 to span the anatomy of interest. As a result, the repetition time and flip angle were increased to 1000 msec and 62°, respectively.

Increases in response times and error rates between performing a saccade in one direction and switching from that direction to the opposite direction were termed *switch costs*. Similar switch costs were detected in both directions, that is, whether the initial instruction was to perform a pro- or an antisaccade. Activity detected in the caudate nucleus with the first fMRI described above, however, was detected only during switching from a pro- to an antisaccade. In the second experiment it was again observed that switch trials involved greater BOLD signal in the caudate nucleus than nonswitch trials. The authors therefore propose that basal ganglia are important for switching of planned responses by over-riding biases in the response system toward a particular action, and that this is necessary to enable changes in responses to occur when necessary, as new stimuli and information are received.

The relevance of these two preceding examples to clinical assessments is demonstrated by recent studies using the same method to study a range of neurological conditions and normal development and aging. The studies include the effects of Parkinson's disease (52,53), attention-deficit hyperactivity disorder (ADHD) (54), changes with brain development (55,56) and with normal aging (66). One example of results from a study of the effects of normal aging is shown in Figure 9.2, with a comparison between activity involved with pro- and antisaccade eye-movement tasks.

In one example, Cameron et al. (52,53) carried out a study of patients with Parkinson's disease, both on and off their medication, and compared results with age-matched healthy control subjects. Patients with Parkinson's disease show impairments in performing antisaccades, presumably because this action requires suppression of the more automatic prosaccade response. Having the patients attempt to perform antisaccades therefore provides a test of executive function as well as volitional movement initiation. In this study a rapid event-related design was used, which means that individual trials are close enough together in time that the BOLD responses to each event can overlap. As discussed in Chapter 7, the fast event-related method provides higher design efficiency than event-related methods and yet still enables detection of responses to individual trials, including when the participant makes an error. The tasks for the fMRI study were again very similar to those described in the study above and involved having the participants look at a fixation point, which then changed color to indicate where to perform a pro-saccade or an antisaccades. The target point then appeared to the right or left of the fixation point, and the participants' responses were recorded continuously via an MRI-compatible eye-tracking system.

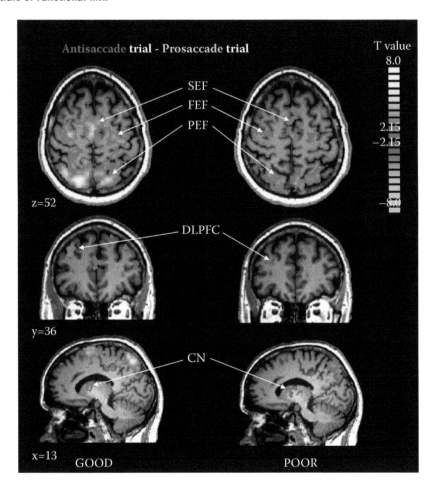

FIGURE 9.2 Examples of fMRI results obtained from a group of healthy volunteers aged 61 to 83 years old. The results show the differences in activity in the brain between the performance of antisaccades and prosaccades and are shown separately for volunteers who performed the task well (<10% error rate) and those who did not. The group results were obtained with a fixed effects analysis (FFX), and show that all subjects had higher activity in the antisaccade task in the dorsolateral prefrontal cortex (DLPFC), frontal eye fields (FEF), supplementary eye fields (SEF), parietal eye fields (PEF), and caudate nucleus (CN). Those who performed the antisaccade task well exhibited increased BOLD signal change in the DLPFC compared with subjects who had a higher rate of errors, suggesting that higher performance might involve increased recruitment of frontal regions. (Figure courtesy of Peltsch, A. et al. Organization for Human Brain Mapping, 15th Annual Meeting, June 18–23, 2009 (57).)

Functional MRI data were acquired at 3 tesla, using imaging with an EPI encoding scheme. The acquisition parameters included an echo time of 30 msec, a repetition time of 1500 msec, and a flip angle of 72°. The spatial parameters included a field of view of 211 mm × 211 mm, with an acquisition matrix of 64 × 64, and 24 axial-oblique slices each 3.3 mm thick. The resulting voxels were cubic with dimensions 3.3 mm × 3.3 mm × 3.3 mm.

The fMRI data were analyzed using a deconvolution-based general linear model (GLM) using BrainVoyager analysis software (see Section 8.2). Regions of interest were identified in the supplementary eye fields (SEF), frontal eye fields (FEF), parietal eye fields (PEF), caudate nucleus (CD), and dosolateral prefrontal cortex (DL-PFC). The peak of BOLD response in each region was detected, and 5 voxels around this peak were selected for characterizing the BOLD responses in each region.

Results from fMRI and behavioral testing showed differences in the magnitude of peak BOLD responses during both preparation and execution phases of the responses. Patients with Parkinson's disease had a greater number of direction errors, compared with control subjects, in which they failed to suppress a prosaccade. This finding corresponded with fMRI results showing decreased cortical activity in areas critical to antisaccade generation, particularly during preparation. This is consistent with a model of increased basal ganglia inhibition on thalamo-cortical circuits. Comparisons of Parkinson's patients on and off their medications also showed differences, including enhanced overall activity in the DL-PFC and CD when off medications. The observed difference in activity also corresponded with differences in task performance.

Within the same research group, another study was done of children with ADHD, using almost identical fMRI methods (54). People with ADHD have difficulty exerting voluntary control over saccade generation, have more variable reaction times, and make more direction errors in antisaccade tasks (58–61). They are also unable to inhibit saccades during prolonged fixation and make saccades prematurely when asked to delay their response after appearance of a visual target (60,62,63). As a result, studies of saccadic eye movements provide a means of observing the regions of the brain that are altered in ADHD as compared with healthy control subjects.

There are three symptom-based subtypes of ADHD: mainly inattentive, mainly hyperactive-impulsive, or both combined. The combined (ADHD-C) and inattentive (ADHD-I) subtypes are the most common. The goal of this study was to investigate the neural correlates of saccade control in ADHD-C and ADHD-I children. Using a rapid event-related design, 15 ADHD-C children, 15 ADHD-I children, and 15 age-matched control volunteers performed pro- and anti-saccade trials that were interleaved with pro- and anticatch trials (i.e., instruction was presented but no target appeared, requiring no response). This method enabled examination of brain activity patterns when each group either prepared or executed correct antisaccades. Eye-tracking was simultaneously recorded during all fMRI experiments.

Functional MRI results demonstrated that when preparing to make a correct antisaccade, control participants had increased activity (higher BOLD signal) in the dorsolateral prefrontal cortex and frontal, supplementary, and parietal eye fields. However, ADHD-C and ADHD-I children only exhibited activity in the frontal and supplementary eye fields, respectively. When executing a correct antisaccade, all three groups showed activity in parietal eye fields. Control and ADHD-I children also had activity in frontal eye fields, while control and ADHD-C children showed activity in supplementary eye fields. Activity in the caudate nucleus and dorsolateral prefrontal cortex was only observed in the ADHD-C group. The authors therefore concluded that both ADHD-C and ADHD-I children are impaired in their flexible, voluntary control of saccadic behavior, producing more direction errors during antisaccade trials. During antisaccade preparation, ADHD-C and ADHD-I groups recruited very few areas of the oculomotor network compared with controls, implying that children with ADHD do not differentially prepare between pro- and antisaccades, thus leading to more direction errors in the antisaccade task. The authors also noted that caudate and dorsolateral prefrontal cortex activity was only observed in ADHD-C children during antisaccade execution, suggesting that these children needed to employ additional brain areas to successfully inhibit an unwanted, automatic response.

The fMRI methods used for all of these studies employed eye tracking and visual display of relatively simple graphics to provide fixation points, colored to provide instructions, and target points offset to the right or left. The acquisition parameters are summarized in Table 9.7. In comparison with other examples of fMRI parameters, as listed above, this application requires higher temporal resolution, and so the authors reduced the number of slices, and therefore the anatomical coverage, to gain this speed. The repetition time (TR) was set to the minimum possible with the number of slices selected, and the flip angle was adjusted accordingly to maximize the signal-to-noise ratio.

Table 9.7 Examples of Imaging Parameters Used for fMRI Studies of Saccadic Eye Movements; Selected Examples Were Taken from a Selection of Published Scientific Papers and Recent Conference Presentations, and All Use a Gradient-Echo Imaging Method with an EPI Encoding Scheme.

Reference	Field Strength (Tesla)	TE (msec)	TR (msec)	Flip Angle (deg)	Resolution (mm)	Time Points	Slices
Connolly et al. 2002	4	28	500	30	3 × 3 × 6	1056	6, axial slices
Cameron et al. 2009a	3	30	750	56	3.3 × 3.3 × 3.3	448	11, axial slices
Cameron et al. 2009a	3	30	1000	62	3.3 × 3.3 × 3.3	336	16, axial slices
Cameron et al. 2009b, 2010	3	30	1500	72	3.3 × 3.3 × 3.3	224	24, axial slices

Sources: Connolly, J.D. et al., *Nat Neurosci* 5, 12, 1345–1352, 2002 (50); Cameron, I.G. et al., *Eur J Neurosci* 29, 12, 2413–2425, 2009a (51); Cameron, I.G.M. et al., Organization for Human Brain Mapping; 15th Annual Meeting, June 18–23, 2009b (52); Cameron, I.G.M. et al., Society for the Neural Control of Movement Annual Meeting 2010 (53).

9.2.7 Spinal Cord Injury and Disease

Functional MRI of the spinal cord (spinal fMRI) has been under development, and its use expanding, almost as long as has fMRI of the brain, with the first paper on spinal fMRI being published in 1996 (64,65). Its use has not expanded as rapidly as brain fMRI, however, partly because it meets a narrower scope of applications for the broad fields of neuroscience, and because of the inherent technical challenges. To date, there have almost 50 papers published on spinal fMRI in humans and animals. The current methods have been proposed to be ready for clinical trials aimed at assessing the effects of spinal cord injury (SCI) and disease and for use as a tool for spinal cord research.

The need for a noninvasive method of mapping neural function in the spinal cord, such as fMRI, is demonstrated by the fact that clinical decisions about the best treatment course to take following trauma to the spinal cord, or after the effects of diseases such as multiple sclerosis, require knowledge of how the spinal cord is functioning. This knowledge is obtained from tests that must be applied without causing additional damage to the cord or undue pain or stress to the patient (66). Current standard clinical tests for spinal cord function include the American Spinal Injury Association (ASIA) International Standards for Neurological Classification (67), which involves pin-prick tests across dermatomes and motor tests of various muscle groups. Electrophysiological tests involving stimulation of cortical areas and recording of motor and sensory-evoked potentials are also used to reveal functional connections. Other assessments are primarily surveys of the patient's abilities and quality-of-life factors, such as functional independence measure, functional assessment measure, the Spinal Cord Independence Measure, and the Walking Index for Spinal Cord Injury. These tests are limited because they do not reveal information about spinal cord function below the most rostral point of damage, thoracic regions of the cord are difficult to assess, or the assessments are subjective. None of these tests reveal information about the causes of bowel or bladder dysfunction or sexual dysfunction, and they can rarely reveal if there are multiple sites of damage. Functional MRI of the spinal cord has been shown to demonstrate activity caudal to sites of SCI at any level of the cord (68,69), the effects of multiple sclerosis (70,71) and peripheral nerve damage (72), and can also demonstrate spinal cord activity related to sexual function (73), and therefore has the potential to provide the needed information that cannot be obtained with current clinical assessments.

The most commonly used method for fMRI of the spinal cord employs SEEP (signal enhancement by extravascular water protons) contrast, as described in Chapter 6. Functional MRI data are acquired with single-shot fast spin-echo imaging to provide good image quality with low spatial distortion in the presence of magnetic susceptibility differences between the bone and cartilage of the spine and the tissues of the spinal cord. Since SEEP contrast can be detected with proton-density weighting, most of the published papers employ a relatively short echo time, TE, of around 30–40 msec. Data are acquired in thin contiguous sagittal slices, 2 mm thick, with in-plane spatial resolution of 1 mm × 1 mm to 1.5 mm × 1.5 mm, and fields of view of 20 cm × 10–28 cm × 21 cm (74,75). Image quality in the spinal cord is improved by the use of spatial saturation pulses to eliminate signal from all structures anterior to the spine, such as the heart, lungs, throat, tongue, and so forth and with the use of flow-compensating gradients to reduce cerebrospinal fluid flow artifacts in the head-foot direction.

9.2.7.1 True Physiological Variation

A key source of variation that has been shown in fMRI results is the true differences in neuronal activity that can occur between repeated studies. While this variability cannot be considered an error, per se, because its detection demonstrates the reliability and sensitivity of the fMRI method, it can present a challenge for repeated or group studies, as well as for assessment of individual patients.

It is well known (76,77) that activity in the spinal cord in both ventral and dorsal regions is modulated by descending input from the brainstem and cortex, and depends on factors of awareness, alertness, and attention as well as control of motor reflex responses. Studies of emotional and cognitive influences on activity in the spinal cord have been carried out by systematically varying participants' attention focus, whether toward a thermal sensation on the hand, toward a movie, or toward mentally challenging tasks (78,79). The results showed that activity in the cord, in response to a thermal sensory stimulus, did indeed depend on the participant's attention focus in each study. A separate study demonstrated that the activity in the spinal cord, in response to thermal stimuli applied to the hand, depends both on the stimulus temperature and on the order of experiments (74). This result again implicates factors of emotion and attention. From these studies it was concluded that a significant source of variation in spinal cord activity, at least in response to thermal sensations, is true physiological variation of neuronal activity.

The overall conclusions from these studies are that emotional and attention factors need to be controlled, as much as possible, in spinal fMRI studies of any specific function. Even changes in anxiety or alertness over time, as participants become accustomed to being inside the MRI system and potentially become bored with the study, were seen to affect spinal fMRI results (74,79). These observations therefore demonstrate that spinal fMRI results are highly sensitive to true neuronal activity, as they were able to detect this unexpected variation in the responses.

9.2.7.2 Human Studies and Clinical Applications

Our understanding of the normal functioning of the healthy spinal cord has also been advanced by the results obtained over the past several years with spinal fMRI of human subjects. Although this understanding is important, the ultimate goal of spinal fMRI research is reported as being to provide a means of observing function in the injured and diseased spinal cord to assist in diagnosis, prognosis, and treatment planning and monitoring. To this end, a number of clinically related studies have been carried out to date in patient populations with spinal cord injury (SCI), neuropathic pain, and multiple sclerosis (MS) using a variety of sensory stimuli and motor or proprioceptive tasks.

Sensory-related neuronal activity in the spinal cord (SC) has been consistently observed with fMRI in healthy and clinical subject populations in a number of studies since 2002 (69,70,74,80–87). Early studies of patients with SCI investigated the functional response in the lumbar SC to thermal stimulation of the fourth lumbar dermatome (69,80). The thermal probe was placed against the inner skin of the calf, and the temperature was ramped from skin temperature (32°C) to 10°C at different rates in repeated experiments. Neuronal activity was consistently observed in the lumbar SC caudal to the site of injury, regardless of whether the subject could consciously feel the stimulus or not. Though the percent signal change detected in images of the spinal cord was similar between healthy controls and SCI patients, the spatial distribution of activity was notably different. While healthy volunteers showed predominantly ipsilateral dorsal gray-matter activity in response to sensory stimuli, absent or diminished activity in dorsal gray matter and enhanced contralateral ventral gray-matter activity was observed in subjects with complete SCI (patients who were unable to feel the stimulus). Results from patients with incomplete SCI were essentially divided based on the degree to which subjects were able to perceive the stimulus. For subjects with preserved sensation, the observed activity pattern was very similar to that of healthy controls (consistent ipsilateral dorsal gray-matter activity, in addition to central and bilateral ventral gray-matter activity). Conversely, patients with decreased sensation exhibited diminished ipsilateral dorsal gray-matter activity (similar to complete SCI), yet bilateral ventral gray-matter activity was similar (in some cases even diminished) compared with healthy controls. The ability of spinal fMRI to distinguish subtle functional differences between well-established classes of SCI lends credence to its capacity to quantitatively assess the function of the SC.

Spinal cord and brainstem fMRI responses to noxious stimuli have been investigated in both healthy and neuropathic pain patient populations (72,80). In healthy controls, when a thermal stimulus was ramped from 29°C to 15°C, the percent signal change ranged only between 2% and 3%. However, when that stimulus is further ramped to 10°C (reported as noxious), the percent signal change more than doubled to approximately 8%, and the neuronal activity observed became concentrated in superficial regions of the ipsilateral dorsal horn (80), corresponding to Rexed's laminae I and II, which is consistent with known pain pathways. A specific and concentrated neuronal response to noxious insult prompted the investigation of the difference between healthy and pathological pain processing. A recent study focusing specifically on pain pathways in neuropathic pain patients yields further insight into the normal and pathological processing of noxious stimuli (72). When painful pressure was applied to the median nerve of both healthy controls and patients diagnosed with carpal tunnel syndrome (CTS, a common neuropathy caused by compression of the median nerve), several differences were observed. While both groups showed consistent activity in the spinal cord, with pressure applied to produce the same pain ratings in both groups, the CTS patients showed more activity in areas of the brainstem (nucleus raphe magnus [NRM]) and in the contralateral dorsal horn of the spinal cord. These areas are known to be involved in the descending modulation of pain responses, indicating that changes in this descending modulation may play an important role in the perceived pain in this condition.

Spinal fMRI has also been used to assess and compare the functional differences in the SC gray matter between healthy controls and patients with relapsing-remitting MS. Neuronal activity in the cervical spinal cord was investigated following tactile stimulation of the palm of the right hand, and was found from C5 to C8 in all patients and controls (78), which corresponds to the expected regions of neuronal recruitment. In general, MS patients showed approximately 20% greater signal intensity changes than controls (3.9% compared with 3.2%), with activity dispersed throughout the dorsal, central, and ventral cord, most likely attributable to the interneuronal systems of the SC (88–90). Interestingly, MS patients tended to show an over-recruitment of dorsal gray matter (i.e., show bilateral dorsal activity, whereas healthy controls showed predominantly ipsilateral dorsal activity), which is indicative of reduced functional lateralization in the SC. This result appears to support SC gray-matter re-organization (as was previously found in the cortex (91), as well as postmortem (92–95) and *in vivo* MRI studies (96) of the spinal cord, which suggest that gray matter is not spared by MS pathology. The purpose of this gray-matter reorganization is not yet clear. However, spinal fMRI could serve as the precise tool needed to assess changes in the functional activity of gray matter throughout the evolution of the disease and may yield insight as to whether these changes are predictive of clinical outcome.

Neuronal activity in the SC related to various motor tasks has been demonstrated by a number of groups (64,68,71,82,97–103). In 2007, 12 patients with SCI, classified as ASIA A (no sensory or motor function preserved, n = 4), ASIA B (sensory but no motor function preserved, n = 3), ASIA C (weak motor function is preserved, n = 3), or ASIA D (motor function preserved in a condition sufficient for near-normal use, n = 2), performed a pedaling motor task (68). All subjects participated in the passive task (researcher manually moved pedals and subjects' feet move in pedaling motion), but only ASIA C and D patients performed the active task (autonomous alternating pedaling). Consistent with results from studies in SCI patients using sensory stimulus, neuronal activity was detected caudal to the site of injury in all subjects, regardless of the extent or level of injury. The number of active voxels in the lumbar SC was greater during active compared with passive participation; however, the overall percent signal change was greater during passive (15.0%) compared with active (13.6%) pedaling. The pattern of distribution of neuronal activity in SCI patients was similar to healthy controls for each task. Active participation resulted in bilateral neuronal activity in both dorsal and ventral horns, corresponding to a neuronal response to motor and sensory stimulation, typical of purposeful movement. Passive participation yielded some ventral horn activity, but most activity was seen in the dorsal horn, typical of a neuronal response to proprioceptive and mechanical information produced by this type of movement. Also, the number of active voxels detected in the SC of each subject population mirrored the severity of the impairment. That is, fewer active voxels were detected in the SC of ASIA C/D SCI patients than in the healthy control group (99). Likewise, still fewer active voxels were observed in the SC of ASIA A SCI subjects compared with ASIA C/D SCI subjects. Perhaps most intriguingly, six subjects were able to use only one limb during active participation (unilateral movement generation), as opposed to typical pedaling with both feet (bilateral movement generation). The latter results in neuronal activity distributed across both sides of the cord. In this study, it was found that during unilateral movement generation, neuronal activity appears to be prominent in the contralateral ventral horn. This corresponds with known physiology (69,104–107) and verifies once again that spinal fMRI is able to reliably and sensitively detect subtle differences in neuronal function. Though spinal fMRI cannot (at this time) determine the cause of the observed activity patterns, it can be used to supplement the ASIA diagnosis with functional activity maps, providing additional insight into SC physiology and enhancing the design of rehabilitation programs. Neuronal function could be investigated before, during, and after rehabilitation to provide a quantitative measure of progress in addition to the qualitative measure provided by ASIA and other subjective (outcome-based) tests.

A study similar to the passive participation pedaling task has been carried out in an MS patient population as well, investigating the extent of cervical SC functional activity in healthy controls and patients with remitting-relapsing (RR) or secondary progressive (SP) course MS (71). A passive, calibrated, 45° flexion/extension was repetitively administered to the relaxed, prone, right hand of the patient by the researcher. Activity was observed in the cervical SC from C5 to C8 in all subjects, but several differences between controls and patients were noted. Approximately a 20% greater signal intensity change was observed in the cervical SC of MS patients (3.4%) compared with controls (2.7%), analogous to results from the study investigating the spinal fMRI effects of tactile stimulation of the palm in MS patients and healthy controls (70). Also, increased bilateral ventral gray-matter activity was observed in MS patients compared with controls. Spinal fMRI can be used not only to detect differences between patient and control populations, but also to investigate functional differences between various classifications of MS severity. One study has shown that patients with less severe MS (RRMS) had a task-related spinal fMRI activity pattern similar to healthy controls, while patients with more severe MS (SPMS) showed a pattern of cord activity more similar to SCI patients. If cervical cord functional activity varies over the course of the disease, spinal fMRI could prove incredibly useful in assessing the nature and evolution of MS within individual patients.

Based on these examples that have been reported to date, spinal fMRI has been proposed to provide invaluable information that can in turn be used for assessing residual function, designing rehabilitation programs, predicting the potential for recovery of function in SCI patients, and also assessing new experimental treatment strategies. A recently reported study employed a custom-made apparatus to apply multipoint thermal stimulation in a patient with a complete (ASIA A) C5 spinal cord injury, with a metallic fixation device spanning the C4 to C6 vertebrae (65). Functional activity was observed in the spinal cord both rostral and caudal to the site of injury, in spite of the fact that the patient did not consciously perceive all of the stimuli. Importantly, functional activity was detected in the SC in close proximity to the fixation device, and the fMRI data took only 7 minutes to acquire. This is a critical step toward implementing spinal fMRI as a routine clinical tool. Perhaps the most significant result that has emerged from this preliminary study is the detection of a significantly greater function on the left side of the body below the injury level, suggesting the presence of some preserved descending pathways on the left side, although clinical assessments indicated only slight right/left differences (Figure 9.3). Although this is a single case study, the results present an intriguing example of the information that may be made available for planning of rehabilitation therapy, and for monitoring progressive changes over time.

9.2.8 Concluding Points about the Use of fMRI for Clinical Diagnosis and Monitoring

As can be seen from the examples above, fMRI has been used to date only for a relatively narrow range of applications in clinical practice, all related to presurgical mapping. However, quite a number of examples show that fMRI has potential for clinical assessment of conditions such as multiple sclerosis, stroke, Parkinson's disease, Alzheimer's disease, ADHD, schizophrenia, mild cognitive impairment, traumatic brain injury, disorders of consciousness, spinal cord injury, and normal development and aging. Many of these examples can reasonably be extended to other neurological disorders as well. The greatest variation in the fMRI methods across all of these conditions is seen to be in the study design, in terms of method and timing of tasks or stimulation, with much more consistency in the fMRI acquisition parameters. However, there are still key differences in the acquisition parameters, particularly in the repetition time, TR, because this impacts directly on the sampling rate of the fMRI data, and in the image resolution because this results from a choice of balance between acquisition speed, anatomical coverage, signal-to-noise ratio, and spatial specificity.

The key areas that appear to be well-suited for clinical use, in addition to the presurgical mapping that is already being done, are the methods involving fMRI of the resting state and fMRI of saccadic eye movements. These methods stand out among the examples listed above because each involves a single method that can be standardized and used to examine a wide range of disorders. Instead of acquiring data only to support or refute one specific diagnosis, these methods could be used to indicate which neurological condition is most likely and could complement conventional tests.

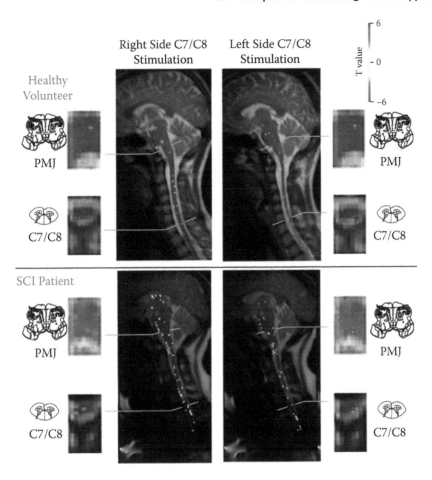

FIGURE 9.3 Comparison of fMRI results obtained from a healthy control volunteer and a person with a cervical spinal cord injury, with a metal fixation plate spanning the injury site, and metal screws in the C4 and C6 vertebral bodies. In each participant thermal stimuli at 44°C were applied to four different dermatomes in distinguishable patterns in time so that the response to each can be identified. Stimulation was applied symmetrically on the right and left sides, to the little-finger side of the palm in both participants, and to the upper arm in the healthy participant and the shoulder in the person with a spinal cord injury. The results are shown only for stimulation of the right and left palms in both people, in selected 2 mm sagittal slices showing the right-side and left-side activity, and in selected 1 mm transverse slices at the level of the ponto-medullary junction (PMJ) and at the level of the C7 or C8 spinal cord segment. The results show clearly symmetric responses to right- and left-side stimulation in the healthy participant with activity in the right and left dorsal horn of the gray matter, respectively. The responses detected in the injured participant show an almost normal-appearing response to left-side stimulation, but a very different response with stimulation of the right side, indicating preserved pathways through the injury site, consistent with clinical assessments using the ASIA standard.

References

1. Schleim S, Roiser JP. FMRI in translation: The challenges facing real-world applications. *Front Hum Neurosci* 2009;3:63.
2. Pillai JJ. The evolution of clinical functional imaging during the past 2 decades and its current impact on neurosurgical planning. *AJNR Am J Neuroradiol* 2010;31(2):219–225.
3. Haller S, Bartsch AJ. Pitfalls in fMRI. *Eur Radiol* 2009;19(11):2689–2706.
4. Tieleman A, Deblaere K, Van RD, Van DO, Achten E. Preoperative fMRI in tumour surgery. *Eur Radiol* 2009;19(10):2523–2534.
5. Voyvodic JT, Petrella JR, Friedman AH. fMRI activation mapping as a percentage of local excitation: Consistent presurgical motor maps without threshold adjustment. *J Magn Reson Imaging* 2009; 29(4):751–759.
6. Laatsch L. The use of functional MRI in traumatic brain injury diagnosis and treatment. *Phys Med Rehabil Clin N Am* 2007;18(1):69–85, vi.
7. Wada J, Rasmussen T. Intracarotid injection of sodium amytal for the lateralization of cerebral speech dominance: Experimental and clinical observations. *J Neurosurg* 1960;17(2):266–282.
8. Yetkin FZ, Mueller WM, Morris GL, McAuliffe TL, Ulmer JL, Cox RW, Daniels DL, Haughton VM. Functional MR activation correlated with intraoperative cortical mapping. *AJNR Am J Neuroradiol* 1997;18(7):1311–1315.
9. Petrella JR, Shah LM, Harris KM, Friedman AH, George TM, Sampson JH, Pekala JS, Voyvodic JT. Preoperative functional MR imaging localization of language and motor areas: Effect on therapeutic decision making in patients with potentially resectable brain tumors. *Radiology* 2006;240(3): 793–802.
10. Fitzgerald DB, Cosgrove GR, Ronner S, Jiang H, Buchbinder BR, Belliveau JW, Rosen BR, Benson RR. Location of language in the cortex: a comparison between functional MR imaging and electrocortical stimulation. *AJNR Am J Neuroradiol* 1997;18(8):1529–1539.
11. Majos A, Tybor K, Stefanczyk L, Goraj B. Cortical mapping by functional magnetic resonance imaging in patients with brain tumors. *Eur Radiol* 2005;15(6):1148–1158.
12. Arora J, Pugh K, Westerveld M, Spencer S, Spencer DD, Todd CR. Language lateralization in epilepsy patients: fMRI validated with the Wada procedure. *Epilepsia* 2009;50(10):2225–2241.
13. Bartos R, Jech R, Vymazal J, Petrovicky P, Vachata P, Hejcl A, Zolal A, Sames M. Validity of primary motor area localization with fMRI versus electric cortical stimulation: A comparative study. *Acta Neurochir* (Wien) 2009;151(9):1071–1080.
14. Xie J, Chen XZ, Jiang T, Li SW, Li ZX, Zhang Z, Dai JP, Wang ZC. Preoperative blood oxygen level-dependent functional magnetic resonance imaging in patients with gliomas involving the motor cortical areas. *Chinese Med J* 2008;121(7):631–635.
15. Berntsen EM, Samuelsen P, Lagopoulos J, Rasmussen IA, Jr., Haberg AK, Haraldseth O. Mapping the primary motor cortex in healthy subjects and patients with peri-rolandic brain lesions before neurosurgery. *Neurol Res* 2008;30(9):968–973.
16. Gasser T, Ganslandt O, Sandalcioglu E, Stolke D, Fahlbusch R, Nimsky C. Intraoperative functional MRI: implementation and preliminary experience. *NeuroImage* 2005;26(3):685–693.
17. Nelson L, Lapsiwala S, Haughton VM, Noyes J, Sadrzadeh AH, Moritz CH, Meyerand ME, Badie B. Preoperative mapping of the supplementary motor area in patients harboring tumors in the medial frontal lobe. *J Neurosurg* 2002;97(5):1108–1114.
18. Buckner RL, Andrews-Hanna JR, Schacter DL. The brain's default network: Anatomy, function, and relevance to disease. *Ann NY Acad Sci* 2008;1124:1–38.
19. Murphy K, Birn RM, Handwerker DA, Jones TB, Bandettini PA. The impact of global signal regression on resting state correlations: Are anti-correlated networks introduced? *NeuroImage* 2009;44(3):893–905.
20. Zou Q, Long X, Zuo X, Yan C, Zhu C, Yang Y, Liu D, He Y, Zang Y. Functional connectivity between the thalamus and visual cortex under eyes closed and eyes open conditions: A resting-state fMRI study. *Hum Brain Mapp* 2009;30(9):3066–3078.
21. Chang C, Glover GH. Relationship between respiration, end-tidal CO_2, and BOLD signals in resting-state fMRI. *NeuroImage* 2009;47(4):1381–1393.
22. Shehzad Z, Kelly AM, Reiss PT, Gee DG, Gotimer K, Uddin LQ, Lee SH, Margulies DS, Roy AK, Biswal BB, Petkova E, Castellanos FX, Milham MP. The resting brain: Unconstrained yet reliable. *Cereb Cortex* 2009;19(10):2209–2229.

23. Qi Z, Wu X, Wang Z, Zhang N, Dong H, Yao L, Li K. Impairment and compensation coexist in amnestic MCI default mode network. *NeuroImage* 2010;50(1):48–55.

24. Greicius MD, Srivastava G, Reiss AL, Menon V. Default-mode network activity distinguishes Alzheimer's disease from healthy aging: evidence from functional MRI. *Proc Natl Acad Sci USA* 2004;101(13): 4637–4642.

25. Wang K, Liang M, Wang L, Tian L, Zhang X, Li K, Jiang T. Altered functional connectivity in early Alzheimer's disease: A resting-state fMRI study. *Hum Brain Mapp* 2007;28(10):967–978.

26. Morgan VL, Gore JC, bou-Khalil B. Functional epileptic network in left mesial temporal lobe epilepsy detected using resting fMRI. *Epilepsy Res* 2010;88(2–3):168–178.

27. Cauda F, Sacco K, Duca S, Cocito D, D'Agata F, Geminiani GC, Canavero S. Altered resting state in diabetic neuropathic pain. *PLoS One* 2009;4(2):e4542.

28. Helmich RC, Derikx LC, Bakker M, Scheeringa R, Bloem BR, Toni I. Spatial remapping of cortico-striatal connectivity in Parkinson's disease. *Cereb Cortex* 2010;20(5):1175–1186.

29. Bluhm RL, Miller J, Lanius RA, Osuch EA, Boksman K, Neufeld RW, Theberge J, Schaefer B, Williamson P. Spontaneous low-frequency fluctuations in the BOLD signal in schizophrenic patients: Anomalies in the default network. *Schizophr Bull* 2007;33(4):1004–1012.

30. Craddock RC, Holtzheimer PE, III, Hu XP, Mayberg HS. Disease state prediction from resting state functional connectivity. *Magn Reson Med* 2009;62(6):1619–1628.

31. Broyd SJ, Demanuele C, Debener S, Helps SK, James CJ, Sonuga-Barke EJ. Default-mode brain dysfunction in mental disorders: A systematic review. *Neurosci Biobehav Rev* 2009;33(3):279–296.

32. Meindl T, Teipel S, Elmouden R, Mueller S, Koch W, Dietrich O, Coates U, Reiser M, Glaser C. Test–retest reproducibility of the default-mode network in healthy individuals. *Hum Brain Mapp* 2010;31(2):237–246.

33. Franco AR, Pritchard A, Calhoun VD, Mayer AR. Interrater and intermethod reliability of default mode network selection. *Hum Brain Mapp* 2009;30(7):2293–2303.

34. Filippi M, Rocca MA. Functional MR imaging in multiple sclerosis. *Neuroimaging Clin N Am* 2009; 19(1):59–70.

35. Rocca MA, Riccitelli G, Rodegher M, Ceccarelli A, Falini A, Falautano M, Meani A, Comi G, Filippi M. Functional MR imaging correlates of neuropsychological impairment in primary-progressive multiple sclerosis. *AJNR Am J Neuroradiol* 2010.

36. Wegner C, Filippi M, Korteweg T, Beckmann C, Ciccarelli O, De SN, Enzinger C, Fazekas F, Agosta F, Gass A, Hirsch J, Johansen-Berg H, Kappos L, Barkhof F, Polman C, Mancini L, Manfredonia F, Marino S, Miller DH, Montalban X, Palace J, Rocca M, Ropele S, Rovira A, Smith S, Thompson A, Thornton J, Yousry T, Matthews PM. Relating functional changes during hand movement to clinical parameters in patients with multiple sclerosis in a multi-centre fMRI study. *Eur J Neurol* 2008;15(2):113–122.

37. Lowe MJ, Beall EB, Sakaie KE, Koenig KA, Stone L, Marrie RA, Phillips MD. Resting-state sensorimotor functional connectivity in multiple sclerosis inversely correlates with transcallosal motor pathway transverse diffusivity. *Hum Brain Mapp* 2008;29(7):818–827.

38. Page SJ, Harnish SM, Lamy M, Eliassen JC, Szaflarski JP. Affected arm use and cortical change in stroke patients exhibiting minimal hand movement. *Neurorehabil Neural Repair* 2010;24(2):195–203.

39. Riecker A, Groschel K, Ackermann H, Schnaudigel S, Kassubek JR, Kastrup A. The role of the unaffected hemisphere in motor recovery after stroke. *Hum Brain Mapp* 2010;31(7):1017–1029.

40. Menke R, Meinzer M, Kugel H, Deppe M, Baumgartner A, Schiffbauer H, Thomas M, Kramer K, Lohmann H, Floel A, Knecht S, Breitenstein C. Imaging short- and long-term training success in chronic aphasia. *BMC Neurosci* 2009;10:118.

41. Bernat JL. Chronic consciousness disorders. *Annu Rev Med* 2009;60:381–392.

42. Owen AM, Coleman MR, Boly M, Davis MH, Laureys S, Pickard JD. Detecting awareness in the vegetative state. *Science* 2006;313(5792):1402.

43. Coleman MR, Davis MH, Rodd JM, Robson T, Ali A, Owen AM, Pickard JD. Towards the routine use of brain imaging to aid the clinical diagnosis of disorders of consciousness. *Brain* 2009;132(Pt 9):2541–2552.

44. Coleman MR, Rodd JM, Davis MH, Johnsrude IS, Menon DK, Pickard JD, Owen AM. Do vegetative patients retain aspects of language comprehension? Evidence from fMRI. *Brain* 2007; 130(Pt 10):2494–2507.

45. Eickhoff SB, Dafotakis M, Grefkes C, Stocker T, Shah NJ, Schnitzler A, Zilles K, Siebler M. fMRI reveals cognitive and emotional processing in a long-term comatose patient. *Exp Neurol* 2008;214(2): 240–246.

46. Slobounov SM, Zhang K, Pennell D, Ray W, Johnson B, Sebastianelli W. Functional abnormalities in normally appearing athletes following mild traumatic brain injury: A functional MRI study. *Exp Brain Res* 2010;202(2):341–354.

47. Kohl AD, Wylie GR, Genova HM, Hillary FG, Deluca J. The neural correlates of cognitive fatigue in traumatic brain injury using functional MRI. *Brain Inj* 2009;23(5):420–432.

48. Nakamura T, Hillary FG, Biswal BB. Resting network plasticity following brain injury. *PLoS One* 2009;4(12):e8220.

49. Bullmore E, Sporns O. Complex brain networks: Graph theoretical analysis of structural and functional systems. *Nat Rev Neurosci* 2009;10(3):186–198.

50. Connolly JD, Goodale MA, Menon RS, Munoz DP. Human fMRI evidence for the neural correlates of preparatory set. *Nat Neurosci* 2002;5(12):1345–1352.

51. Cameron IG, Coe BC, Watanabe M, Stroman PW, Munoz DP. Role of the basal ganglia in switching a planned response. *Eur J Neurosci* 2009;29(12):2413–2425.

52. Cameron IGM, Pari G, Alahyane N, Coe BC, Stroman PW, Munoz DP. Deficits in Anti-Saccade Performance in Parkinson's Disease: A Rapid Event-Related fMRI study. Organization for Human Brain Mapping 2009; 15th Annual Meeting, June 18–23, San Francisco, CA.

53. Cameron IGM, Pari G, Alahyane N, Coe BC, Stroman PW, Munoz DP. Correlates of Voluntary Saccade Impairment and Automatic Saccade Bias in Parkinson's Disease. Society for the Neural Control of Movement Annual Meeting 2010; Naples FL.

54. Haakvoort-Schwerdtfeger RM, Alahyane N, Coe BC, Stroman PW, Munoz DP. Children with different subtypes of Attention Deficit/Hyperactivity Disorder (ADHD) Recruit Distinct Neural Networks During a Task of Executive Function. Organization for Human Brain Mapping 2009; 15th Annual Meeting, June 18–23, San Francisco, CA.

55. Alahyane N, Coe BC, Stroman PW, Munoz DP. Functional Neural Circuitry of Voluntary and Inhibitory Control Processes in the Developing Brain. Organization for Human Brain Mapping 2009; 15th Annual Meeting, June 18–23, San Francisco, CA.

56. Alahyane N, Coe BC, Stroman PW, Munoz DP. Function of Brain Networks Subserving Saccade Control in the Developing Brain. Society for the Neural Control of Movement Annual Meeting 2010; Naples FL.

57. Peltsch A, Alahyane N, Coe BC, Stroman PW, Garcia A, Munoz DP. Saccade characteristics in healthy aging assessed using fMRI. Organization for Human Brain Mapping 2009; 15th Annual Meeting, June 18–23, San Francisco, CA.

58. Karatekin C. Improving antisaccade performance in adolescents with attention-deficit/hyperactivity disorder (ADHD). *Exp Brain Res* 2006;174(2):324–341.

59. Mostofsky SH, Lasker AG, Singer HS, Denckla MB, Zee DS. Oculomotor abnormalities in boys with tourette syndrome with and without ADHD. *J Am Acad Child Adolesc Psychiatry* 2001;40(12): 1464–1472.

60. Munoz DP, Armstrong IT, Hampton KA, Moore KD. Altered control of visual fixation and saccadic eye movements in attention-deficit hyperactivity disorder. *J Neurophysiol* 2003;90(1):503–514.

61. O'Driscoll GA, Depatie L, Holahan AL, Savion-Lemieux T, Barr RG, Jolicoeur C, Douglas VI. Executive functions and methylphenidate response in subtypes of attention-deficit/hyperactivity disorder. *Biol Psychiatry* 2005;57(11):1452–1460.

62. Rommelse NN, van der Stigchel S, Witlox J, Geldof C, Deijen JB, Theeuwes J, Oosterlaan J, Sergeant JA. Deficits in visuo-spatial working memory, inhibition and oculomotor control in boys with ADHD and their non-affected brothers. *J Neural Transm* 2008;115(2):249–260.

63. Ross RG, Hommer D, Breiger D, Varley C, Radant A. Eye movement task related to frontal lobe functioning in children with attention deficit disorder. *J Am Acad Child Adolesc Psychiatry* 1994;33(6):869–874.

64. Yoshizawa T, Nose T, Moore GJ, Sillerud LO. Functional magnetic resonance imaging of motor activation in the human cervical spinal cord. *NeuroImage* 1996;4(3 Pt 1):174–182.

65. Leitch JK, Figley CR, Stroman PW. Applying functional MRI to the spinal cord and brainstem. *Magn Reson Imaging* 2010.

66. Committee on Spinal Cord Injury BoNaBH. *Spinal Cord Injury: Progress, Promise, and Priorities.* Washington, DC: The National Academies Press; 2005.

67. American Spinal Injury Association Neurological Standards Committee. *International Standards for Neurological Classification of Spinal Cord Injury.* 2000.

68. Kornelsen J, Stroman PW. Detection of the neuronal activity occurring caudal to the site of spinal cord injury that is elicited during lower limb movement tasks. *Spinal Cord* 2007;45(7):485–490.

69. Stroman PW, Kornelsen J, Bergman A, Krause V, Ethans K, Malisza KL, Tomanek B. Noninvasive assessment of the injured human spinal cord by means of functional magnetic resonance imaging. *Spinal Cord* 2004;42(2):59–66.

70. Agosta F, Valsasina P, Caputo D, Stroman PW, Filippi M. Tactile-associated recruitment of the cervical cord is altered in patients with multiple sclerosis. *NeuroImage* 2008;39(4):1542–1548.

71. Agosta F, Valsasina P, Rocca MA, Caputo D, Sala S, Judica E, Stroman PW, Filippi M. Evidence for enhanced functional activity of cervical cord in relapsing multiple sclerosis. *Magn Reson Med* 2008;59(5):1035–1042.

72. Leitch J, Cahill CM, Ghazni N, Figley CR, Stroman PW. Spinal cord and brainstem activation in carpal tunnel syndrome patients in response to noxious stimuli: A spinal fMRI study. International Society for Magnetic Resonance in Medicine 2009; 17th Annual Meeting, Honolulu, HI, May 18–25.

73. Kozyrev N, Sipski-Alexander M, Richards JS, Figley CR, Stroman PW. Mapping a neural model of sexual responses in the human spinal cord with functional magnetic resonance imaging. Society for Neuroscience 2008; 2008 Neuroscience Meeting Planner:577.12/QQ2.

74. Stroman PW. Spinal fMRI investigation of human spinal cord function over a range of innocuous thermal sensory stimuli and study-related emotional influences. *Magn Reson Imaging* 2009;27:1333–1346.

75. Stroman PW, Figley CR, Cahill CM. Spatial normalization, bulk motion correction and coregistration for functional magnetic resonance imaging of the human cervical spinal cord and brainstem. *Magn Reson Imaging* 2008;26(6):809–814.

76. Gebhart GF. Descending modulation of pain. *Neurosci Biobehav Rev* 2004;27(8):729–737.

77. Hoffman GA, Harrington A, Fields HL. Pain and the placebo: What we have learned. *Perspect Biol Med* 2005;48(2):248–265.

78. Stroman PW, Coe BC, Leitch J, Figley CR, Munoz DP. Attentional modulation of thermal sensory responses in the human spinal cord. International Society for Magnetic Resonance in Medicine 2009; 17th Annual Meeting, Honolulu, HI, May 18–25.

79. Stroman PW, Coe BC, Munoz DP. Influence of attention focus on neural activity involved with thermal sensations in the human spinal cord. *Magn Reson Imaging* 2010;29:9–18.

80. Stroman PW, Tomanek B, Krause V, Frankenstein UN, Malisza KL. Mapping of neuronal function in the healthy and injured human spinal cord with spinal fMRI. *NeuroImage* 2002;17(4):1854–1860.

81. Stroman PW, Krause V, Malisza KL, Frankenstein UN, Tomanek B. Functional magnetic resonance imaging of the human cervical spinal cord with stimulation of different sensory dermatomes. *Magn Reson Imaging* 2002;20(1):1–6.

82. Komisaruk BR, Mosier KM, Liu WC, Criminale C, Zaborszky L, Whipple B, Kalnin A. Functional localization of brainstem and cervical spinal cord nuclei in humans with fMRI. *AJNR Am J Neuroradiol* 2002;23(4):609–617.

83. Li G, Ng MC, Wong KK, Luk KD, Yang ES. Spinal effects of acupuncture stimulation assessed by proton density-weighted functional magnetic resonance imaging at 0.2 T. *Magn Reson Imaging* 2005; 23(10):995–999.

84. Zambreanu L, Wise RG, Brooks JC, Iannetti GD, Tracey I. A role for the brainstem in central sensitisation in humans. Evidence from functional magnetic resonance imaging. *Pain* 2005;114(3):397–407.

85. Ng MC, Wong KK, Li G, Lai S, Yang ES, Hu Y, Luk KD. Proton-density-weighted spinal fMRI with sensorimotor stimulation at 0.2 T. *NeuroImage* 2006;29(3):995–999.

86. Wang WD, Kong KM, Xiao YY, Wang XJ, Liang B, Qi WL, Wu RH. Functional MR imaging of the cervical spinal cord by use of electrical stimulation at LI4 (Hegu). *Conf Proc IEEE Eng Med Biol Soc* 2006;1:1029–1031.

87. Xie CH, Kong KM, Guan JT, Chen YX, Wu RH. Functional MR imaging of the cervical spinal cord by use of 20Hz functional electrical stimulation to median nerve. *Conf Proc IEEE Eng Med Biol Soc* 2007; 2007:3392–3395.

88. Williams PL, Warwick R. *Gray's Anatomy.* 36th ed. Edinburgh: Churchill Livingstone; 1980.

89. Brodal A. *Neurological Anatomy in Relation to Clinical Medicine*. New York: Oxford University Press; 1981.

90. Kandel E, Schwartz JH, Jessell TM. *Principles of Neural Science*. New York: Elsevier Science; 1991.

91. Filippi M, Rocca MA. Cortical reorganisation in patients with MS. *J Neurol Neurosurg Psychiatry* 2004;75(8):1087–1089.

92. Bot JCJ, Blezer ELA, Kamphorst W, Lycklama à Nijeholt GJ, Ader HJ, Castelijns JA, Ig KN, Bergers E, Ravid R, Polman C, Barkhof F. The spinal cord in multiple sclerosis: Relationship of high-spatial-resolution quantitative MR imaging findings to histopathologic results. *Radiology* 2004;233(2):531–540.

93. Gilmore CP, DeLuca GC, Bo L, Owens T, Lowe J, Esiri MM, Evangelou N. Spinal cord atrophy in multiple sclerosis caused by white matter volume loss. *Arch Neurol* 2005;62(12):1859–1862.

94. Gilmore CPM, Bo LM, Owens TD, Lowe JM, Esiri MMD, Evangelou ND. Spinal cord gray matter demyelination in multiple sclerosis: A novel pattern of residual plaque morphology. *Brain Pathol* 2006;16(3):202–208.

95. Nijeholt GJL, Bergers E, Kamphorst W, Bot J, Nicolay K, Castelijns JA, van Waesberghe JHTM, Ravid R, Polman CH, Barkhof F. Post-mortem high-resolution MRI of the spinal cord in multiple sclerosis: A correlative study with conventional MRI, histopathology and clinical phenotype. *Brain* 2001;124(1):154–166.

96. Agosta F, Pagani E, Caputo D, Filippi M. Associations between cervical cord gray matter damage and disability in patients with multiple sclerosis. *Arch Neurol* 2007;64(9):1302–1305.

97. Backes WH, Mess WH, Wilmink JT. Functional MR imaging of the cervical spinal cord by use of median nerve stimulation and fist clenching. *AJNR Am J Neuroradiol* 2001;22(10):1854–1859.

98. Govers N, Beghin J, Van Goethem JW, Michiels J, van den HL, Vandervliet E, Parizel PM. Functional MRI of the cervical spinal cord on 1.5 T with fingertapping: To what extent is it feasible? *Neuroradiology* 2007;49(1):73–81.

99. Kornelsen J, Stroman PW. fMRI of the lumbar spinal cord during a lower limb motor task. *Magn Reson Med* 2004;52(2):411–414.

100. Madi S, Flanders AE, Vinitski S, Herbison GJ, Nissanov J. Functional MR imaging of the human cervical spinal cord. *AJNR Am J Neuroradiol* 2001;22(9):1768–1774.

101. Maieron M, Iannetti GD, Bodurka J, Tracey I, Bandettini PA, Porro CA. Functional responses in the human spinal cord during willed motor actions: Evidence for side- and rate-dependent activity. *J Neurosci* 2007;27(15):4182–4190.

102. Stroman PW, Nance PW, Ryner LN. BOLD MRI of the human cervical spinal cord at 3 tesla. *Magn Reson Med* 1999;42(3):571–576.

103. Wilmink JT, Backes WH, Mess WH. Functional MRI of the spinal cord: Will it solve the puzzle of pain? *JBR—BTR* 2003;86(5):293–294.

104. Kawashima N, Nozaki D, Abe MO, Akai M, Nakazawa K. Alternate leg movement amplifies locomotor-like muscle activity in spinal cord injured persons. *J Neurophysiol* 2005;93(2):777–785.

105. Kautz SA, Patten C, Neptune RR. Does unilateral pedaling activate a rhythmic locomotor pattern in the nonpedaling leg in post-stroke hemiparesis? *J Neurophysiol* 2006;95(5):3154–3163.

106. Ferris DP, Gordon KE, Beres-Jones JA, Harkema SJ. Muscle activation during unilateral stepping occurs in the nonstepping limb of humans with clinically complete spinal cord injury. *Spinal Cord* 2004;42(1):14–23.

107. Jankowska E, Edgley SA. How can corticospinal tract neurons contribute to ipsilateral movements? A question with implications for recovery of motor functions. *Neuroscientist* 2006;12(1):67–79.

Glossary of Terms

A/D converter: Analog-to-digital converter; takes the continuous (analog) MR signal from a receiver coil and converts it to digital form for computer analysis and storage.

Acquisition time: The time taken to acquire a complete set of image data; also called the imaging time; not to be confused with the sampling time, which includes only the actual time taken to sample the data points.

Action potential: A brief change in electrical potential between the inside and outside of a nerve when it is stimulated, serving to transmit nerve signals.

ADC: Apparent diffusion coefficient; for diffusion-weighted MRI this typically refers to diffusion of water.

Affine transformation: A linear transformation including rotation, scaling or shear, and a translation or shift.

AFNI: Functional MRI analysis software, Analysis of Functional Neuroimages.

Aliasing: An error or artifact that occurs when a signal at one frequency becomes indistinguishable from another frequency; in MR images this typically results in a "wrap-around" effect where features extending out one side of an image appear to continue into the opposite side.

Amplifier: Electronic components used to increase the power or magnitude of electronic signals.

Amplitude: Magnitude, absolute size, or intensity.

Amplitude envelope: The time-dependent magnitude of the magnetic field used to produce a radio-frequency (RF) pulse. The "shape" of the RF pulse.

Anaerobic glycolysis: Transformation of glucose to pyruvate to provide energy for cellular processes, typically when limited amounts of oxygen (O_2) are available.

Analog-to-digital: Conversion of electronic signals from analog (continuous) to digital (discrete).

Analysis of Functional Neuroimages: Functional MRI analysis software, AFNI, developed at the National Institutes of Health in the United States.

ANALYZE: A software package for multidimensional display, processing, and measurement of multimodality biomedical images, developed at the Mayo Clinic.

Analyze format: The image format that was first developed for use in the ANALYZE software and was adopted for use in SPM, Statistical Parametric Mapping, for fMRI analysis.

Anatomical imaging: Specific application of MRI for visualization of anatomical features.

Angular momentum: Similar to linear momentum, which is the product of the velocity and mass of an object, except that it refers to the momentum of rotation and is the product of rotational velocity and rotational moment of inertia.

ANIMAL: Automatic nonlinear image matching and anatomical labeling, a method for spatial normalization and coregistration of images, developed at the Montreal Neurological Institute.

Artifact: In MRI an artifact is an error that results in incorrect signal being displayed in one or more pixels in an image, altering the apparent structure of objects, and can include spurious signals that did not arise from the object being imaged, or MR signals that appear at incorrect locations.

ASCII: American Standard Code for Information Interchange, a method of representing characters with numeric codes that can be used in electronic files; there are 95 ASCII characters defined, which are assigned numbers 32 to 126.

Astrocyte: One of the glial cells that are components of neural tissue, expected to play a role in neurovascular coupling and contrast mechanisms used for fMRI.

Automatic Nonlinear Image Matching and Anatomical Labeling: ANIMAL, a method for spatial normalization and coregistration of images, developed at the Montreal Neurological Institute.

Axial (or transverse): Image slice orientation with the plane of the image orthogonal to the long axis of the body; image plane extending in the right–left and anterior–posterior directions.

B_0: The static magnetic field of an MRI system.

B_1: The rotating magnetic field of a radio-frequency (RF) pulse.

Balanced acquisitions: Imaging methods with magnetic field gradients applied to cancel out and produce zero net effect at the end of the acquisition—typically used for steady-state imaging.

Balanced FFE: Balanced fast field-echo, also known as balanced fast gradient-echo, a steady-state imaging method.

Band-pass filter: Filter designed to exclude frequency components of a signal that are below a lower threshold and above an upper threshold, therefore leaving only those frequency components within a middle band.

Bandwidth, receiver: The frequency range selected for sampling the MR signal—inversely related to the sampling rate or *dwell time* (DW), also inversely related to the signal-to-noise ratio (SNR).

Bandwidth, RF pulse: The range of Larmor frequencies that is affected by a radio-frequency (RF) pulse—related to the slice thickness via the slice gradient strength.

Baseline: Typically refers to the signal intensity of a voxel during the reference conditions, or rest periods, of an fMRI time series.

Basis functions: The model time courses that are used in a general linear model (GLM) to fit to measured data.

Basis set: The complete set of basis functions used in a general linear model (GLM).

B-FFE: A balanced fast-field echo or gradient echo.

Binary, or base 2: Numbers represented using only the digits 0 and 1, as when recorded electronically.

Bit: One digit of a binary number.

Block design: Functional MRI study design in which the task or stimulus conditions each last for several seconds so that the BOLD response reaches approximately a steady state during each period, or block.

Blood: In addition to being a specialized bodily fluid that delivers necessary substances to the body's cells, blood is the source of the BOLD contrast used for fMRI, via the magnetic properties of hemoglobin.

Blood flow: One of the major potential sources of image artifacts in MRI and fMRI, and a significant component of neurovascular coupling; also referred to for fMRI as cerebral blood flow, or CBF.

Blood oxygenation–level dependent: BOLD, the primary contrast mechanism used for fMRI.

Body temperature: Normal body temperature measured under the tongue is 36.8°C or 98.2°F at the core of the body.

BOLD: Blood oxygenation–level dependent; the primary contrast mechanism used for fMRI.

Bonferroni correction: A method used to address the problem of multiple comparisons—to test n dependent or independent hypotheses (such as with n voxels in a set of fMRI data), use a statistical significance level of 1/n times what it would be if only one hypothesis were tested.

Bootstrapping: In statistics, bootstrapping is a resampling technique used to obtain estimates of summary statistics—for fMRI, the range or variance of measured values can be estimated by repeating the measures with subsets of data created by substituting data from some participants with duplicates of data from others, over a large number of possible resampling combinations.

Bootstrap ratio: The ratio of a measured value to its variance determined with a bootstrapping method—used as a test of significance.

Bore: The open space inside an MR magnet—where people or objects are placed for imaging.

Brain voyager: Functional MRI analysis software.

Bulk motion: Movement of a person during imaging, distinct from physiological motion.

b-value: In diffusion-weighted imaging, the b-value expresses the degree of diffusion weighting in the resulting image (analogous to TE in T_2-weighted imaging); the b-value is determined by the magnitude and duration of the gradients that are applied to impose the sensitivity to water diffusion (not to be confused with β-value used in fMRI analysis).

Byte: An 8-digit binary number (*two bits* used to refer to 25 cents, or one quarter of a dollar; so there are 8 bits in a dollar, and 8 bits in a byte).

Capillaries: The smallest blood vessels in the body, 5–10 μm in diameter, connecting arterioles and venules, the primary location of gas exchange with tissues to supply oxygen and remove carbon dioxide.

Center the data: To subtract the average values from each data set, time series, and so forth.

Chemical shift: The shift in Larmor frequency due to nearby magnetically active nuclei in a molecule.

$CMRO_2$: Cerebral metabolic rate of oxygen consumption, commonly expressed in units of μmol/sec.

CMYK: Color representation used in printing and digital images, expressed with values for color components cyan (C), magenta (M), yellow (Y), and black (K).

CNS: Central nervous system; consists of the brain, brainstem, and spinal cord.

Coil: In magnetic resonance, a *coil* refers to a loop of conducting material (metal wire, foil, etc.) designed to produce a magnetic field, such as for the static magnetic field, gradient fields, or radio-frequency pulses, or designed as a receiving antenna to detect the MR signal.

Color scale: The order and range of colors used to express measured quantities in images or other graphics.

Compass: An instrument for determining direction relative to the Earth's magnetic poles.

Complete: In the context of data analysis, *complete* refers to a basis set being sufficient to fit to any continuous data set.

Component: A constituent part; with respect to the MR signal, a component is one coordinate of the magnetization vector, or the amount projected onto each axis of a coordinate system; in data analysis, a component of a signal is the contribution from one of multiple sources, or one characteristic feature of the overall signal.

Computer program: A sequence of instructions written to perform a specified task for a computer.

Confounds: In fMRI data analysis, *confounds* refer to those components of signal variance that are not related to neural activity.

Conjunction analysis: A group analysis method by taking the minimum T-value across all of the people or studies, for each voxel.

Connectivity: Detection of regions of the CNS that are interconnected and function as a network.

Continuous: A series of numbers that progresses smoothly from one value to the next, or values that can take any numerical value, as opposed to being discrete.

Continuum: Anything that goes through a gradual transition from one condition to a different condition without any abrupt changes.

Contrast: The difference between values; for MRI, *contrast* refers to the difference in pixel values between two points or two objects in the image, or differences in pixel values between two different points in time.

Contrast mechanisms: The physical and/or physiological processes that produce signal intensity differences between points in space or in time, in MR images.

Control room: The area of an MRI facility where the control console is located.

Convolution: A mathematical operation on two functions, or two sets of data values, to produce a third function or set of data values; the result is a modified combination of the original two inputs.

Coordinates: Any of the magnitudes of values that serve to define the position of a point, line, plane, and so forth in reference to a fixed set of axes or reference lines.

Coronal: Slice orientation with the plane extending in the head-foot direction and the right–left direction.

Correlation: A measure of dependence between two quantities.

Correlation time: A time used to characterize random processes, such as Brownian motion of water molecules, as the largest time interval over which the position of a particle at the end of the interval is correlated with its position at the start of the interval; used in MR relaxation theory to characterize the speed of movement and frequency of random interactions between magnetically active particles.

Cosine: Trigonometric function; in a right triangle, the cosine of the angle of one corner (not the right angle corner) is equal to the ratio of the side adjacent to the given corner, to the length of the hypotenuse.

Cross-excitation: In MRI, *cross-excitation* refers to the edge of one slice being affected by the excitation pulse for the adjacent slice. Cross-excitation also refers to when neurotransmitters diffuse from one synapse to another.

Cryostat: A vessel, similar to a vacuum flask, or dewar, used to maintain cold cryogenic temperatures; in MR it refers to the vessel that contains liquid helium to maintain the temperature of the superconducting coils that produce the static magnetic field.

CSF: Cerebrospinal fluid.

dB/dt: The rate of change of magnetic field with respect to time; typically refers to the field changes when magnetic field gradients are turned on and off; excessive values can induce peripheral nerve stimulation in a person being imaged, and also eddy currents in conducting materials within the MRI system.

DCM: Dynamic causal modeling.

Decompose: In data analysis, refers to breaking down a data set into its constituent components.

Deformation field: The map of displacements needed at each point of an image data set to transform the images and map them to match a reference data set, such as in spatial normalization.

Delta function: An instantaneous function of time or position having a value of zero everywhere except at time or position equal to zero, where it has a value greater than 1, and the continuous integral of the function is equal to 1.

Dendrites: The branching process of a neuron that conducts impulses toward the cell.

Deoxyhemoglobin: Deoxygenated hemoglobin, the paramagnetic source of blood oxygenation–level dependent contrast in MRI.

Dephase: To become distributed in phase or out of phase.

Dewar: A container with an evacuated space between two walls that are highly reflective, capable of maintaining its contents at a near-constant temperature over relatively long periods of time.

DICOM: Digital Imaging and Communications in Medicine, a standard image format and data transfer and storage standard for communications between medical imaging devices.

Diffusion: The movement of molecules by means of random thermal motion.

Diffusion anisotropy: The measure of how diffusion values depend on direction; specifically a ratio of the differences of diffusion rates in three different spatial directions to the overall magnitude of the diffusion rate.

Diffusion tensor imaging: DTI—an extension of diffusion-weighted imaging in which the three-dimensional nature of diffusion is characterized by means of a 3×3 matrix for each voxel; commonly but mistakenly used interchangeably with the term *fiber-tracking*, which is just one use of DTI data.

Diffusion-weighted imaging: An MR imaging method in which the signal intensity of each voxel is weighted according to the water self-diffusion rate in a particular direction, with the direction determined by the diffusion-weighting gradients that are applied.

Digital: Consisting of discrete values, as opposed to the continuous values represented by analog signals.

Dipole–dipole: Refers to magnetic interactions between magnetic dipoles, this form of interaction is a dominant source of energy exchange contributing to relaxation of the MR signal from biological tissues.

Discrete: Discontinuous, consisting of distinct or individual parts.

Distortions: Spatial errors in images, inaccurate representations of spatial features.

DTI: Diffusion tensor images or imaging.

DW: Dwell time.

Dwell time: DW, the time taken to sample each point of MR data, inversely related to the receiver bandwidth.

DWI: Diffusion-weighted images or imaging.

Dynamic causal modeling: A method for characterizing and modeling connectivity between brain regions.

e: The transcendental number, approximately equal to 2.7183.

Echo: In MR, an echo is produced by bringing the magnetic moments back into phase, thereby momentarily recovering the net signal.

Echo time: The time from the center of the radio-frequency (RF) excitation pulse to the peak of the echo.

Echo-planar imaging: A method for rapid acquisition of the MR image data, consisting of sweeping the spatial encoding across all two-dimensional (2D) combinations of values to span an entire 2D plane in *k-space*, which can then be transformed into an image.

Echo train length: The number of echoes acquired in a fast spin-echo acquisition.

Eddy currents: Electrical currents induced by rapid switching of magnetic fields.

Edema: Accumulation of fluid in interstitial spaces of tissues.

Effective connectivity: The influence one region has over another.

Effective transverse relaxation time: T_2^*, the combined effect of true transverse relaxation characterized with T_2, and the effect of spatial magnetic field variations, $1/T_2^* = 1/T_2 + \Delta B$.

Efficiency: For fMRI study design, the effectiveness or sensitivity per amount of time taken.

Eigenvectors: Characteristic vectors of a matrix, the vectors that when multiplied by a matrix result in scalar multiples of the same vectors.

Eigenvalues: Characteristic values of a matrix, the scalar multiples of the eigenvectors mentioned in the definition for *eigenvector*, above.

Electronics: In MRI, refers to the electrical components of the MR system hardware, such as preamplifiers, amplifiers, A/D converters, computers, and so forth.

Energy metabolism: The biochemical reactions that take place within cells to provide energy for biosynthesis of cellular and extracellular components, transport of ions and organic chemicals against concentration gradients, and the conduction of electrical impulses in the nervous system.

EPI: Echo-planar images or imaging.

Equilibrium: The balance of forces including magnetic, electrical, osmotic, mechanical, and so forth, resulting in a steady state.

Equilibrium magnetization: The steady-state magnetization parallel to the static magnetic field after relaxation processes have occurred.

Equilibrium state: In MRI, this term is used interchangeably with *equilibrium magnetization*.

Equipment room: The area of an MRI facility containing the electronic equipment such as high-power amplifiers, pumps for water distribution for cooling, cold-head pumps, and so forth, typically located in a room separated from the control room.

Ernst angle: The flip angle of a radio-frequency excitation pulse that produces the greatest MR signal magnitude in the least amount of time when repeated excitations are being applied.

Event-related design: Functional MRI study design consisting of very brief tasks or stimuli, separated in time sufficiently for the MRI signal to return to baseline values between each event, so that the responses to individual events can be detected.

Excitation: In MRI, refers to applying a radio-frequency pulse to tip the magnetization away from its equilibrium orientation.

Excitatory: A neuronal signal that causes the release of neurotransmitters such as glutamate, glycine, and so forth at a synapse, causing ion channels to open, allowing positively charged ions to flow into the postsynaptic cell, increasing the probability of an action potential being initiated.

Exponentially decaying function: A function that decreases at a rate proportional to its value; can be expressed as a negative exponent of the transcendental number e, such as a function of time, t, can be expressed as $S = S(0)e^{-at}$, where $S(t)$ is the value at any time, $S(0)$ is the value at $t = 0$, and a reflects the rate of decay over time; in some forms this equation could also be written as $S = S(0)\exp(-at)$, with the same meaning.

Extracellular: Outside of cells and blood vessels; between the cells.

Extracellular fluid: Fluid within the extracellular space.

FAIR: Flow-sensitive alternating inversion-recovery, a perfusion imaging method.

False color: An image that represents objects in colors that differ from those that a color photograph would show; typically used to represent measured values or characteristics of objects within colors instead of indicating the actual physical colors of the objects.

False-discovery rate: The rate or incidence of false positives in multivoxel fMRI results; also refers to a statistical method used in multiple hypothesis testing to correct for multiple comparisons.

False negative: A result that incorrectly fails to reject the null hypothesis; in fMRI, a result that indicates no activity in a voxel when there truly is; also called a type II error.

False positive: A result that incorrectly rejects the null hypothesis; in fMRI, a result that indicates activity in a voxel when there truly is none; also called a type I error.

Family-wise error: The probability of there being a false positive within multivoxel fMRI results.

Fast event-related design: Also called a *mixed* design; an fMRI study design that employs brief stimuli or tasks, as in an event-related design, but with insufficient time between events to allow the MRI signal to return to baseline, and so individual responses can overlap in time.

Fast imaging: Any MR imaging method designed specifically for rapid acquisition of images at the expense of image quality.

Fast spin echo: A spin-echo imaging method in which multiple successive echoes are produced and a different spatial encoding is given to each echo, thereby allowing multiple lines of k-space to be sampled with each excitation, reducing the total time taken to acquire an image.

FastSPGR: Fast, spoiled gradient-echo imaging method; consists of a gradient echo applied with short repetition time and low flip angle to gain speed at the expense of signal intensity, and in which all transverse magnetization is dephased at the end of each acquisition so that it does not contribute to subsequent data acquired to produce the same image.

FCA: Fuzzy clustering algorithm, a nonparametric method of fMRI data analysis.

FDR: False-discovery rate.

FFE: Fast field-echo; also called a fast gradient-echo.

Fiasco/FIAT: Functional MRI analysis software.

Fiber tracking: A method of using diffusion tensor imaging data to estimate the path of large white-matter tracts.

FID: Free induction decay.

Field of view: The spatial extent covered by an image.

FIESTA: Fast imaging employing steady-state acquisition, a gradient-echo imaging method.

File: A computer file; block of digital information stored electronically.

Filter: A function that processes data to remove unwanted characteristics or components; in MRI, both hardware and software filters are used, as frequency components are excluded from the MR signal while it is sampled, and high-frequency noise, physiological effects, and the like are often removed during fMRI analysis.

Finite: Having limits, not infinite, measurable.

First-level analysis: In fMRI, refers to analysis of individual data sets, as opposed to a group analysis.

FISP: Fast imaging with steady-state precession, a gradient-echo imaging method.

Fitting: The process of applying regression analysis to data, sometimes called line fitting or curve fitting, but can also be extended to fitting a set of patterns to data, such as with a general linear model (GLM).

Fixed effects: A group analysis method with statistical tests based on the average response across the group relative to the standard error of the response in each individual.

FLAIR: Fluid-attenuating inversion recovery.

FLASH: Fast low-angle shot, a fast gradient-echo imaging method.

Flip angle: The angle that magnetization is rotated through when a radio-frequency (RF) pulse is applied.

Fluid-attenuating inversion recovery: FLAIR, an imaging variant typically used to null the signal from cerebrospinal fluid (CSF).

fMRI: Functional magnetic resonance imaging, a specific application of MRI.

FMRIB Software Library: Functional MRI analysis software.

FMRlab: Functional MRI analysis software.

Format: The consistent structure or organization that is adopted for the storage of data.

Fourier convolution theorem: The Fourier transform of a convolution between two functions is the pointwise product of the separate Fourier transforms of each of the two functions.

Fourier transform: The decomposition of a signal into its frequency components represented by sine and cosine functions.

FOV: Field of view.

Free induction decay: FID, the decay of the MRI signal due to relaxation and spatial field variations.

Free water: Refers to the component of water in a solution, tissues, and so forth that is not bound to surfaces, macromolecules, ions, and the like but rather is free and has the behavior of pure water.

Frequency: The number of cycles per unit time of a wave or oscillation; has units of hertz (Hz) indicating the number of cycles per second, or units of radians per second, where 2π radians is one complete cycle.

Frequency encoding: Spatial encoding method used in MRI consisting of recording the MR signal while a magnetic field gradient is applied, so that the frequency of the signal reflects the position of the signal source along the gradient direction.

Frequency-encoding gradient: The magnetic field gradient that is applied while the signal is recorded so that spatial information is encoded into the frequency of the MR signal.

FSE: Fast spin echo.

FSL: FMRIB Software Library, fMRI analysis software.

FT: Fourier transform.

Functional connectivity: The fMRI responses detected in two regions are correlated.

Functional imaging: The general term used for any imaging method (MRI, PET, etc.) that is used to detect or characterize any physiological function, including neural function.

Functional MRI: Magnetic resonance imaging (MRI) method for detecting neural function.

Fuzzy clustering: A nonparametric data analysis method consisting of identifying consistent patterns within a data set and creating clusters of similar patterns and determining the probability of each individual pattern being a member of each cluster.

FWHM: Full width at half maximum.

Gauss: Unit of magnetic field strength, equal to 0.0001 tesla.

Gaussian function: Symmetric *bell curve* shape that quickly falls off toward plus or minus infinity; a Gaussian function of time, t, centered at $t = 0$, has the form e^{-t^2/σ^2}, where σ controls the width.

GE: Gradient-echo.

General linear model: GLM; a linear statistical method consisting of regression with a set of basis functions.

Ghosts: In MRI, *ghosts* are low-intensity replications of objects within an image, displaced from the actual positions of the objects, typically caused by movement during the image acquisition.

GLM: General linear model.

Glutamate: An excitatory neurotransmitter.

Gradient: In MR, refers to a magnetic field that is directed parallel to the static magnetic field, B_0, but has a magnitude that varies linearly as a function of position.

Gradient coils: Wire or other conductor wrapped onto a rigid form in a pattern designed to produce a linear magnetic field gradient at the center of the coil when an electrical current is passed through the conductor.

Gradient direction: The direction in which the magnitude of a magnetic field varies when a gradient is applied, not necessarily the same as the direction of the magnetic field itself.

Gradient echo: An echo produced by applying a magnetic field gradient and then reversing the gradient direction to bring the magnetization back into phase to produce a brief signal peak.

Gradient moment nulling: The practice of applying magnetic field gradients so that the cumulative net effect is zero; the result is that the transverse magnetization is left in phase at the end of an imaging sequence.

GRASS: Gradient recall acquisition using steady states, a gradient-echo imaging method.

Gray matter: The component of neural tissue consisting predominantly of neuronal cell bodies and their dendrites.

Gray scale: A color scale consisting of shades of gray spanning from white to black.

GRE: Gradient-recalled echo, gradient-echo.

Group analysis: In fMRI, refers to the combined analysis of data from a number of people and/or repeated experiments in order to characterize consistent features of the neural activity that was detected.

G_x: Magnetic field gradient along the x-axis direction, commonly generalized to refer to the frequency-encoding direction.

G_y: Magnetic field gradient along the y-axis direction, commonly generalized to refer to the phase-encoding direction.

Gyromagnetic ratio: The ratio of the magnetic dipole moment to angular momentum; the ^{1}H (hydrogen) nucleus that is most commonly detected to produce the MRI signal has a gyromagnetic ratio of 42.576 MHz/tesla.

ħ: Planck's constant divided by 2π (equal to approximately 1.055×10^{-34} Joule · seconds (J · s)).

Half-Fourier: Spatial encoding scheme in which only slightly more than half of k-space is sampled in order to reduce the imaging time.

Half-Fourier single-shot turbo spin echo: HASTE, a fast spin-echo imaging method incorporating half-Fourier sampling, in which all of the data required to produce an image are sampled in one acquisition.

HARDI: High angular-resolution diffusion imaging.

HASTE: Half-Fourier single-shot turbo spin echo.

Header: Information included in an image data file in addition to the image; typically provides details of the image acquisition parameters.

Hematocrit: The proportion of blood volume occupied by red blood cells.

Heme: The iron-containing component of hemoglobin, $C_{34}H_{32}N_4O_4Fe$, that binds to oxygen.

Hemodynamic response function: HRF; the temporal pattern of the BOLD response to a brief stimulus, which can be convolved with any timing pattern of a stimulus to predict the time course of the BOLD response.

Hemoglobin: The iron-containing oxygen-transport component of red blood cells.

High angular-resolution diffusion imaging: HARDI, a method of diffusion-weighted imaging that employs data acquired over a very large number of different diffusion directions in order attain high precision.

Homogeneity: Uniformity; in MRI refers to the consistency of the static magnetic field strength over a region being imaged.

HRF: Hemodynamic response function.

Hydration layer: Refers to water that is attracted or adsorbed to the surface of macromolecules, ions, and so forth in solution, such as in biological fluids.

Hydrogen nucleus: The nucleus of the hydrogen atom, (^{1}H), a single proton, the source of the signal in most biological applications of magnetic resonance, including MRI.

i: The imaginary number, the square root of negative one, -1.

ICA: Independent components analysis.

Image: A physical likeness or representation of a person, animal, or thing.

Image quality: In MRI, refers generally to the cumulative effects of spatial accuracy, signal-to-noise ratio, and contrast between features in order to clearly depict the anatomical and/or functional features of interest.

Imaging method: The method used to elicit an MR signal and impose contrast between structures or objects based on one or more properties of proton density, relaxation properties, flow, diffusion, and the like.

Incoherent: As opposed to coherent or steady-state methods; MR acquisition methods in which the transverse magnetization is spoiled or dephased at the end of each signal acquisition period.

Independent components analysis: ICA, a method of data analysis based on decomposing data into a number of statistically independent components that occur consistently across the data sample.

Induction: The process by which a body having electric or magnetic properties produces magnetism, an electric charge, or an electromotive force in a neighboring body without contact.

Inhibitory: A neuronal signal that causes the release of neurotransmitters such as γ-aminobutyric acid (GABA) and others at a synapse, causing ion channels to open, allowing negatively charged ions to flow into the postsynaptic cell, decreasing the probability of an action potential being initiated.

Interleaved acquisition: Multi-slice image acquisition in which the slices are acquired out of order to reduce effects of cross-excitation.

Interpolate: To estimate values that are between known or measured values in a sequence of numbers.

Interstimulus interval: In fMRI study design, the time span between successive applications of stimuli.

Intracellular: Inside a cell.

Inversion: In MRI, refers to rotating magnetization by 180°.

Inversion pulse: A radio-frequency (RF) pulse that rotates magnetization by 180° around some chosen axis.

Inversion recovery: A method of imposing T_1-weighting into MR signal by applying an inversion (180°) pulse and then allowing time for longitudinal relaxation to occur before measuring the signal or applying an imaging sequence.

IR: Inversion recovery.

ISI: Interstimulus interval.

Isocenter: The physical center of an MRI system, the point at which the static magnetic field and gradient fields are centered.

Isochromat: A sufficiently small volume of space that the magnetic field within the volume can be assumed to be perfectly uniform.

Keypad: A device consisting of one or more buttons that can be pressed to register a response.

k-space: The coordinate space spanned by the MR data that is acquired to construct an image, representing the Fourier transform of the image.

Larmor equation: The equation relating the frequency of precession of magnetic moments, and therefore the frequency of the MR signal, to the magnetic field strength; $\omega = \gamma B$, where ω is the frequency, γ is the gyromagnetic ratio of the nuclei being detected (typically ^{1}H nuclei), and B is the magnetic field strength.

Larmor frequency: The frequency, ω, in the Larmor equation; the frequency of precession of magnetic moments, and therefore the frequency of the MR signal.

LFP: Local field potential.

Linear: Created or occurring in straight lines; mathematically refers to functions that can be added according to $f(x + y) = f(x) + f(y)$.

Linear combinations: A point-by-point summation of patterns or sequences of numbers, typically with each pattern or sequence multiplied by a weighting factor.

Linearly independent: A set of patterns of sequences of numbers is linearly independent if none of them can be expressed as a linear combination of other patterns or sequences in the set; two such patterns are linearly independent if the vector dot product of the two (i.e., the sum of the point-by-point product of the numbers in the two patterns) is equal to zero.

Lipids: A broad group of naturally occurring molecules that includes fats, phospholipids, and the like and are biologically important for energy storage and as structural components of cell membranes.

Local field potentials: A type of electrophysiological signal that is related to the sum of all dendritic synaptic activity within a volume of tissue.

Longitudinal axis: The coordinate axis defined to be parallel to the static magnetic field of the MRI system, B_0.

Longitudinal magnetization: The component of magnetization that is parallel to the longitudinal axis.

Longitudinal relaxation: Recovery of the longitudinal magnetization toward its equilibrium value via processes of relaxation.

Lorentzian function: A Lorentzian function of time, t, centered at $t = 0$, has the form $1/(1 + t^2\sigma^2)$, where σ controls the width.

LPH: Left-posterior-head; listing the positive anatomical directions, or directions of increasing coordinate values, in one system of displaying anatomical images.

M_0: The magnitude of the equilibrium magnetization; depends on the proton density, temperature, and magnetic field strength.

Magnet: In general, a magnet is a material or object that produces a magnetic field; in MR it refers to the device used to produce the intense static magnetic field, B_0.

Magnet room: The area of an MRI facility that contains the magnet, typically isolated from surrounding areas for safety, and shielded to exclude radio-frequency signals from external sources.

Magnetic dipole: The fact that every magnetic field has a north and a south pole, indicating the direction of magnetic field, which points always from the south pole to the north pole; in MR *magnetic dipole* is occasionally used as jargon to refer to the nuclei that produce the MR signal.

Magnetic energy: Also called Zeeman energy, equal to the dot product of the magnetic dipole moment and the magnetic field vector, the work of magnetic force on realignment of the vector of the magnetic dipole moment.

Magnetic field: The continuous spatial distribution of the force that acts on magnetic materials and on the movement of electric charges that extends around a magnet, in terms of both magnitude and direction of the force at every point in space.

Magnetic field gradient: See *gradient*.

Magnetic moment: Magnetic dipole moment; a measure of the tendency of a magnet, electric current, or any object producing a magnetic field to align with another magnetic field.

Magnetic resonance: MR; also called nuclear magnetic resonance (NMR); the phenomenon of absorption of certain frequencies of radio and microwave radiation by atoms placed in a magnetic field; the same phenomenon used as the basis of magnetic resonance imaging (MRI).

Magnetic susceptibility: The most commonly used form in MR is the volume magnetic susceptibility, the ratio of magnetization of a material to the applied magnetic field strength; values can be expressed in SI and CGS units, and the two will differ by a factor of 4π, so it is important to check which units you need when using these values.

Magnetization: The net magnetic moment per unit volume.

Magnetization transfer contrast: Image contrast obtained by preferentially saturating the signal from "bound" water, thereby reducing the total signal intensity according to the proportions of bound and "free" water and the exchange rate between the two.

Magnitude: Size, extent, dimension, amplitude.

MATLAB: A software package that provides a numerical computing environment, developed and distributed by The Mathworks Inc., Natick, Massachusetts.

Matrix: A rectangular array of numbers.

Maxwell-Boltzmann statistics: The statistical distribution of material particles over various energy states in thermal equilibrium; used in MR relaxation theory.

Mean: Average; the sum of a set of numbers divided by the number of values in the set.

Membrane potential: The voltage difference between the interior and exterior of a cell, typically generated by differences in ion concentrations.

Metabolism: The set of chemical and physical processes that happen in living organisms to maintain their material substance and to provide energy.

Metabolites: A substance that is used in a metabolic process.

Mixed design: Also called fast event-related design, an fMRI study design consisting of brief stimuli or tasks, as in event-related designs, but with only short intervals between events so that the responses to successive events overlap.

MNI coordinate system: Normalized coordinate system defined for mapping brain anatomical structures, as adapted from the Talairach coordinate system by researchers at the Montreal Neurological Institute, MNI.

Montreal Neurological Institute: MNI.

Motion: The action of changing place or position; for MRI and fMRI, motion is a significant source of image artifacts and time-varying signal intensity changes that do not reflect neural activity.

Motion correction: The process of realigning images acquired in a time series to reduce the effects of motion by ensuring that each voxel in the image data represents the same volume of space at all time points in the series; this process, however, does not remove image artifacts or distortions that may have been created by the motion.

MR signal: The electrical signal that is induced in a receiver coil by magnetic moments precessing in phase and at the same frequency, thereby producing a detectable time-varying magnetic field; for biological MRI, this signal is typically produced by the nuclei of hydrogen atoms in water and lipids.

MRI: Magnetic resonance imaging.

Myelin sheaths: The extensions or processes of oligodendrocytes (in the CNS) or Schwann cells (in the peripheral nervous system) that wrap around axons in white matter to produce an electrically insulating barrier that contributes significantly to the propagation speed of action potentials along the axons.

Nearest neighbor: A method of applying mathematical operations on images in which the value for the pixel nearest the point of interest is used, as opposed to interpolating values between existing pixel values.

Neural: Pertaining to a nerve or the nervous system.

Neural function: The action or operation of the nervous system, which may include the combined action of neurons and glial cells.

Neuronal activity: Refers specifically to the action or operation of neurons within the nervous system.

Neuronal cell body: The soma of a neuron; the bulbous end of a neuron, containing the cell nucleus where the bulk of metabolic activity occurs; the major constituent of gray matter.

Neurotransmitter: Any of several chemical substances that transmit signals between a neuron and another cell.

Neurovascular coupling: The close link between neuronal activity and local blood flow modulated by vascular reactivity; this coupling is an essential component of the blood oxygen-level dependent MR signal that is exploited for fMRI.

NIfTI: Neuroimaging Informatics Technology Initiative.

NIfTI-1: A data format developed by the NIfTI Data Format Working Group to make it easier to use different functional MRI data analysis software.

NITRC: Neuroimaging Informatics Tools and Resources Clearinghouse, an Internet resource for structural and functional image analysis and processing software.

Noise: In MR, random electrical noise causing relatively low-intensity random signal fluctuations, or in some uses also includes physiological "noise" consisting of MR signal fluctuations arising from cardiac- and respiratory-related motion including flow of blood, cerebrospinal fluid (CSF), and so forth.

Nonlinear interactions: Processes whose individual effects do not sum to equal their combined effects when they occur simultaneously.

Nonparametric methods: Functional MRI analysis methods that do not rely on the assumption that the measured responses can be matched to one or a combination of a set of predetermined patterns.

Normal: (1) orthogonal, at right angles; (2) defined so that the sum of the squares of the absolute values is equal to 1.

N_{sample}: The number of data points between measured or sampled.

Nuclear magnetic resonance: NMR; identical to magnetic resonance (MR) except the term *nuclear* was dropped; the term NMR is more commonly used today in reference to high-field MR spectroscopy.

Null hypothesis: In statistical analysis, the hypothesis that a measured value, or difference between a measured value and some reference value, is equal to zero; statistical tests on sets of measurements may be done to determine the probability that the null hypothesis is correct; in fMRI, statistical tests often determine the probability that the magnitude of a response could equal zero.

Nyquist frequency: The highest frequency that can be determined accurately at a given sampling rate; equal to half of the sampling rate.

Oxidative metabolism: Metabolic reactions that occur in cells to convert biochemical energy from nutrients into adenosine triphosphate (ATP).

Oxidative phosphorylation: A metabolic process that takes place in mitochondria to produce adenosine triphosphate (ATP).

Oxygen consumption: The rate of oxygen usage; in fMRI it is also often referred to as the cerebral metabolic rate of oxygen consumption ($CMRO_2$).

Oxygen extraction fraction, E: The proportion of oxygen that is taken up by the tissues out of the total amount of oxygen available in the blood.

Parallel imaging: A method of making use of the spatial information provided by detecting the MR signal with multiple receiver coils in order to reduce the number of phase-encoded lines that must be applied and reduce the total acquisition time.

Paramagnetic: A material that is magnetized only in the presence of an externally applied magnetic field and has a magnetic susceptibility that is greater than zero.

Parametric methods: Analysis methods that rely on the assumption that data come from a type of probability distribution; in fMRI, analysis methods that typically rely on regression to a set of basis functions, or predicted response patterns.

Partial-Fourier: Spatial encoding scheme in which not all k-space is sampled in order to reduce the imaging time; a more general form of half-Fourier.

Partial least-squares: PLS; a nonparametric group analysis method for fMRI.

Partial-volume effects: The result of having image voxel span more than one type of tissue or more than one structure, producing an average or sum of the MR signals from all of the tissues in the voxel.

PCA: Principal components analysis.

Perfusion: The rate of blood flow per volume of tissue.

Perfusion-weighted imaging: MR imaging method designed so that pixel intensities have a dependence on the local rate of tissue perfusion in each corresponding voxel.

Peripheral nerve stimulation: In MR, stimulation of impulses in peripheral nerves by rapid switching of magnetic field gradients; produces slight feelings of tingling, pressure, or muscle twitches; can be avoided by reducing gradient strengths or by eliminating closed loops of body parts by skin-on-skin contact, such as keeping hands apart, not crossing legs at the ankles, and so forth.

PET: Positron emission tomography.

Phase: Direction or angle from a reference direction.

Phase encoding: Spatial-encoding method used in MRI consisting of recording the MR signal after a magnetic field gradient has been applied for a fixed duration and then turned off, so that the phase of the signal reflects the position of the signal source along the gradient direction.

Phase-encoding gradient: The magnetic field gradient that is applied prior to recording the signal so that spatial information is encoded into the phase of the MR signal.

Physiological motion: Movement caused by physiological processes such as the heartbeat and breathing.

Pixel: One point of data in an image; *picture element*; in MRI a pixel represents a *voxel*, or volume element, of real space.

Planck's constant: h, a physical constant, 6.626×10^{-34} Joule · seconds (J · s)).

PLS: Partial least squares.

Polynomial: A function consisting of the sum of two or more terms, each of which is a constant multiplied by a variable raised to the power of a whole number (integer), such as $f = a + bx + cx^2$, where a, b, and c are constants and x is a variable.

Preamplifier: Electronic component of an MR system that amplifies the analog signal detected by an MR coil before it is transmitted to other components for digitization, processing, and storage.

Precession: A change in the orientation of the rotation axis of a rotating body, such as a spinning top wobbling around so that its rotation axis describes a cone; typically caused by a torque being applied to a rotating body.

Preprocessing: Steps taken prior to data analysis to remove confounding effects of noise, motion, and the like, or to subtract mean values, whiten the data, and so forth.

Principal components analysis: PCA; a process of reducing a data set into uncorrelated components, with the first component describing the largest amount of variance in the data, and each successive component describing the largest amount of remaining variance after the previous components have been removed, and so on.

Proton density: The number of hydrogen nuclei (protons) within a given volume that can contribute to the MR signal.

PSIF: Reversed FISP; a steady-state gradient-echo imaging method.

Pulse sequence: The precise timing sequence of radio-frequency pulses, magnetic field gradients, and receiver on/off states that are applied to acquire an MR image or other MR signal; includes the RF pulse amplitudes, phase, and frequency, and the gradient amplitudes and directions, at every point in time during the sequence.

p-value: The statistical probability, p, that the null hypothesis is valid.

PWI: Perfusion-weighted imaging.

Q-ball imaging: A method of estimating white matter fiber pathways (i.e., Tractography) based on diffusion-tensor imaging with data acquired with a very large number of different diffusion-encoding directions and amplitudes (HARDI).

Radio-frequency: The range of the electromagnetic spectrum between 3 kHz and 300 GHz.

Ramp: (1) to increase or decrease in a linear manner; (2) the changing gradient strength over time when a magnetic field gradient is applied or removed, given that the field does not change instantaneously.

Random effects: A group analysis method with statistical tests based on the average response across the group relative to the standard error of the responses across the group.

Random walk: The pathway or trajectory taken by a random process such as thermal motion of a molecule, in which each successive step has a random direction; characterizes processes of diffusion.

Rapid exchange: The case in which hydrogen nuclei are able to exchange quickly between two or more relaxation environments, on the time scale of the relaxation, so that the net observed relaxation rate is a weighted average of the relaxation rates in each of the environments.

RARE: Rapid acquisition with relaxation enhancement, a fast spin-echo imaging method.

RAS: Right–anterior–superior, listing the positive anatomical directions, or directions of increasing coordinate values, in one system of displaying anatomical images.

Readout: Frequency encoding or sampling of the MRI data.

Receiver coil: An MR coil used specifically for detecting the MR signal, as opposed to the transmission of radio-frequency pulses.

Rectangular function: A function having only two values, being at the higher value for a period of time and at the lower value otherwise, also called a *boxcar* function.

Refocusing: In MRI, a radio-frequency pulse or magnetic field gradient used specifically to bring magnetization back into phase.

Register: To coalign two images.

Relaxation: The process by which magnetization in materials in a magnetic field reaches the steady-state equilibrium magnetization; in biological tissues occurs predominantly via interactions between magnetic moments mediated by random thermal motion.

Repetition time: TR; the time between successive radio-frequency excitations of a given volume of tissue or other material.

Residual: Any unaccounted for variance, such as any deviation in measured values from a fit to a specific pattern.

Resolution: The capability of distinguishing between two separate but adjacent objects; the minimum distance between two objects that can be detected as being separate; in MRI, the upper limit of resolution is equal to the dimensions of a voxel, although smoothing effects or artifacts may reduce the actual resolution.

RF: Radio-frequency.

RF coils: Any coil designed for transmitting or receiving radio-frequency signals.

RF pulse: A magnetic field rotating at a frequency within the radio-frequency range that is applied in a brief (~ milliseconds) pulse.

RGB: Red-green-blue; color representation used in many common image formats and visual displays.

Rotating frame: A frame of reference, or coordinate system, that rotates at the Larmor frequency in the same direction as magnetization precessing within a magnetic field, used for describing the interactions between magnetic moments and a magnetic field, as in MR.

Sagittal: Slice orientation with the plane extending in the head-foot and anterior–posterior directions.

Sampling time: The total time taken specifically for recording the data points for an image; does not include time for gradients, RF pulses, and so on and is therefore less than the acquisition time.

SAR: Specific absorption rate.

Saturation: Reduction of an MR signal due to T_1-weighting, in some cases the complete elimination of the signal.

SE: Spin echo.

Second-level analysis: Group analysis based on results of individual first-level analyses.

SEEP: Signal enhancement by extravascular water protons.

Segment: A portion of k-space or of image data.

Segmented EPI: An echo-planar imaging method in which not all of k-space is sampled in a single acquisition, but rather requires multiple acquisitions; also called multishot EPI.

Sensitivity: For fMRI, the ability to detect signal changes corresponding to neural activity.

SEM: Structural equation modeling.

Serial correlations: In data analysis, successive points of a time-series not being independent; for fMRI this can affect the efficiency of the study design and can also lead to underestimates of the variance of results and over-estimates of the significance level.

Sharp: Spatially selective RF pulses being designed to provide slice profiles with very narrow transition edges, usually at the expense of taking more time to apply.

Shim: In MR, (1) the uniformity or homogeneity of the static magnetic field, or (2) the action of making the field more uniform by adjusting currents in a set of coils designed specifically for shimming, which are incorporated into all commercially available MRI systems.

Signal: The electrical impulses that are detected at the receiver coil.

Signal enhancement by extravascular water protons: SEEP, a contrast mechanism for fMRI; the primary contrast mechanism used for fMRI of the spinal cord.

Signal-to-noise ratio: SNR, the ratio of the magnitude of the MR signal to the standard deviation of the random signal fluctuations that arise due to noise (although several variations of this definition exist).

sinc function: A function of the form $\text{sinc}(x) = \text{sine}(x)/x$.

Sine: Trigonometric function—in a right triangle, the sine of the angle of one corner (not the right angle corner) is equal to the ratio of the side opposite the given corner to the length of the hypotenuse.

Single-shot: Refers to any imaging method in which all of the data used to produce a two-dimensional image are acquired after a single excitation.

Single-shot fast spin echo: SSFSE; HASTE; RARE; a fast-spin echo imaging method that acquires all of the required two-dimensional image data after a single excitation pulse (followed by a string of refocusing pulses).

Slab: A thick slice.

Slice: A region with small spatial extent in one direction that approximates a plane.

Slice profile: The distribution across the slice thickness of the effect of a spatially selective RF pulse.

Slice rewind gradient: The gradient that is applied after a spatially selective pulse to bring the magnetization across the slice thickness back into phase in order to produce a detectable MR signal.

Slice thickness: The spatial extent of a slice across the smallest dimension.

Slice timing correction: The process in fMRI analysis of accounting for the fact that multiple slices in a data set are acquired at slightly different points in time and therefore may reflect different points in the BOLD response.

Slow exchange: The case in which hydrogen nuclei are not able to exchange quickly between two or more relaxation environments, on the time scale of the relaxation, so that the net observed relaxation rates in each of the environments may be separated.

Smoothing: The process of applying a weighted average across groups of adjacent points in an image or time series in order to emphasize features or remove features such as high-frequency fluctuations.

SNR: Signal-to-noise ratio.

Spatial encoding: (1) imposing position-dependent frequency and phase into MR signal in order to retrieve spatial information; (2) the method used to sample k-space points or the order of sampling, as in conventional sampling, echo-planar imaging, spiral, and so forth.

Spatial frequency: The number of cycles per unit space of a wave or oscillation of a signal that depends on position, such as MR image data; has units of (1/cm).

Spatial normalization: The process of warping images to match features, such as anatomical regions, to a standardized reference image, in order to provide a common shape for comparison of result across studies of different people.

Spatial resolution: See *resolution*.

Specific absorption rate: SAR, a measure of energy deposited in tissues by radio-frequency pulses during MRI; clinical MRI systems include monitoring to ensure that safety guidelines for SAR are not exceeded.

SPGR: Spoiled gradient-recalled echo.

SPICE: Acronym listing the key features of a spin echo.

Spin: In MR jargon, used interchangeably with *magnetic moment* and *nucleus*, referring to the source of the MR signal; more accurately, a fundamental characteristic property of elementary particles, composite particles, and atomic nuclei that is related to the angular momentum and magnetic moment of such particles, although not the only determining factor.

Spinal fMRI: Functional MRI specifically of the spinal cord.

Spin echo: An echo produced by applying a radio-frequency excitation pulse, followed a short time later by a 180° refocusing pulse, having the effect of reversing the influence of any static magnetic field variations and bringing the magnetization back into phase to produce a brief signal peak.

Spiral: A method of sampling k-space data in a spiral pattern; a fast image acquisition method.

SPM: Statistical Parametric Mapping.

Spoiled: Indicates that an acquisition method employs steps to dephase all transverse magnetization after each acquisition, as opposed to the steady-state methods.

Spoiled GRASS: Refers to a steady-state imaging method, gradient recall acquisition using steady states, except with spoiling so that it is not a steady-state method, resulting in a standard gradient-echo imaging method.

Spoiling gradients: Magnetic field gradients applied after an acquisition intended to dephase all remaining transverse magnetization.

SSFP: Steady-state free precession.

SSFSE: Single-shot fast spin echo.

Starling Law of Capillaries: Water passage through blood capillary walls is the net effect of the balance of hydrostatic and osmotic forces and can result in water moving out of vascular spaces into the interstitial space, or in the opposite direction.

Statistical Parametric Mapping: SPM, a software package for analyzing functional neuroimaging data, including fMRI.

Statistically significant: Unlikely to have occurred by chance.

Statistical tests: Methods of making statistical decisions using experimental data.

Statistical threshold: The level of significance chosen to infer that a result is statistically significant, that is, adequately significant.

Steady state: Steady-state methods, in contrast to spoiled methods, return the transverse magnetization to being in phase prior to each subsequent RF pulse, and contribute to increasing the signal detected in subsequent acquisitions.

Stimulated echo: Similar to a spin echo but produced by a sequence of three RF pulses, with the echo occurring at a time delay after the third pulse equal to the interval between the first two pulses.

Stimulus: In fMRI, any effect producing a sensory response (tactile, visual, auditory, etc.).

STIR: Short tau inversion recovery; an inversion-recovery method that can be used with either a spin echo or a gradient echo.

Structural equation modeling: SEM, a method of analyzing connectivity in fMRI data.

SUGAR: Acronym listing the key features of a gradient-echo.

Superconducting: Materials having zero electrical resistance, typically only at very low temperatures, as with the coils used to produce the static magnetic field in most MRI systems.

Suppression: Action of an RF pulse intended to eliminate MR signal, often spatially selective.

T_1: Longitudinal relaxation time; spin-lattice relaxation time.

T_1-weighted: Having MR signal intensity that is a function of the T_1-value of the material.

T_2: Transverse relaxation time; spin–spin relaxation time.

T_2^*: Effective transverse relaxation time, including effects of both transverse relaxation and spatial field variations within each voxel.

Talairach coordinate system: Normalized coordinate system defined for mapping brain anatomical structures.

Tesla: Unit of magnetic field strength, equal to 10,000 gauss; particle passing through a magnetic field of 1 tesla at 1 meter per second carrying a charge of 1 coulomb experiences a force of 1 newton.

TFE: Turbo field-echo; fast gradient-echo.

Theorem of reciprocity: Theorem that the magnetic field intensity produced at any point in space by an MR coil when carrying an electric current is proportional to the sensitivity of the coil to an MR signal originating from that same point, when the coil is used as a receiver.

Thermal energy: Heat; temperature.

Thermal motion: Random Brownian motion due to thermal energy.

TI: Inversion time.

Time point: One point in a time series; a volume represented by images in an fMRI time series.

Time series: A sequence of measurements spanning a period of time.

Tissue probability maps: Spatial maps indicating the probability at each point of being a member of each of several tissue types, such as gray matter, white matter, and cerebrospinal fluid.

Tissues: Biological materials such as organs, muscle, and neural tissues, as opposed to fluids, bone, or air spaces.

TM: Mixing time; the time between the second and third pulses of a stimulated echo sequence.

Transition edge: The zone affected by a spatially selected RF pulse between the range that receives the full effect of the pulse and the surrounding area that is left unaffected.

Transverse component: The component of magnetization that is within the transverse plane, or orthogonal to the static magnetic field, B_0; the component of magnetization that induces a signal in a receiver coil.

Transverse plane: The plane that is orthogonal to the static magnetic field, B_0.

Transverse relaxation: Refers to the decay of the transverse magnetization toward its equilibrium value of zero via processes of relaxation.

TrueFISP: True fast imaging with steady-state precession, a steady-state gradient-echo imaging method.

True SSFP: True steady-state free precession, a gradient-echo imaging method.

T-statistic: A statistic whose sampling distribution is a Student's t-distribution; equal to the mean value of a sample with a normal distribution minus the expected value, and the result is divided by the standard error of the sample.

Tune: In MR, typically refers to the adjustment of an MR coil to have high sensitivity to the specific Larmor frequency of interest.

Turbo FLASH: Fast gradient-echo imaging method.

Turbo factor: The number of spin echoes acquired with each excitation in fast spin-echo imaging.

Turbo SE: Turbo spin echo; fast spin echo.

Turbo spin echo: Fast spin echo.

T-value: See *T-statistic*.

Vascular space occupancy: VASO; a functional imaging method based on detecting changes in blood volume related to changes in neural activity.

VASO: Vascular space occupancy.

Volume: (1) in general, the amount of space occupied by a three-dimensional object or region of space; (2) in fMRI, often used to refer to one set of images spanning a volume, as one time point of an fMRI time series.

Voxel: Volume element; the three-dimensional region represented by one pixel in an image.

Voxel-by-voxel: Mathematical operations or comparisons applied individually to each voxel in fMRI data.

Water: The dominant source of MR signal in biological tissues, such as for MRI and fMRI.

Whisper: Used to indicate a mode of applying magnetic field gradients with lower acoustic sound levels produced, typically at the expense of speed.

White matter: The component of neural tissue consisting predominantly of myelinated axons.

White noise: Random noise with a uniform frequency spectrum.

Whiten: To remove structure in the noise component of data and make it closer to a purely random process.

x-axis: (1) one of the spatial axes in the transverse plane, orthogonal to the static magnetic field, B_0; (2) often used to indicate the frequency-encoding direction in general descriptions of imaging methods.

x–y plane: The transverse plane.

y-axis: (1) one of the spatial axes in the transverse plane, orthogonal to the static magnetic field, B_0; (2) often used to indicate the phase-encoding direction in general descriptions of imaging methods.

z-axis: (1) the spatial axis parallel to the static magnetic field, B_0; (2) often used to indicate the slice direction in general descriptions of imaging methods.

Zeeman energy: Equal to the dot product of the magnetic dipole moment and the magnetic field vector; the work of magnetic force on realignment of the vector of the magnetic dipole moment.

Zero filling: The process of filling regions of k-space with values of zero, instead of sampling the data, in order to reduce imaging time or for interpolation.

ΔB: Change in magnetic field, B.

ΔCBF: Change in cerebral blood flow.

$\Delta CMRO_2$: Change in cerebral metabolic rate of oxygen consumption.

ΔR_2: Change in transverse relaxation rate ($1/T_2$).

ΔR_2^*: Change in effective transverse relaxation rate ($1/T_2^*$).

$\Delta S/S$: The fractional signal change, typically used in fMRI, equal to the change in signal intensity between two conditions divided by the baseline signal intensity.

β-values: The results of a general linear model analysis, indicating the magnitude of each component of a basis set fit to a data set.

γ: The gyromagnetic ratio.

π: The physical constant 3.141592654.

Appendix

Decision Tree for BOLD fMRI Study Design

Stimulus or Task Is Determined by the Neural Function of Interest

Form a clear hypothesis to be tested

What is the simplest set of stimuli or tasks that will answer the question(s) being asked?

One task/stimulus and baseline condition?

Two tasks/stimuli to be compared/contrasted?

More than two tasks/stimuli, and so forth? (One can be considered the baseline.)

1 Stimulus or Task Paradigm Design

1.a Block, Event, or Something in Between?

Function(s) of interest can be sustained or applied/performed repeatedly	→	**Block Design** (the most efficient design)
Some function(s) of interest cannot be sustained and some function(s) cannot be studied as brief events, and both are needed	→	**Mixed Design**
Function(s) of interest cannot be sustained, and/or It is necessary to compare/contrast the subject's responses to events, or task performance, within the fMRI acquisition (such as correct and incorrect responses)	→	**Event-Related Design** Needed if: Responses to individual events must be observed Responses must be grouped or re-ordered based on responses
	→	**Fast Event–Related Design** Needed if: Interactions between successive stimuli/ tasks are to be observed Event-related acquisitions take too much time

1.b How Will Responses to Stimuli/Tasks Be Contrasted or Compared?

Subtraction—Contrast between trial types (A and B)

Factorial—Interactions between trial types (A, B, A + B, not A or B)

Parametric—Differentiate cognitive components of responses (varying intensities of A and B)

Conjunction—Determine common features of trial types (AB, AC, AD, ABC, ABD, ACD)

1.c Timing of Tasks/Stimuli?

Acquire as many time points (volumes) as possible	↔	Limits of person's comfort, fatigue, ability to maintain cognitive/attentional state, etc.
Keep sampling rate as fast as possible (repetition time as short as possible) (lower limit of 1 sec for most applications)	↔	Limits of time needed to image the essential range of the anatomy

Block Designs:
   Equalize amount of time spent in each trial type
   Use blocks that are ~15 seconds long or longer

Event-Related Designs:
   Use shorter inter-stimulus intervals (ISIs) when possible (fast event-related) (down to 4 sec)
   Use variable ISIs when possible, with slowly varying probability of events occurring (stochastic design)

Use an fMRI simulator to help with design planning because there is no one optimal design for each set of tasks or stimuli and each effect of interest.

2 Data Acquisition

2.a Work with the MRI System Available, and Start with Standard Acquisition Parameters

2.a.1 Optimal BOLD Sensitivity

Fast T_2^*-weighted images—single-shot gradient-echo EPI (echo-planar imaging).

Echo time (TE) = 60 msec, 30 msec, 15 msec for field strengths of 7 T, 3 T, 1.5 T, respectively.

2.a.2 Balance Image Resolution, Field of View (FOV), and Speed

64×64 acquisition matrix, set FOV to 211 mm × 211 mm in axial slices (to get 3.3 mm × 3.3 mm resolution), set slice thickness to 3.3 mm for cubic voxels, and set number of slices to 36 to span the cortex, with zero slice gap between the slices.

2.a.3 Minimize T_1-Weighting and Optimize Speed

Set repetition time (TR) = minimum possible (preferably 3 sec or less), and set flip angle = $\cos^{-1}(TR/T1)$ (the Ernst angle).

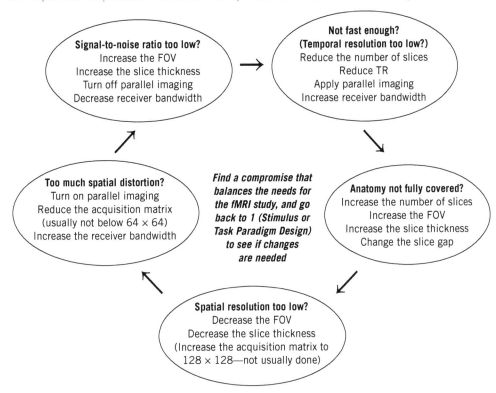

3 Data Analysis

3.a Preprocessing—Preparing the Data for Analysis

Choose which preprocessing steps to include:	
Conversion of Data Format to a Form Accepted by the Analysis Software	
Global normalization	• Removes trends that occur across the entire volume • Includes mean intensity correction, removal of initial time points
Motion correction	• Removes/reduces effects of bulk movement of the body
Slice timing correction	• Accounts for the fact that all slices in a volume are not imaged at the same time • May impose poor assumptions for interpolation if repetition time (TR) is long
Spatial normalization	• Reshapes the data or the results of analysis or both to a standardized shape and size for all participants

(continued on next page)

Conversion of Data Format to a Form Accepted by the Analysis Software (continued)	
Spatial smoothing	• Increases signal-to-noise reduction (SNR), may reduce structure in the noise, may improve alignment of results across participants, reduces the number of tests and decreases the multiple comparisons problem • Reduces spatial resolution, can reduce detection of small areas of activity • Optimal smoothing width should match the expected extent of active regions
Temporal filtering	• Can reduce low-frequency trends and high-frequency noise • Optimal choice is a band-pass filter with a tow-frequency cutoff of 0.008 Hz or fluctuations with a period of 120 sec, and a high-frequency cutoff depending on the TR and timing of stimulus blocks or events

3.b Choice of Analysis Method(s)

Model-Driven, Univariate Methods		
Require estimation or prediction of expected patterns of signal change in active voxels Each voxel is treated as an independent test	**Correlation**	• Sensitive to the shape of the expected response • Does not account for confounding signal changes
	General linear model (GLM)	• Most widely used analysis method for fMRI • Accounts for confounding signal changes if they are modeled or predicted

Data-Driven, Multivariate Methods		
Do not require estimation or prediction of expected patterns of signal change in active voxels Common components of responses are detected across all voxels Results indicate functional connectivity between active regions	**Independent components analysis (ICA)**	• Determines independent signal intensity components • Most widely used data-driven method
	—	—
	Principal components analysis (PCA)	• Determines uncorrelated signal intensity components
	Fuzzy clustering analysis (FCA)	• Determines "clusters" of voxels with similar components • Estimates the probability of a voxel fitting within any cluster

3.c Choice of Statistical Thresholds

The problem of multiple comparisons must be addressed with a choice of statistical threshold, and typically with adjustment based on a choice of methods:

Family-Wise Error (FWE)—the probability of there being a false positive result within all of the voxels

False Discovery Rate (FDR)—the proportion of voxels in the data set that are expected to be false positive results

Bonferroni Correction—divide the desired p-value threshold by the number of independent voxels being tested

Cluster Site Threshold—require active regions to exceed some minimum volume

Index

Index

Bandwidth
  acquisition parameter adjustment, 275
  BOLD optima, 132
  image creation from k-space, 98
  noise generated by human body, 103
  optimization and trade-offs in image acquisition, 107
  signal-to-noise ratio, 104, 106
Basal ganglia, 139
Baseline
  BOLD signal change quantification, 128–129
  fMRI, 123
  rectangular functions, 33
  study design, 162, 273; *See also specific design
    approaches*
Basis functions
  Fourier transform, 26
  GLM, 22, 24, 25
Behaviorally driven designs, 165, 168–169
beta-values, 197, 198
b-FFE, 71, 83
Bile, 81
Binary numbers, 18
Biological materials; *See also* Tissue magnetization
  magnetization transfer contrast, 81
  signal sources, 37
Bits, defined, 18
Bloch equation, 69
Block designs, 164, 165, 182, 273, 274
  clinical studies
    consciousness disorders/coma, 233
    MS, 229
    preoperative mapping, 221
  number of blocks, 180–182
  order and timing, 173–179
Blood
  magnetization transfer contrast, 81
  oxygen saturation effects on magnetic susceptibility, 55, 56
Blood flow
  alternative contrast methods, 143
  and artifacts/distortions, 116, 117, 119
  BOLD signal change response, 125
  DWI, 152–153
  PWI, 143, 150
  SEEP, 144
Blood oxygen; *See* BOLD
Blood plasma, 48
Blood vessel size, 125–126
Blood volume
  BOLD, 122–123
  hemodynamic response, 126
  VASO, 143, 150–151
Blurring
  artifacts and distortions, 120
  convolution operation, 36, 116, 117
BOLD
  alternative contrast methods, 143–156
    diffusion-weighted imaging (DWI), 151–156
    perfusion-weighted imaging (PWI), 146–150
    SEEP, 144–146, 147

    vascular-space occupancy (VASO), 150–151
  contrast mechanisms, 122–130
    oxygen saturation and blood volume, 122–123
    physiological origins, 123–128
    signal change, quantification of, 128–130
  DWI and, 155
  methods, 130–134
  MS studies, 228
  perfusion-weighted imaging (PWI), 149
  preoperative mapping, 221
  PWI, 150
  resting-state studies, 226
  saccadic eye movement studies, 238, 240–241
  study design, 273–277; *See also* Study design
    comparison of responses, 274
    data acquisition, 274–275
    data analysis, 275–277
    neural function of interest, 273
    stimulus or task paradigm, 273–274
    timing of tasks/stimuli, 274
  tissue differences in magnetic susceptibility, 56
  traumatic brain injury, 236, 237
Bone
  BOLD, causes of distortion and signal loss, 136
  and magnetic field variations, 55, 61
  SEEP, 146
Bonferroni correction, 204, 277
Bootstrap method, 207–208
Boundaries; *See* Interfaces/boundaries
BPP expression, 49–50
Brain imaging
  air spaces, effects of, 55
  artifacts and distortions, 119
  clinical studies; *See* Clinical applications
  instrumentation, 10–11, 12
  inversion-recovery methods, 79–80
  relaxation times for various regions, 50
Brainstem, 244
  air spaces, effects of, 55
  air/tissue interface effects, 119
  BOLD
    causes of distortion and signal loss, 136
    method modifications for special regions, 139
BrainVoyager, 188, 240
Brownian motion, 103
b-values, DWI, 153
Bytes, defined, 18

C

Canonical HRF, 123
Cardiac stimulation, rapid gradient switching and, 108
Carpal tunnel syndrome (CTS), 244
CASL (continuous arterial spin labeling), 148
Causal modeling, dynamic, 208–211
Cellular volume changes, 143, 144
Center frequency, spatially selective RF pulses, 89
Cerebral metabolic rate of oxygen consumption (CMRO), 123, 125, 126, 147
Cerebral slices, SEEP, 145, 146

Milton Keynes UK
Ingram Content Group UK Ltd.
UKHW052029141024
449569UK00017B/746